Sustainable Consumption and Production Patterns: Policy Design and Evaluation

Sustainable Consumption and Production Patterns: Policy Design and Evaluation

Editors

Yasuhiko Hotta
Tomohiro Tasaki
Shunsuke Managi

MDPI • Basel • Beijing • Wuhan • Barcelona • Belgrade • Manchester • Tokyo • Cluj • Tianjin

Editors
Yasuhiko Hotta
Institute for Global
Environmental Strategies
Japan

Tomohiro Tasaki
National Institute for
Environmental Studies
Japan

Shunsuke Managi
Kyushu University
Japan

Editorial Office
MDPI
St. Alban-Anlage 66
4052 Basel, Switzerland

This is a reprint of articles from the Special Issue published online in the open access journal *Sustainability* (ISSN 2071-1050) (available at: https://www.mdpi.com/journal/sustainability/special_issues/cpp_sus).

For citation purposes, cite each article independently as indicated on the article page online and as indicated below:

LastName, A.A.; LastName, B.B.; LastName, C.C. Article Title. *Journal Name* **Year**, *Volume Number*, Page Range.

ISBN 978-3-0365-4299-7 (Hbk)
ISBN 978-3-0365-4300-0 (PDF)

Contents

About the Editors

Yasuhiko Hotta

Yasuhiko Hotta (DPhil.) is the Director in the Sustainable Consumption and Production Area, Institute for Global Environmental Strategies. Hotta holds a Dphil in International Relations from the University of Sussex (2004). He has been involved in both policy initiatives and research projects in relation to sustainable resource circulation such as the G8's 3R Initiative, Working Group for 3R Policies for Southeast and East Asia at Economic Research Institute for ASEAN and East Asia (ERIA), and the OECD's Working Party for Resource Productivity and Waste. From 2016 to 2021, Hotta served as a theme leader of Theme 3 of the PECoP-Asia research project for SCP in Asia. He is also a part-time lecturer for the Tokyo Institute of Technology. In 2021, Hotta was appointed as the Vice President of Asia Pacific Roundtable on Sustainable Consumption and Production (APRSCP).

Tomohiro Tasaki

Tomohiro Tasaki (PhD) is the head researcher of the Material Cycles and Social Systems Research Section in the Material Cycles Division, and a researcher in the Social Systems Division at the National Institute for Environmental Studies (NIES), Japan. He was the leader of Theme 2 of the PECoP-Asia research project. His academic background includes both systems engineering and policy science. His current work is mainly focused on the following three themes: establishing a regime inclusive of future generations and sustainability indicators, transition to sustainable consumption and production patterns and sustainable lifestyles, and evaluation and analysis of 3R (reduce, reuse, and recycle)/circular economy policies and waste treatment systems in the era of population decline/aging. He holds a Ph.D. from the Graduate School of Yokohama National University, Japan.

Shusuke Managi

Shunsuke Managi (PhD) is the Distinguished Professor of Technology and Policy and Director of the Urban Institute at the Kyushu University, Japan. He has been awarded several national research grants on topics such as urbanization, transportation, energy, climate change, sustainability, urbanization, and population change. He has received several research fellowships from organizations such as the Helmholtz Association and Cheney Senior Fellowship and has served as an expert on energy and environmental policy. He is an editor of Environmental Economic and Policy Studies and on the editorial board for six journals including Resource and Energy Economics, and a lead author for the Intergovernmental Panel on Climate Change. He is the author of 12 books and 140 academic journal papers.

Preface to "Sustainable Consumption and Production Patterns: Policy Design and Evaluation"

From 2016 to 2021, a five-year research project on sustainable consumption and production titled "S-16 Project on Policy Design and Evaluation to Ensure Sustainable Consumption and Production Patterns in Asian Region (PECoP-Asia)" was conducted as a collaborative research project among 11 leading research universities and institutes in Japan: University of Tokyo, National Institute of Environmental Studies (NIES), Institute for Global Environmental Strategies (IGES), Keio University, Osaka University, National Institute of Advanced Industrial Science and Technology (AIST), Ritsumeikan University, Kobe University, Kyushu University, Nanzan University, United Nations University/Institute for the Advanced Study for Sustainability. This collaborative project (JPMEERF16S11600) was funded by Environmental Research and Technology Development Fund of Environmental Restoration and Conservation Agency (ERCA) of Japan. As core members of this research project, the guest editors decided to publish a Special Issue in Sustainability (MDPI journal), contributing to the research and knowledge development on policy design and evaluation for sustainable consumption and production. This book is a reprint of the Special Issue Sustainable Consumption and Production Patterns: Policy Design and Evaluation.

Under the era of the Paris Agreement and SDGs, the agenda for sustainability has shifted from regulation to designing fundamental changes in the socio-technical system towards a decarbonized and circular society. This shift would largely depend on taking advantage of emerging forces in the socio-technical system. This includes new business models, drivers of wealth, wellbeing and human development, urbanization, disruptive technologies, and digitization. Socio-technical innovations are crucial to ensure that these forces do not divert our society away from sustainability. Sustainable Consumption and Production (SCP) policy design and evaluation encounter a fundamental shift in its focuses. Firstly, SCP policies are expanding from the environmental policy domain to socio-technology policy domain. Secondly, strengthening linkages between consumption and production is a key emerging trend. Thirdly, the transition to SCP is a socio-technical regime shift requiring successive changes in social practices, technology use in daily life, and associated infrastructure. Fourth, bottom-up approaches are necessary to enhance the effectiveness and acceptance of SCP policies as well as to enable new business models and lifestyles for SCP.

Based on this recognition, this book is intended to highlight why SCP policy design and evaluation needs to overcome conventional environmental policy framework. Emerging SCP policy design and evaluation do not focus on individual products or behaviors/or improving efficiency in management system in relation to environmental sustainability; instead, they address a more socio-economic system and target collective efforts for the transition. This is fundamentally different from environmental policy design responding to pollution. It is vital to identify and develop communication tools for sharing visions among stakeholders to facilitate collective actions towards sustainable lifestyles. Emerging SCP policy design under the era of SDGs include communication/planning tools as well as those expecting multiple effect/unintended effect contributing to social well-being. At the same time, for sustainability transition and to achieve global targets for sustainability such as SDGs, it is vital to identify gaps and opportunities for sustainable consumption in the specific context of rapidly emerging economies such as China, India, Vietnam, and Thailand, as presented in this book. Effort has been made for this book/Special Issue to feature studies contributing to policy design and evaluation in this direction. The papers in this book suggest that SCP policy design and evaluation needs to pay more attention to social aspects of sustainability such as social infrastructure and well-being and socio-technical systems and societal empowerment to ensure an effective and just transition to sustainability. Thus, we are very pleased to compile this book based on the Special Issue together. The guest editors would like to express their sincere appreciation to the authors for their contributions as well as all collaborators of the PECoP-Asia project for their support.

Yasuhiko Hotta, Tomohiro Tasaki, and Shunsuke Managi
Editors

 *sustainability*

Article

Expansion of Policy Domain of Sustainable Consumption and Production (SCP): Challenges and Opportunities for Policy Design

Yasuhiko Hotta [1,*], **Tomohiro Tasaki** [2] **and Ryu Koide** [1,2]

[1] Sustainable Consumption and Production Area, Institute for Global Environmental Strategies, Kamiyamaguchi 2108-11, Hayama 240-0115, Kanagawa, Japan; koide.ryu@nies.go.jp

[2] Material Cycles Division, National Institute of Environmental Studies, Japan, 16-2, Onogawa, Tsukuba 305-8506, Ibaraki, Japan; tasaki.tomohiro@nies.go.jp

* Correspondence: hotta@iges.or.jp

Abstract: Since 2015, the international policy community has started to agree on international agreements with ambitious middle-term and long-term goals, highly relevant to sustainable consumption and production (SCP) such as those seen in the Paris Agreement, SDGs, and the plastic-related agreements at the G7 and G20 processes. Along with this trend, there has been growing attention given to socio-technical system change or "transition". Policy debate is putting more focus on the need to change consumption and production patterns and deal with various ecological consequences within planetary boundaries such as decarbonization, absolute reduction in material throughput, or creation of a plastic-free society. This paper examines the expansion of the policy domain of SCP in three phases; SCP focusing on pollution control and cleaner production (SCP 1.0), SCP from the perspective of product lifecycle (SCP 2.0), and SCP focusing on systematic changes in socio-technical systems driving consumption and production (SCP 3.0). The potential impact of a wider SCP policy domain can be comparable to the historical shift in discourse related to ecological modernization theory from pollution prevention to efficiency. This emerging trend corresponds to the need for a fresh approach to policy design which can facilitate transition to sustainability.

Keywords: sustainable consumption and production; sufficiency; efficiency; transition; discourse analysis; policy design

Citation: Hotta, Y.; Tasaki, T.; Koide, R. Expansion of Policy Domain of Sustainable Consumption and Production (SCP): Challenges and Opportunities for Policy Design. *Sustainability* **2021**, *13*, 6763. https://doi.org/10.3390/su13126763

Academic Editor: Paulo Peças

Received: 15 April 2021
Accepted: 11 June 2021
Published: 15 June 2021

Publisher's Note: MDPI stays neutral with regard to jurisdictional claims in published maps and institutional affiliations.

1. Introduction

One of the key policy concepts on the global sustainability agenda since the 1990s has been sustainable consumption and production (SCP). This was triggered by economic globalization and the increasing distance between where goods are produced and where they are consumed [1]. SCP was a key component of Agenda 21, the action plan adopted at the UN Conference on Environment and Development (Rio Summit) in 1992 [2]. Then, at the United Nations Conference on Sustainable Development in 2012 (Rio+20), the UN 10 Years Framework Programme of Sustainable Consumption and Production (10YFP) was adopted, and in 2015, SCP became one of the 17 Sustainable Development Goals (SDGs). In a report published in 2015, the United Nations Environment Programme (UNEP) defined SCP as "a holistic approach to minimizing the negative environmental impacts from consumption and production systems while promoting quality of life for all" [3]. However, the focus of SCP has widened over the last few years due to a shift in emphasis of environmental policy and sustainability agenda.

Since 2015 in particular, several important international agreements with strong links to SCP have been adopted including the Paris Agreement, and the 2030 Agenda on Sustainable Development (SDGs). Furthermore, agreements on marine plastic issues emerged from the G20 and G7 processes, including the G20 Osaka Blue Ocean Vision and the Ocean

Plastic Charter launched by Canada and agreed in 2018 by EU, UK, France, Italy, Germany and Canada. Although these international agreements are based on voluntary efforts, they have some common features in terms of setting very ambitious and unprecedented middle-term and long-term goals. These goals include decarbonization or achieving a net zero society, elimination of marine plastic discharge, and reviewing lifestyles and economies within planetary boundaries, which are central to our economy, which depends on material and energy consumption and production. To achieve these goals, we must make fundamental changes to our social and economic structures including business models and lifestyles.

However, SCP as a policy concept is often used in an intermixed and confused manner due to existence of the different versions as described in this paper. This occurs because the scope of SCP policy has been expanded over years. For example, SCP policies can target individual industrial processes, products and services, or the entire system and subsystems. Systemic solution and individual solution respectively call for different policy instruments and discussion. To make a policy process effective, it is necessary to summarize these different versions of SCP and facilitate the policy discussions on sustainability transition. The objective of this paper is therefore to summarize the different policy backgrounds on how SCP as a policy domain evolved from SCP 1.0 to SCP 2.0 and then onto the recent interest over sustainability transition (SCP 3.0) by applying environmental policy discourse analysis. In this way, we will characterize the emerging trend of SCP that has developed along with very ambitious and unprecedented middle-term and long-term goals for sustainability such as a net zero society, elimination of marine plastic discharge, and lifestyles and economies within planetary boundaries. The paper posits that a new approach for policy design is required to tackle emerging opportunities and challenges caused by this recent trend (SCP 3.0).

In this paper, we look at how the focus of SCP policy has shifted and widened over the last 30 years, and pay particular attention to the development of SCP policy discourse. Our main methodology is an analysis of environmental policy discourse, recognizing that framing, interpretation, and generation of storylines play key roles in the politics of sustainability. We summarize and describe this shift in three phases—SCP 1.0 to SCP 3.0. Building on ecological modernization theory, this paper provides a streamlined narrative on how the SCP-related policy domain first expanded from pollution prevention and cleaner production (SCP 1.0) to increasing efficiency throughout of life-cycle of materials, products, and services (SCP 2.0). Then, in the late 2000s, policy discussions on the SCP-related domain expanded to include the systematic transition of socio-technical systems, lifestyles, and infrastructure driving consumption and production (SCP 3.0). This expansion can be observed in the recent international policy agendas of sustainability such as the Paris Agreement, and agreements emerging from the G7 and G20 processes. Finally, the paper argues that the long-term and middle-term sustainability goals in SCP 3.0 require a future-oriented policy design going beyond conventional evidence-based policymaking.

2. Methodology: Environmental Policy Discourse Analysis

In our streamlined narrative for SCP policy discourse development, we highlight the emergence of a new version of SCP which requires a fresh approach to policy design. Policy debate on sustainability is strongly linked to framing, interpretation, and generation of storylines. As Fairhead and Leach point out in their political studies of science with regard to deforestation [4], it is the framing and interpretation of issues on a global level that guides policy and strategy on national and local levels of politics and science.

Environmental sustainability encompasses not just physical issues related to deteriorating environmental quality surrounding human settlements, but also includes a complex set of scientific data, interpretation of this data into political and social implications, and then the distribution of values based on that interpretation. Thus, as Hajer [5], Litfin [6] and Dryzek [7] have stated, the persuasiveness of the story or discourse is important for effective consensus building and policymaking. The storyline or narrative should assist the scientist, environmentalist, or politician, to show how their work fits into the whole picture of social and environmental issues [5,7].

For example, Hajer argues that "(e)nvironmental politics is only partially a matter of whether or not to act, it has increasingly become a conflict of interpretation in which a complex set of actors can be seen to participate in a debate in which the terms of environmental discourse are set" [5]. In a similar way, Litfin argues in her study of the international negotiations over ozone layer protection that "(e)nvironmental policy is heavily dependent on such cognitive factors as scientific knowledge, philosophical ideas, and public opinion" [6]. This kind of discursive approach appears to focus only on conceptual, philosophical and symbolic levels of conflict. However, framing certain issues within climate change, decarbonization or building a circular economy has political, social and economic implications for the daily routines of every single sector in a modern industrialized society [8]. For example, by framing an individual citizen in the context of climate change or waste problems, we transform them into a major consumer of energy and a source of greenhouse gas emissions. Similarly, a household would be transformed into a source of waste as well as a processing site for waste separation and recycling. A farming family who cultivate rice paddies in Southeast Asia, if framed in a certain context, would be transformed into a source of methane gas emissions. This type of framing and interpretation of issues brings about a change in attitude for governments, think-tanks, the academic community, industry, and NGOs so that they then take particular policies and strategies [4].

Sustainability issues provoke a process of re-framing and reconsidering of various "structured" daily activities that are part of a modern socio-technical system. However, we must question the values and worldviews behind how environmental issues are understood and interpreted. Otherwise, as Lipschutz and Conca argue, we would end up simply reproducing rather than restructuring (in the sense of transforming and altering or transition) the modern socio-technical system [9].

It is also important to remember that although proper interpretation of the issues is important for policy debate on environmental sustainability, it does not mean that one particular interpretation or a specific application of scientific knowledge is based on a clear-cut political position. When mediating between conflicting positions, environmental discourse can be used to construct coalitions. These coalitions can be merged to contextualize certain political positions in the politics of sustainability rather than the collective activities of individual actors [5]. Furthermore, discourse is not always consistent. In addition, indeed, even vagueness and contradiction are important elements of policy discourse on sustainability when trying to mediate between conflicting positions. Hajer calls this aspect of policymaking 'discursive closure'. This means that the interpretation and definition of a certain problem can set the target for policymaking (it is also important to see where the focus of the problem lies and what is left out) [5].

The power of discourse for framing, interpretation, and generation of certain storylines is important to provoke and continue sustainable transition as a way to transform socio-technical systems. Ganz argues that, when looking to provoke societal change, generating a storyline can be a key instrument for relating, motivating, renewing motivation, and sustaining collaborative actions among stakeholders [10].

Thus, this paper uses environmental policy discourse analysis to demonstrate the changing emphasis in SCP policy domains by reviewing research papers relevant to SCP policy, sustainable lifestyles and sustainability transition. We also reviewed policy reports and outcome documents of international processes relevant to SCP policy domains, especially those of the UN (especially UNEP), G7, and G20.

3. Three-Phase Development of SCP Policy Domain

Environmental policy and business strategies have gradually shifted their focus from pollution prevention and cleaner production, through lifecycle-based efficiency, to the systematic change of socio-technical systems. Based on environmental policy discourse analysis [5,7], this section will provide a streamlined narrative of how the SCP policy domain has expanded over the past few decades, using phases which we call SCP 1.0, SCP 2.0, and SCP 3.0.

3.1. SCP 1.0: Pollution Prevention and Cleaner Production

The rise of the first phase of SCP, or SCP 1.0, corresponded with the need to prevent environmental pollution resulting from rapid industrialization in developed economies in the 1960s and 1970s. There was also a need for more energy-efficient and cleaner production in response to the energy crisis of the 1970s.

SCP 1.0 typically emerged in Japan with the development of policies and industrial initiatives on pollution prevention and cleaner production. Indeed, theorists and advocates of ecological modernization initiatives agree that Japan's experience in pollution prevention in the 1970s was a model case that successfully implemented environmental policy in harmony with economic competitiveness [5,7,11–13]. Industrial pollution became the focus of environmental issues, and from the late 1960s to the early 1970s, Japan enacted strong environmental regulation. Rapid urbanization and motorization resulted in worsening urban air pollution due to an increase in automobile use in the 1970s. At the same time, as Jänicke and Weidner point out, the energy crisis also prompted a restructuring of production systems [11]. The Japanese government promoted structural changes in Japanese industry in response to the oil crisis. According to Mitsuhashi, business investment for this restructuring was supported by strong governmental regulations as well as low-interest loans from governmental financial bodies [14].

This governmental-oriented restructuring of Japan's production system in the 1970s contributed to savings in both energy and resources. These cost-cutting measures in production were a response to the energy crisis, but they contributed to the subsequent improvements in environmental technology. Japan's industrial sector was almost fully dependent on imported oil as an energy source, so the energy crisis resulted in a shift from an economy depending on heavy and chemical industries with high-energy consumption to an economy dependent on assembly-based industries such as car and electronic manufacturing. This also helped Japanese industry make the transition to a cleaner production system.

As Nakanishi suggested, changes in manufacturing processes due to energy-saving, technological innovation and industrial restructuring, brought about greater reductions in environmental pollution than the end-of-pipe measures stemming directly from strict environmental regulations [15]. For example, wastewater pollution (COD; chemical oxygen demand) from the pulp industry was reduced by 84% using in-process technology. Water treatment using end-of-pipe technology only contributed to 16% reduction in pollution. Therefore, a shift in the production process motivated by efficient production results in less environmental pollution than end-of-pipe treatment technology. Nakanishi argues that changes in resources and materials used in final products also served to cut costs, as well as protect the environment [15]. Thus, it is highlighted that initial strong environmental regulation not only motivated the installation of end-of-pipe technologies, but also triggered energy-saving, technological innovation and industrial restructuring, resulting in better environmental and economic performance.

A similar argument can be seen in the justification of cleaner production (CP) as an environmental measurement promoted by UNEP. As an industrial and technological concept for pollution prevention, the history of CP goes back to the 1970s and emphasizes the efficient use of resources while minimizing waste and pollution. Thus, CP can be used to achieve a sustainable society through efficient production processes in industry and distinguishes itself from simply promoting the 'end-of-pipe' approach. In this respect, CP tends to focus on the re-engineering of production processes and the international transfer of environmental technology. In 1989, UNEP launched the Cleaner Production Program under the UNEP Industry and Environment (now UNEP DTIE) section in Paris. Then, in 1992, under Agenda 21 [2], "Promoting cleaner production" (Chapter 30 of Agenda 21) became one way to "strengthen the role of business and industry(Chapter 30 of Agenda 21)" in activities for sustainable development. Thus, throughout 1990s, CP became one of the most important principles underpinning the relationship between environment and industry in the UN system. Another major UN organization aside from UNEP promoting cleaner production was the United Nations Industrial Development Organization (UNIDO) (especially the establishment of National Cleaner Production Centers).

Notably, CP was presented as a "win-win strategy". According to UNEP, " … [CP] protects the environment, the consumer and the worker while improving industrial efficiency, profitability, and competitiveness" [16]. In this sense, UNEP understands CP as a development strategy in its own right [17]. Discussions on CP should focus on the need for industry and business to make changes in their production and service processes to achieve sustainable development. UNEP argues that more efficient production processes and services through the application of environmental technology and appropriate financial mechanisms will contribute to pollution prevention [18]. In a similar way, UNIDO emphasizes that "(Cleaner Production) refers to the approach (the mindset and way of thinking) of how goods and services are produced with the minimum environmental impact within present technological and economic limits. Cleaner Production does not deny economic and industrial growth, but it insists that growth be ecologically sustainable" [19].

CP also takes pollution prevention strategies that have achieved success in industrialized society and applies them to developing countries. In fact, UNIDO and UNEP cooperated to establish national CP centers mainly in developing countries. CP working groups were formed and created a network to accumulate academic knowledge on specific technology and engineering issues. UNEP is also working to set up financial mechanisms promoting cleaner production projects in developing countries.

As UNEP indicates [3], CP is a starting point for SCP policy development focusing more on technology and engineering, as well as on the management side of sustainability. This means that CP serves to mainstream pollution prevention as one way to achieve sustainability. At the same time, CP was a prelude to SCP 2.0, as described below. UNEP also emphasizes that focus of CP has shifted over the years from "single-issue, reactive, site-specific and end-of-pipe" to a "systems approach", and from "production-oriented" to "life-cycle" orientation [20]. CP is increasingly linked to resource and energy efficiency [3]. Also, the concept of SCP and policy emerged out of CP. For example, UNEP and UNIDO established a regional network of CP centers originally called Roundtables on Cleaner Production and Sustainable Consumption. These networks later developed into Regional Roundtables on SCP, including those for Africa and the Asia Pacific.

3.2. SCP 2.0: Lifecycles

The SCP policy domain expanded from pollution prevention and CP in the 1970s and 1980s, to efficiency of the product lifecycle in the 1990s, as indicated in Table 1. As discussed in the previous section, UNEP's development of CP also shifted from SCP 1.0 to SCP 2.0. The 1980s and 1990s saw the emergence of global environmental challenges such as climate change, loss of biodiversity, destruction of rainforest, as well as waste issues and food safety, stemming from globalization, mass production, mass consumption and improper waste management. Thus, we could observe the co-evolution of SCP2.0 (with its focus on product lifecycles) with the globalization of production and distribution. With the globalization of production and consumption, environmental policy put less focus on direct and site-specific issues and began to target indirect and lifecycle-based ones such as GHGs, resource depletion, and waste. Environmental impacts associated with trade and the globalized production system provoked more awareness on the importance of footprint indicators [21–23] for better understanding of indirect environmental and material costs.

Ecological modernization theory [5,7,13,24–26] states that there was a significant shift in the discourse of policymaking and corporate strategy on the environment and industry from the 1970s to 1990s, mainly in advanced industrialized societies [5,7,27,28]. The theory claims that industrialization can be achieved in harmony with environmental conservation with no economic disadvantages. In other words, the relationship between environmental protection and economic competitiveness has been reconfigured from "contradictory and conflicting" to "compatible." The theory argues that there has been a shift in the discourse on environmental policy and corporate strategy or descriptions of "social and institutional transformation" rather than there being a "physical improvement" of environmental conditions [13]. The focus of policy domain has expanded from preventing environmental pollution to increasing efficiency. It also now includes lifecycle thinking, represented by the concepts of material cycles, lifecycle analysis and industrial ecology.

In other words, SCP 2.0 shows that environmental sustainability can be achieved by more efficient use of materials and energy throughout the product lifecycle and at the value-chain level [29]. SCP 2.0 aims at harmonizing the environment and economy towards sustainability by increasing efficiency throughout the product lifecycle at the product level or facility-unit level. Chemical management, waste management, and recycling can now be seen as a part of product management system from lifecycle perspective.

We can see the concept of SCP 2.0 in Toyota's marketing strategy for its hybrid vehicle, the Prius, released on 10 December 1997, to coincide exactly with the timing of the third Conference of the Parties to the UN Framework Convention on Climate Change (COP3) held in Kyoto, Japan from 1 to 11 December 1997. The Prius was marketed as environmentally friendly, incorporating technological innovation in response to global warming. The main marketing strategy for the Prius was its low environmental impact and high energy-saving potential. The vehicle's internal computer automatically adjusts the driving mechanism to use its electric motor at low speeds, and a combination of the motor and the gasoline engine at normal speed, with electricity stored to the battery. The Prius has low CO2 emissions and high fuel efficiency throughout its lifecycle. However, critics alleged that, while the Prius might have less environmental impact, production methods are potentially more damaging to the environment than for conventional vehicles. In response to this criticism, Toyota used lifecycle assessment (LCA) to calculate the CO2 emissions over the vehicle's lifecycle and proved that the Prius actually emitted 36% fewer emissions over its lifecycle (Toyota, Environmental Report 1999, 1999). The launch of the Prius came four years earlier than the Japanese government's introduction of a green tax mechanism for automobiles in 2001.

SCP 2.0 mainstreamed the efficient use of energy and resources throughout the product lifecycle. In the product manufacturing and utilization stage (upstream), SCP 2.0 aimed to improve energy efficiency and product recyclability, as well as to pursue environmentally friendly design and efficient production. After the product is used (downstream), it is more important to focus on waste reduction, reuse, recycling, and environmentally appropriate treatment. The lifecycle of products and materials can be divided into: resource extraction, production/manufacturing, distribution, consumption, recycling, and waste management. Lifecycle thinking for policy intervention usually place policies regulating environmental impacts in each lifecycle stage, or incorporates environmental externalities in each lifecycle stage or in a combination of different lifecycle stages [30,31]. However, a lifecycle-based approach is considered to be part of policy interventions for individual products, services, or material streams such as plastics, home appliances, automobiles, or food. In addition, the role of consumption tends to be downplayed as a driving force for product and service system. Instead, consumption represents one stage of the whole system.

Nevertheless, SCP 2.0 follows the lines of conventional economic development and incremental changes. For example, Steinberger and Krausmann symbolically criticize resource productivity as a well-referred indicator to measure the policy progress of a resource-efficient sustainable economy [32] and argue that this indicator is correlated with GDP as income, so richer countries tend to benefit more from a higher resource productivity indicator [33]. However, there are increasing concerns about the so-called "rebound effects", which can undermine the environmental gains made in resource efficiency. It is widely acknowledged that improvements in unit-level and product-level resource and energy efficiency do not necessarily result in an overall reduction in energy and resource consumption [34–37]. In this context, the limitations of efficiency discourse have been gradually recognized.

3.3. SCP 3.0: Transition and Sufficiency

We are now observing a significant expansion in the SCP policy domain, comparable to the expansion from SCP 1.0 to SCP 2.0 described by ecological modernization theory. In light of concerns regarding the rebound effect [34–36,38] and the implied limitations of efficiency improvement in products and services, there is growing awareness of the need to create socio-technical systems that enable or constrain behavior and reduce consumption of non-renewable resources/materials [39–42] such as fossil fuels and plastics.

Socio-technical systems require urgent policy intervention to encourage the demand side/civil sector to transition to a sustainable regime, by modifying business models, lifestyles, and infrastructure. The current mainstreamed SCP 2.0 tends to be technically oriented and primarily focuses on the individual product, unit and facility, or on the behavior of individuals [43,44], rather than on societal and technical dynamics beyond a single product system. Indeed, the study on social practices and climate change emphasizes the need for change in terms of the interdependence of social practices on consumption rather than individual behavior [45]. Only by pushing forward with these dynamics can we hope to make the systemic changes to physical and social infrastructures necessary for sustainable business models and lifestyles.

A focus on efficiency has its limitations if we want to control ever-increasing demand and consumption. Trying to find a way to overcome these limitations since the 2000s, the expert community has been examining the principle of sufficiency [9,14,28,29], which has been seen as a new concept for decision-making on sustainability [46]. Huber defines the policy approach on sufficiency as "a strategy of self-limitation of material consumption within the boundaries of low-level production and consumption". In contrast, efficiency is defined as "a strategy to allow further economic growth and ecological adaptation of industrial production by improving the environmental performance, i.e., improving the efficient use of material and energy, thus increasing resource productivity in addition to labour and capital productivity". This definition has been discussed as a main feature of SCP 2.0 in this paper.

In contrast to efficiency, Princen defines sufficiency as a principle which is "a sense of enoughness and too muchness" and "social restraint as the logical analog to ecological constraint" [17,18]. Boulanger also makes a contrast between sufficiency strategy and efficiency strategy for sustainable consumption, stating that sufficiency strategy "striv[es] to get the maximum wellbeing from each unit of material service consumed" as well as "minimize[s] the role of material services in the definition and production of wellbeing" [44]. Spengler emphasizes that environmental sustainability based on sufficiency is motivated by the limits of environmental capacity, by the cumulative consequences of modern consumption patterns, and by the promotion of an additional solution besides a "technical fix" [43].

In the second half of the 2010s, sufficiency discourse has been implicitly or explicitly emphasized in international policy processes including climate change and sustainable development, as discussed in the next section. This idea of sufficiency in mainstream policy discussion has emerged in line with need for a transition of lifestyles and infrastructure as summarized in Table 1.

Studies and discussions on SCP [49–54] point out that sustainable consumption patterns as well as social practices including lifestyles cannot be detached from infrastructure setting. For example, studies by Shove on the relationship between social practices and consumption emphasizes that consumption is not a sum of individual behavior but a socio-technical system of daily routines based on various infrastructures controlled by cultural, economic and technological drivers [55]. By using LCA methodology, Tukker et al. identifies provision systems associated with final consumption domains such as housing, mobility, and food as the key drivers of material and energy consumption and production [56]. Similarly, UNEP's "Sustainable Consumption and Production: A Handbook for Policy Makers" [3] clearly articulates the need to make a shift in lifestyle domain and policy focus to food, housing, and transportation. Strategic investment is also required for sustainable infrastructure. Akenji [57], as well as UNEP [3], argues that it is necessary to focus on the shift in infrastructure of energy production and provisions, housing, and the transport system or urban development for more fundamental changes.

Focusing on lifestyle and infrastructure is one way to overcome limitations and increase the efficiency of individual products and their lifecycles. At the same time, there is a growing awareness that environmental education and consumer awareness are not enough to change the behavior of consumers because of the so-called lock-in effect [58]. This means that consumer behavior is determined by infrastructure and selection of products and services. Raising awareness is not enough to change behavior—there also needs to be a change in social design. In another example, Akenji emphasizes that adoption of a sustainable lifestyle is influenced by the design of provision systems and infrastructure which predetermine the level of flexibility, and appropriate infrastructure can enable sustainable lifestyles [57].

By reviewing major SCP-related studies from 2000 to 2010, Cohen et al. points out that policy debate on SCP had expanded to include "the prospects of transitions toward sociotechnical regimes that could enable more sustainable modes of consumption" [59]. Soler et al. argues that material and institutional infrastructures, which are systems providing the operational basis for products and services, enable and support sustainable consumption by making sustainable products and services accessible and convenient while penalizing unsustainable ones [52]. The concept of one planet living by BioRegional is one example of how sustainable lifestyles are conceptualized [60]. By using the headline indicators of ecological footprint and carbon footprint, BioRegional suggests ten principles for one planet living—health and happiness, equity and local economy, culture and community, land use and wildlife, sustainable water, local and sustainable food, sustainable materials, sustainable transport, zero waste, and zero carbon [61]. In a similar way, Spengler [43] positions the sufficiency concept as a central policy concept for emerging areas of SCP such as mobility, housing, appliances, and products, emphasizing that "[s]ufficiency as policy would . . . shift the idea from being an individualistic strategy to reduce the environmental impacts of one's personal consumption pattern to the collective level, in search of policy options that support or even require such changes in consumption patterns" [62].

Thus, SCP 3.0 emphasizes the importance of socio-technical system change and innovation, as well as the transition towards sustainability. Policy debate under SCP 3.0 is characterized by ambitious policy goals which require socio-technical transition as well as a roadmap to achieve such goals.

Policy discussions under SCP 3.0 are strongly linked to increasing awareness on planetary boundaries [63,64]. The concept of planetary boundaries is also linked to our carbon footprint and ecological footprint. These indicate the indirect environmental impacts of the supply chain and trade [65]. As the rebound effect suggests [34,35,38,66], we must make changes to our socio-technical systems. Otherwise, however much effort we make on an individual/unit-level to limit material needs and unnecessary consumption, we will not be able to control our current levels of ever-increasing consumption and all the associated ecological impacts. A systematic literature review on sustainable transition by Savaget et al. reveals that literature on sustainable transition mainly deals with how to trigger socio-technical system change [67]. Kohler et al. emphasizes that future challenges for sustainability transition research include widening the scope from focusing on single systems such as energy, mobility, water, food and health to also looking at multi-sector transitions. This could be done by reconfiguring existing socio-technical systems as well as triggering deeper changes on the demand side [68].

Table 1. Three phases of SCP policy discourse and domain.

Approaches	SCP 1.0	SCP 2.0	SCP 3.0
Major concepts	Pollution prevention Cleaner Production (as an intermediate between SCP 1.0 and 2.0)	Industrial ecology Resource efficiency Product lifecycles	One planet living, Sufficiency, Decarbonization Transition
Key issues	Industrial pollution	Climate change, waste, environmental issues associated with consumption	Well-being, Life-style Socio-technical system
Environment-economy relationship	Separate, contradictory, confrontational	Compatible, industrialization harmonized with environmental conservation	Inclusion of social consideration, Sustainability as a key for next socio-technical innovation
Approaches	Installation of end of pipe technologies Technology and management for cleaner production	Increasing material and energy efficiency	Consensus building Changes in infrastructure Changes in lifestyles New business models
Major actors and stakeholders	Government vs. industry	Collaboration of government and market agents	Social entrepreneurship Multi-stakeholder Lifestyles of people
Attitude of policies	React and cure	Anticipate and prevent	Long-term goal setting, investment, creating business environment, creation and communication

Source: Authors partly adopting ecological modernization theory [5,7,27,28].

4. Emergence of SCP 3.0 in the Recent International Policy Discussions

As discussed in Section 3.3, in the late 2000s, policy discussions on SCP-related domain expanded to include the systematic transition of socio-technical systems, lifestyles, and infrastructure driving consumption and production (SCP 3.0). We will use this section to highlight how recent international policy discussions on sustainability have begun to express policy discourse similar to SCP 3.0. Progress is being made to set very ambitious long-term and middle-term goals requiring sustainable transition of modern socio-technical systems.

Major international agreements since 2015 have often set ambitious mid-term and long-term goals, including goals on decarbonization or transition to a net zero society, goals to eliminate marine plastic discharge, or goals to achieve one planet living. However, to meet these goals, we need to make fundamental changes to our society, in terms of our lifestyles and infrastructure. In the following subsections, we examine the discourse that has taken place in major international policy discussions in the context of SCP 3.0.

4.1. The Paris Agreement

The Paris Agreement (PA) was adopted in December 2015 to strengthen global efforts to tackle climate change impacts. It aims to keep "the global temperature rise this century well below 2 degrees Celsius above pre-industrial levels and to pursue efforts to limit the temperature increase even further to 1.5 degrees Celsius." Article 4 of the PA also sets out to "undertake rapid reductions thereafter in accordance with best available science, so as to achieve a balance between anthropogenic emissions by sources and removals by sinks of greenhouse gases in the second half of this century". This is interpreted as net zero emissions of GHGs by the latter half of this century and often labeled as "de-carbonization". Thus, it is argued that the long-term goal of the PA implies the necessity of de-carbonization and de facto zero emissions of GHGs towards the second half of the 21st century [69].

Although one of the limitations of the PA is the non-compulsory nature of its main approach to achieving NDCs (Nationally Determined Contributions), many countries are now setting ambitious mid-term and long-term goals at the national level.

Several European countries including the UK and France have included net zero emission goals in national legislation. The European Union, Canada, Republic of Korea, Spain and Chile put net zero emission goals in their proposed legislation. Major countries including the US, Japan, South Africa, Germany, and China included net zero emission goals by 2050 or 2060 in their official policy documents [70].

Although the scope is limited to carbon emissions, the implementation of the PA can be understood as operationalizing the shift in policy discourse to SCP 3.0 [71,72]. Indeed, Rockstrom et al. argues that the decarbonization target of the PA necessitates a rapid transition including changes in innovation, institutions, infrastructures, and investment [73]. Moreover, the 1.5-degree target requires a significant change in consumption and production patterns including the transition of lifestyles and provision systems [71,74]. The implication of such goal setting is a massive decarbonization of consumption and production systems towards a net zero carbon emission society. Even with extensive application of carbon sequestration technologies, there is no way to avoid making major transitions in energy demand and lifestyles [71,75].

4.2. G7 Toyama Framework on Material Cycles in 2016

In 2016, the G7 Environmental Ministers' Meeting in Toyama adopted the G7 Toyama Framework on Material Cycles [76]. This has one vision, three goals, and eight actions. The three goals are (1) Leading domestic policies for resource efficiency and the 3Rs, (2) Promotion of global resource efficiency and the 3Rs, and (3) Steady and transparent follow-up process. Specific actions in the framework include (1–1) Integration of policies and policy mix, (1–2) Efficient and maximized use of resources, (1–3) Initiatives in cooperation with diverse local actors (industrial and community symbiosis), (1–4) Actions to final demands/consumers, (2–1) Cooperation with other countries, (2–2) Cooperation across the global supply chain, (3–1) G7's domestic efforts, and (3–2) Global efforts. Eight years ago in 2008, G8 adopted the G8 Kobe 3R Action Plan which focuses on 3R policy promotion and environmentally sound waste management in the context of developing countries. Conceptually, 2008 Action Plan embodied SCP 2.0 by emphasizing resource productivity, efficient use of resources, environmentally sound waste management, decoupling between economic development and environmental impacts associated with material uses and waste generation. On the other hand, the 2016 Framework emphasized a more holistic policy approach for sustainable resource management, decentralized actions, attention to consumers, global value chains, and harmonized actions among the G7 member states.

The 2016 Framework emphasizes that the "common goal is to realize a society which uses resources including stock resources efficiently and sustainably across the whole lifecycle, by reducing the consumption of natural resources and promoting recycled materials and renewable resources so as to remain within the boundaries of the planet, respecting relevant concepts and approaches." The framework clearly mentions "reduction of the consumption of natural resources" to "remain within the boundaries of the planet". At the same time, in action 1–4 on actions to final demands, the framework clearly stated that G7 will "promote increased consumer awareness of the environmental and economic advantages of sustainable consumption; "awareness of sufficiency"—an idea that we should not be greedy, but be satisfied with appropriate amounts; smart purchasing; green public procurement; new services involving reuse, repair, and sharing; and eco-labeling." Although it is stated in the context of sufficiency at behavioral level, this is one case where SCP 3.0 is highlighted in high-level political documents on sustainability.

This expansion from SCP 2.0 to a more integrated approach including the concept of SCP 3.0 in the document on resource efficiency for the G8/G7 is a good indication of how SCP 3.0 is being further mainstreamed at the discursive level in major international policy processes.

4.3. Marine Plastic Issues: Ocean Plastic Charter and G20 Osaka Blue Ocean Vision

One of the most widely discussed agendas in relation to SCP is plastic pollution and marine plastic litter including those related to micro- and nano-plastics. Second only to climate change, it has been a focal issue for discourse in policy circles since the late 2000s International policy attention on marine plastics is not limited to mere pollution prevention. Plastic as a material is deeply integral to every aspect of the current global economy of production and consumption from food and beverage packaging, logistics, textiles, cosmetics, mobility (tires), and construction materials [77]. The New Plastic Economy report by the Ellen Macarthur Foundation describes the symbolic characteristic of plastics in the modern economy as follows: "Plastics have become the ubiquitous workhorse material of the modern economy—combining unrivalled functional properties with low cost" [78], and also emphasizes that "[p]lastic packaging is an iconic linear application with USD 80–120 billion annual material value loss". Thus, it is an issue deeply embedded in our daily lifestyles and in the social infrastructure of our throw-away culture. At the same time, compared to GHGs, plastic pollution is widely more visible both in terms of sources as well as ecological consequences.

Since 2015, the G7 has taken up this issue on its sustainability agenda. In 2015, the G7 Elmau Summit adopted the G7 Action Plan to Combat Marine Litter. The plan mentions the prevention of plastics entering the marine environment as a priority action but did not set any numerical targets. Again in 2016 and 2017, the G7 Ise-Shima Summit and the G7 Bologna Environmental Ministers' Meeting confirmed the importance of priority actions set in the 2015 Action Plan. Then, in June 2018, at the Charlevoix Summit held in Canada, the Ocean Plastics Charter was proposed and adopted by all G7 members, apart from Japan and the US. The preface to the charter stated that "the current approach to producing, using, managing and disposing of plastics poses a significant threat to the environment, to livelihoods and potentially to human health. It also represents a significant loss of value, resources and energy". Interestingly, despite an emphasis on working with industry, the Charter sets several numerical targets such as "100% reusable, recyclable, or, where viable alternatives do not exist, recoverable, plastics by 2030", "increasing recycled content by at least 50% in plastic products where applicable by 2030", or "recycle and reuse at least 55% of plastic packaging by 2030 and recover 100% of all plastics by 2040". However, there is not much reference to a reduction in plastic consumption. Having failed to sign the Ocean Plastics Charter, Japan took urgent action to develop a resource circulation strategy for plastics by the time of its G20 Presidency in 2019, including several numerical targets—total 25% reduction of generation of single use plastics by 2030, 60% of packaging and containers to be reused and recycled by 2030, among others. This national strategy placed reduction in the use of single-use plastics as one of the main objectives.

In June 2019, G20 leaders adopted the G20 Osaka Blue Ocean Vision. This Vision aims "to reduce additional pollution by marine plastic litter to zero by 2050". If G20 leaders are serious about this vision/goal, they need to make a fundamental socio-technical system change in their societies and economies, which are currently dependent on plastic use and consumption.

4.4. COVID-19 and Green Recovery

In this section, we argue that policy discussion on green recovery from the COVID-19 pandemic provides an unprecedented opportunity to make the transition to sustainability [79], at least at the discursive level. For example, the OECD presents green recovery from the pandemic as the greatest opportunity for "building back better" including "alignment with long-term emission reduction goals, factoring in resilience to climate impacts, slowing biodiversity loss and increasing circularity of supply chains" [80]. OECD's policy recommendations include: (1) "screening all elements of stimulus packages for their longer terms implications" for net-zero GHG emissions, strengthening climate resilience, reduction in biodiversity loss, innovation based on behavioral change, and improvement in circularity, (2) "building pipelines of "shovel-ready" sustainable infrastructure projects", and (3) maintaining and increasing ambition of long-term environmental objectives. There has also been more policy research emphasizing the importance of a green recovery plan in line with long-term climate goals [81–83]. We have seen unprecedented changes in lifestyles, business models and infrastructure during this COVID-19 pandemic, and a green recovery from this global crisis would encourage further experimental socio-technical system changes along with other long-term sustainability goals [84].

4.5. Conceptual Change in International Policy

These examples of recent major international agreements and policy discussions on sustainability show that there have been ambitious goals set on absolute reduction. These goals have been for primary material and energy consumption, or for the associated environmental impacts requiring fundamental socio-technical system change. This ambitious goal-setting based on planetary boundaries echoes the sufficiency concept, but also requires a radical shift in our socio-technical system (transition). Indeed, as Tukker et al. rightly predicted in 2008, we can see that "a rough agreement on goals exists, but where change is radical, or means are uncertain … .planning [is] difficult" [85]. Thus, policy design should contribute to foster "visioning, experimentation, and support [85]" for a sustainable transition.

5. Discussion: Emerging Challenges and Opportunities for Policy Design for SCP 3.0

Policy domain and policy design interact with each other. For example, the goals and objectives of SCP policy design under SCP 1.0 aimed at preventing environmental pollution by introducing end-of-pipe technologies, as well as altering production processes by integrating externalities. SCP 1.0 was about how to harmonize separate, contradictory, and confrontational relationships between the environment and the economy, as highlighted in the concept of CP. For SCP 2.0, sustainability challenges were widened from direct environmental pollution to include more indirect issues such as climate change and waste which were driven by globalized consumption and production. Different policy instruments have been used at different stages of product lifecycle, services and materials. Efficiency was promoted, making the environment and economy more compatible. Under SCP 3.0, the objectives and goals of SCP policy design expanded beyond environmental policy and strategy to include socio-technical system design. This was necessary to maintain the well-being of society as a whole, encouraging a transition to a sustainable lifestyle through consensus on changes for infrastructure and business. In this final section, we argue that expansion and mainstreaming of the SCP policy domain from SCP 2.0 to SCP 3.0 creates challenges and opportunities for effective policy design, with reference to the arguments in the articles in this special issue.

Firstly, although many recent international agreements and related national action plans on sustainability have set very ambitious long-term and mid-term goals, it is not clear what society will look like after it has achieved those goals—we do not know what kind of industry, infrastructure, business models and lifestyles there will be and how they will operate. Therefore, the first priority for policy design is to envision concrete images of a society that has successfully met its mid-term and long-term goals. We can use a scenario-based approach to trace pathways to the future. Börjeson et al. [86] points out that there are three scenarios: predictive, exploratory, and normative. These correspond to questions about the future: "What will happen?", "What can happen?" and "How can a specific target be reached?". A backcasting approach is typical for a normative analysis and is defined as "generating a desirable future, and then looking backwards from that future to the present in order to strategize and to plan how it could be achieved" [87]. Exploratory analysis refers to what can happen. Policy design for SCP 3.0 requires both perspectives. We must ask what can happen to a society once it has achieved specific targets, and attempt to create a vision of such a society. One of the most famous attempts is the shared socioeconomic pathways (SSPS). This scenario was the successor of SRES that was used for the IPCC 4th assessment report [88]. SSPS will be applied for socio-economic scenario development in the IPCC 6th assessment report [89–91]. The Paris Agreement prompted global discussions on how to achieve net zero carbon emissions and how to change socio-technical systems, including discussions by IEA [92] and UK Climate Change Committee [93]. For example, Kawakami et al. analyzes what a decarbonized society would look like in Japan, under a conventional incremental reform scenario and under a more decentralized transition scenario [94]. In terms of a vision for policy design, Mont et al. developed different extreme scenarios for sustainable lifestyles in 2050 [95]. To facilitate policy dialogue and design, Kishita et al. attempts to combine a backcasting approach with a workshop-style dialogue to create a vision on SCP for developing Asia in 2050 [96]. In this special issue, Mao et al. apply a foresight study approach when designing policy for a long-term transition to a sustainable lifestyle [97]. However, it is still not clear how effective this type of envisioning or scenario approach really is, and more insights are needed. In the era of SCP 3.0, envisioning a future society and lifestyle will be a crucial part of policy design.

Secondly, it is not yet clear how we can turn these visions into reality and how policy can support, promote, and guide us in that endeavor. What is clear is that policymakers need to mobilize investment, they need to create enabling conditions by making changes to regulations and incentive structure, as well as awarding and informing best practices. There also needs to be policy support for learning from model cases, projects, and businesses to achieve a long-term and mid-term vision. In this special issue, Watabe and Gilby attempt to show how local stakeholders can continue to work on model projects through capacity development and learning processes [98]. By examining local context for implementing SDGs, Liu and Nguyen attempted to provide policy recommendation to reduce food waste along the entire supply chain and promote sufficiency strategies for saving food, reducing food waste, and maintaining health and well-being towards SDGs Target 12.3, based on Hanoi's case study [99]. Additionally, for policy-driven sustainability transition, Kohdke et al. emphasizes importance of accounting for the dynamic positioning of stakeholder involved in collaborative efforts for transition [100]. Better understanding of dynamism among stakeholders can deepen the understanding of the challenges of implementation, particularly adaption of timelines for implementation based on changing capacity and needs of stakeholders. More studies should be done to design policy that can turn a vision into reality.

Thirdly, it is important to evaluate whether a realized vision is truly beneficial for the environment and for sustainability. This requires both policy to collect evidence for realizing a vision and then evidence-based policy making (EBPM). EBPM often refers to decision-making on policy that takes account of "multiple sources of information, including statistics, data and the best available research evidence and evaluations" [101]. Experts and stakeholders can inform decision-making by carefully analyzing and demonstrating what the actual impacts would be for a variety of options for a future vision. There is one major criticism of EBPM and that is its over-reliance on rationality [102]. Kano and Hayashi [103] present a wider understanding of EBPM with five perspectives: (1) methodological rigorousness, (2) consistency, (3) proximity, (4) social appropriateness, and (5) legitimacy. However, future-oriented policymaking in the era of SCP 3.0 is a creative process rather than an evaluation of different policy options. Thus, we need to discuss whether or not EBPM in combination with creating and sharing a vision based on these mid- to long-term goals is different from standard EBPM.

Fourthly, when we look to implement and monitor actions to achieve mid-to-long term goals such as decarbonization, a plastic-free society, green recovery from COVID 19, and the SDGs, we must take care not to evaluate sustainability just from economic and environmental points of view. For example, the combined crises of the COVID-19 pandemic and the Paris Agreement can be used to drive policy interests and expectations further on prospects and innovation towards decarbonization, digitalization, and transition to sustainable lifestyles and infrastructure. Additionally, these two combined crises prompted us to reemphasize that equity, safety, welfare, health and education as well as associated public services are fundamental issues for social sustainability. Education has been reconfirmed as being key for social integrity and individual well-being, with consumption to enrich social capital also being important. In this respect, Tsurumi et al. [104] attempts a detailed analysis on the relationship between attitude, consumption behavior, social capital and subjective well-being of people through a case study of rural habitats in Viet Nam. In addition, a study by Piao et al. examines the relationship between a sustainable lifestyle and the essential needs of citizens such as education and health [105]. There are likely to be widespread and fundamental social impacts from a sustainable transition, and policy research on SCP needs to examine the various ways that sustainable lifestyles and associated infrastructure would function.

We tentatively coin the policy design approach responding to the above four challenges of SCP 3.0 as "Envisioning-based Policy Making (EnBPM)" by incorporating the importance of envisioning approach related to SCP 3.0 beyond EBPM as discussed above. Figure 1 illustrates our idea of EnBPM in contrast to EBPM. The coverage of EBPM tends to focus more on direct policy targets, policy implementation and its effectiveness so as to provide a strong evidence with attention to rigorousness and consistency [103]. EnBPM covers wider policy concerns such as future visions of sustainable society, social experimentation of such societal visions before full policy implementation based on long-term goals as well as social sustainability. EnBPM would require a more decentralized and collaborative approach for policy design based on working together to envision and realize future directions of the society among stakeholders because it puts importance on social appropriateness about sustainability rather than EBPM. A vision is created based on prototypes (community and business models) with decentralized coordination among various initiatives. Communication tools and decision support tools that can share the direction of transition would be important as a catalyst for self-sustained, decentralized efforts in various sectors. Again, we observed the importance of framing, interpretation, and generating storylines not only to provide narratives for the past experiences but also to provide future-oriented storylines with research-based evidence.

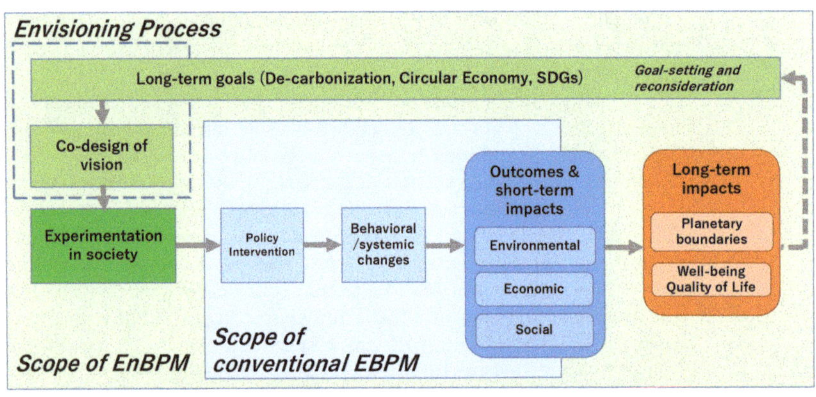

Figure 1. EBPM and EnBPM towards sustainability. Source: Modified and developed from Figure 1 of [106] by authors.

6. Conclusions

This paper argued that policy discussion on SCP is a major expression of the sustainability agenda, and that this discourse has expanded from pollution prevention (SCP 1.0) to lifecycles (SCP 2.0), then to transition (SCP 3.0). For policy discussions on sustainability, it is very important to better understand how SCP is discussed in the different contexts of SCP 1.0, SCP 2.0 and SCP 3.0. Each country and communitytable has different policy priorities [106], and consequently, discussions tend to focus on one of the three SCP phases—SCP1.0, 2.0 and 3.0—or as a combination of these three versions. We could in fact have a hybrid concept, whereby three different but related contexts of environmental policy development endeavor to achieve SCP in a phased manner from SCP 1.0 to SCP 3.0. This could be especially pertinent for emerging economies.

One of the major features of recent international policy agreements is that they have set ambitious long and mid-term goals based on the concept of planetary boundaries. This resulted in an expansion of policy domain from environmental sustainability to socio-technical systems change (SCP 3.0). These changes in emphasis in policy goals and approaches also resulted in the need for a fresh approach for SCP policy design.

The challenge facing SCP policy design in this context is the development of model cases, and a social image or vision. Emerging opportunities and challenges for policy design for SCP 3.0 include: (1) envisioning concrete images of a society that has successfully met its mid-term and long-term goals, (2) policy support for learning from model cases, experimental projects, and new businesses to achieve a long-term and mid-term vision, (3) facilitating creative process among stakeholders, and (4) examination of social implications of innovation towards decarbonization, digitalization, and transition to sustainable lifestyles and infrastructure. To deepen the debate on SCP policies required in the future, it is necessary to change our way of thinking on environmental policies. As a policy design approach in response to these challenges and opportunities, the paper proposed the concept of Envisioning-based Policy Making (EnBPM). In the era of SCP 3.0, policy design and scientific research on SCP can provide rich opportunities and challenges to bring together creative visions, future scenarios, social experimentation, stakeholder engagement, urban and spatial planning, new indicator development, lifestyles and social sustainability, and new business model development. Conventional regulations and economic tools must work to introduce a new approach and innovation into lifestyles and infrastructure [107]. We need to develop a social business model and promote public and private investment to facilitate model cases which can enhance storylines for sustainable transition. We must come up with new business or social model development and social designs within planetary boundaries, and incorporate them into the central agenda of SCP policy. Communication and decision-support tools will also play an important role to promote stakeholder collaboration and dialogue. In that sense, there are rapidly emerging opportunities for collaboration between policy design and scientific research on SCP for envisioning and developing model cases which can generate compelling storylines for SCP 3.0.

Author Contributions: Formal analysis, Investigation, Supervision, Writing—original draft, Y.H.; Conceptualization, Y.H., T.T., and R.K.; Writing—review & editing, T.T. and R.K. All authors have read and agreed to the published version of the manuscript.

Funding: This paper was developed based on the research funded by the Environment Research and Technology Development Fund (S-16-3: JPMEERF16S11630, S-16-2: JPMEERF16S11620) of the Environmental Restoration and Conservation Agency, Japan.

Acknowledgments: The authors would like to express their appreciation to Emma Fushimi of IGES for her careful editing and proof reading of the entire manuscript as well as to Chen Liu of IGES for her review and insightful comments to improve the earlier draft.

Conflicts of Interest: The authors declare no conflict of interest.

References

1. Conca, K. Consumption and Environment in a Global Economy. *Glob Environ. Polit.* **2001**, *1*, 53–71. [CrossRef]
2. United Nations. Agenda 21: The United Nations Programme of Action from Rio. In Proceedings of the United Nations Conference on Environment & Development, Rio de Janeiro, Brazil, 3–14 June 1992.
3. United Nations Environment Programme (UNEP). *Sustainable Consumption and Production: A Handbook for Policy Makers*; Global Edition; United Nations Environment Programme: Nairobi, Kenya, 2015.
4. Fairhead, J.; Leach, M. *Science, Society and Power: Environmental Knowledge and Policy in West Africa and the Caribbean*; Cambridge University Press: Cambridge, UK, 2003.

5. Hajer, M.A. *The Politics of Environmental Discourse*; Oxford University Press: Oxford, UK, 1997; Available online: http://www.oxfordscholarship.com/oso/public/content/politicalscience/019829333X/toc.html (accessed on 25 March 2021).

6. Litfin, K. *Ozone Discourse: Science and Politics in Global Environmental Cooperation*; Basic Books: New York, NY, USA, 1994.

7. Dryzek, J. *The Politics of the Earth: Environmental Discourses*; Oxford University Press: Oxford, UK, 1997.

8. Hawkins, A. Contested Ground: International Environmentalism and Global Climate Change. In *State of Social Power in Global Environmental Politics*; Lipschutz, R.D., Conca, K., Eds.; Columbia University Press: New York, NY, USA, 1993.

9. Lipshcutz, R.D.; Conca, K. *The State and Social Power in Global Environmental Politics*; Columbia University Press: New York, NY, USA, 1993.

10. Ganz, M. Leading Change—Leadership, Organization and Social Movements. In *Handbook of Leadership, Theory and Practice, an HBS Centennial Colloquim on Advancing Leadership*; Harvard Business Press: Boston, MA, USA, 2010.

11. Jänicke, M.; Weidner, H. *Successful Environmental Policy—A Critical Evaluation of 24 Cases*; Sigma: Berlin, Germany, 1995.

12. Jänicke, M.; Mönch, H.; Binder, M. Structural Change and Environmental Policy. In *The Emergence of Ecological Modernisation*; Routledge: London, UK, 2000; p. 133. Available online: https://www.taylorfrancis.com/chapters/structural-change-environmental-policy-martin-jänicke-harald-mönch-manfred-binder/e/10.4324/9781315812540-6 (accessed on 25 March 2021).

13. Mol, A.P.J.; Sonnenfeld, D.A. *Ecological Modernisation Around the World: Perspectives and Critical Debates*; Frank Cass: Portland, OR, USA; Routledge: London, UK, 2000.

14. Mitsuhashi, T. *Kankyo Keizai Nyumon (Introduction to Environmental Economics)*; Nikkei: Tokyo, Japan, 1999.

15. Nakanishi, J. *Mizu no Kankyo Senryaku (Global Water Strategy)*; Iwanami: Tokyo, Japan, 1994.

16. United Nations Environment Programme (UNEP). Understanding Resource Efficient and Cleaner Production. Available online: http://www.unep.fr/scp/cp/understanding/ (accessed on 5 March 2021).

17. United Nations Environment Programme (UNEP). Government Strategies and Policies for Cleaner Production. Available online: https://wedocs.unep.org/handle/20.500.11822/29358 (accessed on 5 March 2021).

18. United Nations Environment Programme (UNEP). Promoting Cleaner Production Investments in Developing Countries. Available online: https://www.unep.org/resources/report/dtie-activities-and-performance (accessed on 5 March 2021).

19. United Nations Industrial Development Organization (UNIDO). Manual on the Development of Cleaner Production Policies—Approaches and Instruments. Available online: https://www.unido.org/sites/default/files/2007-11/9750_0256406e_0.pdf (accessed on 5 March 2021).

20. United Nations Environment Programme. *Global Outlook on SCP Policies*; United Nations Environment Programme: Nairobi, Kenya, 2011.

21. Vanham, D.; Leip, A.; Galli, A.; Kastner, T.; Bruckner, M.; Uwizeye, A. Environmental footprint family to address local to planetary sustainability and deliver on the SDGs. *Sci. Total Environ.* **2019**, *693*, 133642. [CrossRef] [PubMed]

22. Wiedmann, T.; Lenzen, M. Environmental and Social Footprints of International Trade. 2018. Available online: https://www.nature.com/articles/s41561-018-0113-9 (accessed on 1 April 2021).

23. Rudolph, A.; Figge, L. Determinants of Ecological Footprints: What is the role of globalization? *Ecol Indic.* **2017**, *81*, 348–361. [CrossRef]

24. Christoff, P. Ecological modernisation, ecological modernities. *Env Polit.* **1996**, *5*, 476–500. [CrossRef]

25. Huber, J. Towards industrial ecology: Sustainable development as a concept of ecological modernization. *J. Environ. Policy Plan.* **2000**, *2*, 269–285. [CrossRef]

26. Mol, A.P.J. Ecological modernisation and institutional reflexivity: Environmental reform in the late modern age. *Environ. Polit.* **1996**, *5*, 302. [CrossRef]

27. Weale, A. *The New Politics of Pollution*; Manchester University Press: Manchester, UK, 1992.

28. Jänicke, M.; Weidner, H. *Successful Environmental Policy: An Introduction in Successful Environmental Policy*; Edition Sigma: Berlin, Germany, 1995.

29. Von Weizsacker, E.; Lovins, A.B.; Lovins, L.H. *Factor Four: Doubling Wealth-Halving Resource Use*; Earthscan: London, UK, 1997.

30. Bengtsson, M.; Hotta, Y.; Hayashi, S.; Akenji, L. *Policy Tools for Sustainable Materials Management: Applications in Asia*; Institute for Global Environmental Strategies: Kanagawa, Japan, 2010; pp. 1–37.

31. Aoki-Suzuki, C. Examining Future Implementation of Waste Prevention and Resource Reduction Policies in Asia and the Pacific—Referring Practices in European Countries, Hayama, Japan. Available online: https://www.iges.or.jp/jp/pub/examining-future-implementation-prevention-and/en (accessed on 25 March 2021).

32. Government of Japan. *Basic Act for Establishing a Sound Material Cycle Society*; Government of Japan: Tokyo, Japan, 2000.

33. Steinberger, J.K.; Krausmann, F. Material and Energy Productivity. *Environ Sci Technol.* **2011**, *45*, 1169–1176. [CrossRef]

34. Hertwich, E.G. Consumption and the rebound effect: An industrial ecology perspective. *J. Ind. Ecol.* **2005**, *9*, 85. [CrossRef]

35. Maxwell, D.; Owen, P.; McAndrew, L.; Muehmel, K.; Neubauer, A. Addressing the Rebound Effect, a Report for the European Commission DG Environment. Available online: http://ec.europa.eu/environment/archives/eussd/pdf/rebound_effect_report.pdf (accessed on 25 March 2021).

36. Polimeni, J.M.; Mayumi, K.; Giampietro MAlcott, B. *The Myth of Resource Efficiency: The Jevons Paradox*; Earthscan: London, UK, 2008.

37. Schettkat, R. Analyzing Revound Effects. Wuppertal Papaers. Available online: https://epub.wupperinst.org/frontdoor/index/index/docId/3202 (accessed on 25 March 2021).

38. Herring, H.; Sorrell, S. *Energy Efficiency and Sustainable Consumption: The Rebound Effect*; Palgrave Macmillan: Basingstoke, UK, 2009.
39. Akenji, L.; Bengtsson, M.; Bleischwitz, R.; Tukker, A.; Schandl, H. Ossified materialism: Introduction to the special volume on absolute reductions in materials throughput and emissions. *J. Clean Prod.* **2016**, *132*, 1–12. [CrossRef]
40. Lettenmeier, M.; Liedtke, C.; Rohn, H. Eight Tons of Material Footprint—Suggestion for a Resource Cap for Household Consumption in Finland. *Resources* **2014**, *3*, 488–515. [CrossRef]
41. Bringezu, S. Possible target corridor for sustainable use of global material resources. *Resources* **2015**, *4*, 25–54. [CrossRef]
42. Kalmykova, Y.; Rosado, L.; Patrício, J. Resource consumption drivers and pathways to reduction: Economy, policy and lifestyle impact on material flows at the national and urban scale. *J. Clean. Prod.* **2016**, *132*, 70–80. [CrossRef]
43. Spengler, L. Sufficiency as Policy. In *Sufficiency as Policy—Necessity, Possibilities and Limitations*; Nomos Verlagsgesellschaft: Baden-Baden, Germany, 2018.
44. Boulanger, P.M. Three strategies for sustainable consumption. *Sapiens* **2010**, *3*. Available online: http://journals.openedition.org/sapiens/1022 (accessed on 1 April 2021).
45. Shove, E.; Spurling, N. *Sustainable Practices: Social Theory and Climate Change*, 1st ed.; Routledge: London, UK, 2013.
46. Lamberton, G. Sustainable sufficiency—An internally consistent version of sustainability. *Sustain. Dev.* **2005**, *13*, 53–68. [CrossRef]
47. Princen, T. Principles for Sustainability: From Cooperation and Efficiency to Sufficiency. *Glob. Environ. Polit.* **2003**, *3*, 33–50. [CrossRef]
48. Princen, T. *The Logic of Sufficiency*; The MIT Press: Cambridge, MA, USA, 2005.
49. Geels, F.W.; McMeekin, A.; Mylan, J.; Southerton, D. A critical appraisal of Sustainable Consumption and Production research: The reformist, revolutionary and reconfiguration positions. *Glob. Environ. Chang.* **2015**, *34*, 1–12. [CrossRef]
50. Dong, L.; Wang, Y.; Scipioni, A.; Park, H.S.; Ren, J. Recent progress on innovative urban infrastructures system towards sustainable resource management. *Resour. Conserv. Recycl.* **2018**, *128*, 355–359. [CrossRef]
51. Spaargaren, G.; Oosterveer, P. Citizen-consumers as agents of change in globalizing modernity: The case of sustainable consumption. *Sustainability* **2010**, *2*, 1887–1908. [CrossRef]
52. Solér, C.; Koroschetz, B.; Salminen, E. An infrastructural perspective on sustainable consumption—Activating and obligating sustainable consumption through infrastructures. *J. Clean. Prod.* **2020**, *243*, 118601. [CrossRef]
53. Kennedy, E.H.; Cohen, M.J.; Krogman, N.T. Social Practice Theories and Research on Sustainable Consumption. In *Putting Sustainability into Practice*; Kennedy, E.H., Choen, M.J., Krogman, N.T., Eds.; Edward Elger: Cheltenham, UK, 2015.
54. Shove, E.; Walker, G. Governing transitions in the sustainability of everyday life. *Res. Policy* **2010**, *39*, 471–476. [CrossRef]
55. Shove, E. *Comfort, Cleanliness and Convenience: The Social Organization of Normality*; Berg Publishers: Oxford, UK, 2003.
56. Tukker, A.; Jansen, B. Environmental Impact of Products. *J. Ind. Ecol.* **2006**, *10*, 159–182. [CrossRef]
57. Akenji, L. Consumer scapegoatism and limits to green consumerism. *J. Clean. Prod.* **2014**, *63*, 13–23. [CrossRef]
58. Sanne, C. Willing consumers—Or locked-in? Policies for a sustainable consumption. *Ecol. Econ.* **2002**, *42*, 273–287. [CrossRef]
59. Cohen, M.J.; Brown, H.S.; Vergragt, P.J. Individual consumption and systemic societal transformation: Introduction to the special issue. *Sustain. Sci. Pract. Policy* **2010**, *6*, 6–12. [CrossRef]
60. Francis, A.; Wheeler, J. *BioRegional and World Wide Fund for Nature: One Planet Living*; World Wildlife Fund (WWF): Gland, Switzerland, 2004.
61. Bioregional. Implementing One Planet Living: A manual. Available online: https://www.oneplanetnetwork.org/sites/default/files/implementing_one_planet_living_a_manual.pdf (accessed on 25 March 2021).
62. Spengler, L. Two types of 'enough': Sufficiency as minimum and maximum. *Environ. Polit.* **2016**. [CrossRef]
63. Steffen, W.; Richardson, K.; Rockstrom, J.; Cornell, S.E.; Fetzer, I.; Bennett, E.M. Planetary boundaries: Guiding human development on a changing planet. *Science* **2015**, *347*, 1259855. [CrossRef]
64. Raworth, K. *Doughnut Economics: Seven Ways to Think Like a 21st-Century Economist*; Random House: New York, NY, USA, 2017.
65. O'Neill, D.W.; Fanning, A.L.; Lamb, W.F.; Steinberger, J.K. A good life for all within planetary boundaries. *Nat. Sustain.* **2018**, *1*, 88–95. [CrossRef]
66. Sorrell, S.; Dimitropoulos, J. The rebound effect: Microeconomic definitions, limitations and extensions. *Ecol. Econ.* **2008**, *65*, 636–649. [CrossRef]
67. Savaget, P.; Geissdoerfer, M.; Kharrazi, A.; Evans, S. The theoretical foundations of sociotechnical systems change for sustainability: A systematic literature review. *J. Clean. Prod.* **2019**, *206*, 878–892. [CrossRef]
68. Köhler, J.; Geels, F.W.; Kern, F.; Markard, J.; Onsongo, E.; Wieczorek, A. An agenda for sustainability transitions research: State of the art and future directions. *Env. Innov. Soc. Transit.* **2019**, *31*, 1–32. [CrossRef]
69. United Nations. Framework Convention on Climate Change—Paris Agreement. In Proceedings of the 21st Conference of the Parties, Paris, France, 30 November–11 December 2015; p. 3.
70. Energy & Climate Intelligence Unit. Net Zero Emissions Race. 2020. Available online: https://eciu.net/netzerotracker (accessed on 25 March 2021).
71. Institute for Global Enrvironmental Strategies; Aalto University; D-mat Ltd. *1.5-Degree Lifestyles: Targets and Options for Reducing Lifestyle Carbon Footprints*; Technical Report; Institute for Global Environmental Strategies: Hayama, Japan, 2019; Available online: https://www.aalto.fi/sites/g/files/flghsv161/files/2019-02/15_degree_lifestyles_mainreport.pdf (accessed on 25 March 2021).

72. Akenji, L.; Chen, H. A Framework for Shaping Sustainable Lifestyles Determinants and Strategies—II Acknowledgements. Available online: https://www.iges.or.jp/jp/pub/framework-shaping-sustainable-lifestyles/en (accessed on 25 March 2021).

73. Rockström, J.; Gaffney, O.; Rogelj, J.; Meinshausen, M.; Nakicenovic, N.; Schellnhuber, H.J. A roadmap for rapid decarbonization. *Science* **2017**, *355*, 1269–1271. [CrossRef]

74. Koide, R.; Lettenmeier, M.; Kojima, S.; Toivio, V.; Amellina, A.; Akenji, L. Carbon footprints and consumer lifestyles: An analysis of lifestyle factors and gap analysis by consumer segment in Japan. *Sustainability* **2019**, *11*, 5983. [CrossRef]

75. Paul, D. SDG 12 Review at HLPF Calls for Circular Economies, Sustainable Lifestyles. 2018. Available online: http://sdg.iisd.org/news/sdg-12-review-at-hlpf-calls-for-circular-economies-sustainable-lifestyles/ (accessed on 25 March 2021).

76. Ministry of the Environment Government of Japan. Communique—G7 Toyama Environment Ministers' Meeting. Ministry of the Environment Government of Japan, 2016. Available online: https://www.env.go.jp/earth/g7toyama_emm/english/_img/meeting_overview/Communique_en.pdf (accessed on 25 March 2021).

77. United Nations Environment Programme (UNEP); Technical University of Denmark (DTU). *Mapping of Global Plastics Value Chain and Plastics Losses to the Environment (with a Particular Focus on Marine Environment)*; Technical University of Denmark (DTU): Copenhagen, Denmark, 2018; pp. 1–99.

78. Ellen MacArthur Foundation. The New Plastics Economy: Rethinking the Future of Plastics. 2016. Available online: https://www.ellenmacarthurfoundation.org/publications/the-new-plastics-economy-rethinking-the-future-of-plastics (accessed on 25 March 2021).

79. Cohen, M.J. Does the COVID-19 outbreak mark the onset of a sustainable consumption transition? *Sustain. Sci. Pract. Policy* **2020**, *16*, 1–3. [CrossRef]

80. Organisation for Economic Co-operation and Development (OECD). Building Back Better: A Sustainable, Resilient Recovery after COVID. 2020. Available online: https://www.oecd.org/coronavirus/policy-responses/building-back-better-a-sustainable-resilient-recovery-after-covid-19-52b869f5/ (accessed on 25 March 2021).

81. Stigl, S. *Opportunities of Post COVID-19 European Recovery Funds In Transitioning Towards a Circular and Climate Neutral Economy*; European Parliament: Brussels, Belgium, 2020.

82. University of Cambridge Institute for Sustainability Leadership (CISL). *Maximising the Benefits: Economic, Employment and Emissions Impacts of a Green Recovery Plan in Europe*; Technical Report; CISL: Cambridge, UK, 2020.

83. C40 Knowledge Community. Technical Report: The Case for a Green and Just Recovery. Available online: https://www.c40knowledgehub.org/s/article/The-Case-for-a-Green-and-Just-Recovery?language=en_US (accessed on 25 March 2021).

84. Bodenheimer, M.; Leidenberger, J. COVID-19 as a window of opportunity for sustainability transitions? Narratives and communication strategies beyond the pandemic. *Sustain. Sci. Pract. Policy* **2020**, *16*, 61–66.

85. Tukker, A.; Emmert, S.; Charter, M.; Vezzoli, C.; Sto, E.; Munch Andersen, M.; Geerken, T.; Tischner, U.; Lahlou, S. Fostering change to sustainable consumption and production: An evidence based view. *J. Clean. Prod.* **2008**, *16*, 1218–1225. [CrossRef]

86. Börjeson, L.; Höjer, M.; Dreborg, K.H.; Ekvall, T.; Finnveden, G. Scenario types and techniques: Towards a user's guide. *Futures* **2006**, *38*, 723–739. [CrossRef]

87. Vergragt, P.J.; Quist, J. Backcasting for sustainability: Introduction to the special issue. *Technol. Forecast. Soc. Chang.* **2011**, *78*, 747–755. [CrossRef]

88. Intergovernmental Panel on Climate Change (IPCC). Special Report Emissions Scenarios Special Report Working Group III Intergovernmental Panel Climate Change. 2000. Available online: https://www.cambridge.org/jp/academic/subjects/earth-and-environmental-science/climatology-and-climate-change/special-report-emissions-scenarios-special-report-working-group-iii-intergovernmental-panel-climate-change?format=PB&isbn=9780521804936 (accessed on 25 March 2021).

89. Riahi, K.; van Vuuren, D.P.; Kriegler, E.; Edmonds, J.; O'Neill, B.C.; Fujimori, S.; Bauer, N.; Calvin, K.; Dellink, R.; Fricko, O.; et al. The Shared Socioeconomic Pathways and their energy, land use, and greenhouse gas emissions implications: An overview. *Glob. Environ. Chang.* **2017**, *42*, 153–168. [CrossRef]

90. Van Vuuren, D.P.; Riahi, K.; Moss, R.; Edmonds, J.; Thomson, A.; Nakicenovic, N. A proposal for a new scenario framework to support research and assessment in different climate research communities. *Glob. Environ. Chang.* **2012**, *22*, 21–35. [CrossRef]

91. O'Neill, B.C.; Kriegler, E.; Riahi, K.; Ebi, K.L.; Hallegatte, S.; Carter, T.R. A new scenario framework for climate change research: The concept of shared socioeconomic pathways. *Clim. Chang.* **2014**, *122*, 387–400. [CrossRef]

92. International Energy Agency (IEA). World Energy Outlook. 2020. Available online: https://www.iea.org/reports/world-energy-outlook-2020 (accessed on 1 April 2021).

93. UK Committee on Climate Change. Net Zero Technical Report. 2019. Available online: https://www.theccc.org.uk/publication/net-zero-technical-report/ (accessed on 1 April 2021).

94. Kawakami, T.; Kuriyama, A.; Arino, Y. *A Net-Zero World—2050 Japan—Insight into Essential Changes for a Sustainable Future: Hayama*; Institute for Global Environmental Strategies: Kanagawa, Japan, 2020.

95. Mont, O.; Neuvonen, A.; Lähteenoja, S. Sustainable lifestyles 2050: Stakeholder visions, emerging practices and future research. *J. Clean. Prod.* **2014**, *63*, 24–32. [CrossRef]

96. Kishita, Y.; Kuroyama, S.; Matsumoto, M.; Kojima, M.; Umeda, Y. Designing Future Visions of Sustainable Consumption and Production in Southeast Asia. *Procedia CIRP* **2018**, *69*, 66–71. [CrossRef]

97. Mao, C.; Koide, R.; Akenji, L. Applying foresight to policy design for a long-term transition to sustainable lifestyles. *Sustainability* **2020**, *12*, 6200. [CrossRef]

98. Watabe, A.; Gilby, S. To see a world in a grain of sand-the transformative potential of small community actions. *Sustainability* **2020**, *12*, 7404. [CrossRef]
99. Liu, C.; Nguyen, T.T. Evaluation of Household Food Waste Generation in Hanoi and Policy Implications towards SDGs Target 12. *Sustainability* **2020**, *12*, 6565. Available online: https://www.mdpi.com/2071-1050/12/16/6565 (accessed on 15 April 2021).
100. Khodke, A.; Watabe, A.; Mehdi, N. Implementation of Accelerated Policy-Driven Sustainability Transitions: Case of Bharat Stage 4 to 6 Leapfrogs in India. *Sustainability* **2021**, *13*, 4339. Available online: https://www.mdpi.com/2071-1050/13/8/4339 (accessed on 15 April 2021). [CrossRef]
101. Organisation for Economic Co-operation and Development (OECD). Building Capacity for Evidence-Informed Policy-Making Lessons from Country Experiences. Available online: https://www.oecd-ilibrary.org/governance/building-capacity-for-evidence-informed-policy-making_86331250-en (accessed on 1 April 2021).
102. Cairney, P. *The Politics of Evidence-Based Policy Making*; Palgrave Macmillan: London, UK, 2016.
103. Kano, H.; Hayashi, T.I. A framework for implementing evidence in policymaking: Perspectives and phases of evidence evaluation in the science-policy interaction. *Environ. Sci. Policy* **2021**, *116*, 86–95. [CrossRef]
104. Tsurumi, T.; Yamaguchi, R.; Kagohashi, K.; Managi, S. Attachment to material goods and subjective well-being: Evidence from life satisfaction in rural areas in Vietnam. *Sustainability* **2020**, *12*, 9913. [CrossRef]
105. Piao, X.; Ma, X.; Zhang, C.; Managi, S. Impact of Gaps in the Educational Levels between Married Partners on Health and a Sustainable Lifestyle: Evidence from 32 Countries. *Sustainability* **2021**, *12*, 4623. [CrossRef]
106. Wang, C.; Ghadimi, P.; Lim, M.K.; Tseng, M.L. A literature review of sustainable consumption and production: A comparative analysis in developed and developing economies. *J. Clean. Prod.* **2019**, *206*, 741–754. [CrossRef]
107. Koide, R.; Hotta, Y.; Watabe, A. EBPM towards Lifestyle Innovation. *Rev. Environ. Econ. Policy Stud.* **2020**, *13*, 70–73.

 *sustainability* 

Article

Applying Foresight to Policy Design for a Long-Term Transition to Sustainable Lifestyles

Caixia Mao [1,2,*], Ryu Koide [1,3] and Lewis Akenji [1,4]

[1]   Sustainable Consumption and Production Area, Institute for Global Environmental Strategies, 2108-11 Kamiyamaguchi, Hayama, Kanagawa 240-0115, Japan
[2]   Institute of International Peace Strategy, Waseda University, 1-6-1, Nishi Waseda, Shinjyuku-ku, Tokyo 169-8050, Japan
[3]   Center for Material Cycles and Waste Management Research, National Institute for Environmental Studies, 16-2 Onogawa, Tsukuba, Ibaraki 305-8506, Japan; koide.ryu@nies.go.jp
[4]   Hot or Cool Institute, Manteuffelstrasse 47, 12103 Berlin, Germany; la@hotorcool.org
*   Correspondence: mao@iges.or.jp

Received: 6 June 2020; Accepted: 29 July 2020; Published: 31 July 2020

Abstract: Increasing attention is being paid to lifestyles in sustainability research and policymaking. Applying a foresight approach to sustainable lifestyles supports this increased focus by highlighting possible futures while also empowering citizens through a participatory process. Foresight has its origins in theory and practice to serve the policymaking process by involving diverse stakeholders. In the search to empower various stakeholders in the decision-making process on foresight, this paper analyses the results of a global expert survey to identify factors shaping future lifestyles. Survey results show that in consumption, the reasoning behind increased or reduced consumption matters; in infrastructure, affordability and equal accessibility is a concern; there are some uncertain implications of the changes in work and education, and physical and mental health, which need further exploration in the desired direction. Those factors should be included in public discussions on future sustainable lifestyles through adopting sustainable lifestyles as a foresight topic. Additionally, the survey results on stakeholders' changing roles between now and 2050 illustrate how foresight could empower stakeholders through a bottom-up policymaking approach to realise a long term-transition to sustainable lifestyles.

Keywords: policymaking; multi-stakeholder participation; long-term transition; empowerment; sustainable lifestyles

---

1. Introduction

Scholars and policymakers are increasingly paying attention to current ways of living and their impact on sustainability. The Paris Agreement stated the goal for society to transit towards net zero-carbon by the second half of this century, to hold the global average temperature increase well below 2 °C and pursue efforts to limit it to 1.5 °C above the pre-industrial levels [1]. Within this context, the Intergovernmental Panel on Climate Change's (IPCC's) special report on the impacts of global warming of 1.5 °C recognises the changes in human behaviour and lifestyles as enabling conditions for the 1.5 °C consistent systems transitions' [2]. In Japan's Long-term Strategy under the Paris Agreement, it emphasises "Lifestyles Innovation" for the shift of people's daily living towards long term sustainability [3]. Lifestyle, as an instrument in sustainability, was also included in the Sustainable Development Goals, specifically Target 12.8, which specifies the significance for people to have relevant information and awareness on sustainable lifestyles [4]. Sustainable lifestyles can thus be seen both as an approach to achieve long-term sustainability and as a policy objective to ensure the delivery of high quality of life to citizens.

In sustainability research, future-oriented thinking is critical due to the scale of change needed: Systemic change instead of minor modifications [5–7]. Among the different sets of approaches in future studies, foresight emerges as an inclusive policy formation process in both theory and practice, to invite non-government stakeholders such as industry and expert groups to participate in giving shape to futures [8]. It is an exercise to create an improved understanding of possible future developments and forces that would be shaped by different stakeholders, in order to search for alternative and desirable futures [9]. Foresight contributes to the decision-making process in three ways: Openness to different futures; involvement of various stakeholders; and policy orientation [10]. Thus, in futures research, compared to the predictive forecasting approach, which focuses on the 'most likely' futures, foresight emphasises a better understanding of futures by looking for the forces that shape them; it is an action-oriented approach to shaping futures in the policymaking process. Considering those features, the foresight process has an instrumental value to improve our understanding of futures for strategy formation, and also an intrinsic value as a democratic participation process to invite more stakeholders such as the general public into the policymaking process towards the desired direction [8].

In terms of government policymaking, foresight's focused topics have changed over time in line with the government's shifting societal concerns [11]. Additionally, future-oriented thinking has long been applied to the environmental field, most notably in the 1972 publication of 'Limits to Growth' and the 1988 establishment of the IPCC [11]. Recent foresight research has extended its search for long-term sustainability into transition studies, adopting more holistic thinking to include social, environmental, and economic aspects [12–14]. Due to the involvement of stakeholders with multiple knowledge fields and organisational background in the process, foresight serves as an interdisciplinary approach to generate knowledge for policy support and crosses administrative boundaries of policymaking areas and ministries [15]. Including the general public in developing a shared desirable future is also pivotal for actors to take actions to move toward a shared vision [12]. However, foresight's potential remains untapped in terms of forming alternative, progressive visions toward long-term sustainability. 'Sustainable lifestyles' as a concept can engage the general public in the policymaking process, allowing individuals to contribute to how policy can support their lifestyle aspirations within environmental limits [8]. In this context, incorporating tools from the field of foresight studies, such as a participatory policymaking process, could facilitate bottom-up policy formulation toward a long-term transition to sustainable lifestyles.

Within this context, this paper proposes key factors that could be incorporated into the participatory process with various stakeholders to envision future sustainable lifestyles. We developed these suggestions based on analysis of an expert survey on daily living and the roles of stakeholders between now and 2050. We also carried out a review of literature in the field of foresight and sustainable lifestyles. While respondents to the survey are experts from various stakeholder organisations rather than citizens, this global survey is unique in that it specifically focuses on sustainable lifestyles and enquires about stakeholders' changing roles. This paper analyses the factors related to shaping future lifestyles based on the changes identified by the expert survey. The interpretation of sustainable lifestyles differs from person to person and from group to group, so instead of offering analysis on the sustainability of the discussed changes of future lifestyles from the survey results, we provide different sets of factors for the public to consider when engaging in discussions on sustainable lifestyles. We analysed the foresighted changes in the roles of different stakeholders to examine whether the current mainstream government-industry-experts approach of any given foresight exercise is enough to discuss the topic of sustainable lifestyles. Moreover, foresight is a policy-oriented exercise, so understanding how stakeholders' roles change is pivotal when taking actions over a long-term transition.

Following this introduction, Section 2 describes the survey's methodology, including its design, data collection, and analysis. Section 3 reviews foresight literature to examine how foresight research and practice contribute to the policymaking process. Section 4 analyses the survey results, focusing on how lifestyles and the roles of stakeholders will change between now and 2050. Based on the analysis of the survey results, Section 5 proposes policy formation toward long-term sustainable lifestyles under participatory foresight, and Section 6 concludes by discussing its contribution to this field of research.

2. Methodology

2.1. Overall Methodology

To analyse changes in daily living and in stakeholders' roles, we first conducted a literature review to determine how the foresight approach could be used in the participatory process to transition to sustainable lifestyles. This review employed an interdisciplinary approach combining foresight studies and sustainable lifestyles research. We aimed to demonstrate how the knowledge gained from these two fields could help empower citizens and other stakeholders to engage in policy formation through a participatory foresight process. Second, we analysed the results of a global expert survey in terms of foresighted changes in daily living between now and 2050 to understand why and how citizens' participation and an intrinsic foresight approach, which empowers citizens in a democratic decision-making process [8], are necessary to better use foresight in policymaking toward sustainable lifestyles. Using the results on foresighted changes in consumption, infrastructure, work and education, and physical and mental well-being, the analysis focused on identifying the determining factors that could shape future sustainable lifestyles to be considered in the public discussions. The survey results show a diverse set of possibilities for future lifestyles and they illustrate how participatory foresight exercises could empower people to engage in the topic of future lifestyles to make tangible changes in their daily living. Third, we analysed how stakeholders would change from two aspects: Their changing roles and their future strengths and positiveness as considered by the respondents in terms of the implications of future sustainability. This analysis was used to derive factors for consideration related to stakeholder participation in the foresight process for future sustainable lifestyles and their potential roles in the transition towards desired futures. Finally, we combined the results of the literature review and survey analysis to propose participatory foresight processes that are useful for envisioning future sustainable lifestyles.

2.2. The Global Survey and Its Analysis

This study analysed primary data collected in the *Global Foresight Survey of Potential Changes in Society by 2050: Perspectives of Research Institutes and NGOs*, which was designed by this paper's authors. The Institute for Global Environmental Strategies and the United Nations Environment Programme jointly implemented the survey between 25 January and 28 February 2018. The free-text survey targeted 1200 authors of selected journals in the field of future studies: *Futures (300), Journal of Futures Studies (300), Technological Forecasting and Social Change (300), International Journal of Foresight and Innovation Policy (100), World Futures: The Journal of New Paradigm Research (100), International Journal of Forecasting (50), and Risk Analysis (50)*. An additional 300 authors were identified through a search of the following keywords on ScienceDirect and JSTOR: 'Foresight Study', 'Foresight 2050', 'Future Foresight', 'Future Scenarios', and 'Futures Studies'. The survey received 137 valid responses, with the valid response rate of about 9%. Based on the online survey software, it showed that each valid respondent spends an average of approximately 30 min filling out the survey.

Looking at affiliation, 84% of respondents were from research institutes or universities; 9% were from non-governmental/not-for-profit organisations, 5% were from the private sector or consultants, and 2% were from the government sector. The respondents covered a wide scope of knowledge/expertise. There were more respondents with knowledge on the regions of Europe (55%) and Asia-Pacific (31%). A substantial number of respondents were knowledgeable on the remaining regions, including North America (26%), Africa and the Middle East (20%), and Latin America and the Caribbean (17%). A large proportion of respondents either did not focus on specific regions or they focused on the global level (40%). For respondents' research and project areas, they covered a wide range of areas, with respondents allowed to select multiple answers: environment (55% selected this), economy (46%), energy and resource (43%), governance (35%), social policy (33%), natural science and technology (30%), education (28%), international development (21%), health (18%), foreign policy and international affairs (10%), defence and national security (4%), and others (32%). Additional details on the respondents are published in a discussion paper [14].

The survey questions were divided into three parts focusing on changes between now and 2050. Part 1 looked at overall society (culture and social norms, demography, economy, the environment and natural resources, governance structure, and technology and innovation), Part 2 was on the nine domains of lifestyles, and Part 3 focused on the roles of selected stakeholders. The nine domains in Part 2 included five consumption-based domains—food, consumption of manufactured goods, mobility, housing, and leisure [16]—and four non-consumption–based domains measuring people's well-being in terms of work, education, health, and social connections and relationships [17]. In Part 3, the survey provided six stakeholders that are frequently discussed in the literature on sustainability transition: Civil society, governments, households and individuals, local communities, the private sector, and research communities. A comprehensive analysis of the results from Parts 1 and 2 has been published by the authors of this paper [8,14]. This paper focuses on the analysis in Parts 2 and 3.

For this paper, the survey results were analysed using a combination of quantitative and qualitative analysis. The free-text answers on foresighted changes to daily living and stakeholders' roles were first manually labelled by categories of change. To ensure consistency, the given labels were mutually confirmed by two of the authors. Then, the frequency of some labels' presence was summarised in tables for changes in daily living (Sections 4.1.1–4.1.4) and in bar charts for the strengths of stakeholders and the positiveness of futures related to stakeholders' changing roles (Sections 4.2.1 and 4.2.2). Additionally, the foresighted changes in the strengths of stakeholders and positiveness of futures were manually labelled into three levels (stronger, neutral, weaker for strengths; positive, neutral, negative for positiveness) and were cross-analysed and visualised as a mosaic plot (Section 4.2.2). The results of the quantitative analysis were interpreted and supplemented using qualitative analysis, referring to examples of free-text answers from the survey to understand both overall trends and specific context among the survey responses.

3. Literature Review on Foresight's Role in the Policy Process

Future-oriented thinking is pivotal to achieve sustainable lifestyles by 2050. Existing scholarly discourse on sustainable lifestyles tends to focus on a consumption-based domain approach, which has the advantage of measuring environmental impacts such as carbon footprint, as demonstrated by [5–7,18]. Analysis of consumption-based environmental impacts or the footprint perspective provides quantitative evidence for progressive policy actions targeting long-term sustainability [6,19,20]. However, there is more to a person's lifestyle than just consumption. Aspects of day-to-day life—such as education, work, health, and social connections and relationships—also contribute significantly to determining one's level of well-being [14]. Additionally, emerging trends and disruptive changes in society are likely to shape future lifestyles, such as an increase in displaced populations due to climate change, the role of social networking services in human connections and relationships, and the disruption caused by robotics and artificial intelligence (AI) in employment. These should be incorporated into long-term policy design targeting a transition to sustainable lifestyles [8,14].

Calof and Smith [21] describe foresight's policy impacts include value statements by key actors, foresight's roles in the public arena (such as awareness-raising), project design to meet stakeholder needs, outputs of new knowledge in strategy-making, and policy formulation and delivery. The different phases of foresight—design, implementation, result generation, interpretation, and results and policy formulation—are all vital steps to impact policy [21]. Applying foresight in policymaking has two major areas of value: Instrumental and intrinsic [8]. Earlier foresight research and practices weigh more on the instrumental side, focusing on 'strategic' and 'scientific' aspects and allowing the decision-making process to be supported by scientific knowledge with insights from interdisciplinary experts in policymaking. By contrast and from an intrinsic perspective, foresight enables policymaking to be a more democratic process that engages diverse actors such as industry, civil society, and ordinary citizens to influence policy directions before implementation. This inclusive participatory element is important in shaping the decision-making process.

Foresight has been used in different countries for policymaking through government-initiated foresight programmes, particularly on technology and innovation. Such programmes were initiated in the United States and Japan, followed by Western European countries, in the 1990s [22]. Recently, foresight initiatives have been implemented by more countries in different regions; these initiatives have also been initiated by non-government actors, such as United Nations Industrial Development Organization (UNIDO)'s Technology Foresight Initiative in Latin America [23] providing those countries with support on technology and innovation-led policy formulation. Government-led foresight programmes have an advantage in that they are embedded within the government agencies and, thus, directly feed into the policy process. However, agenda-setting on foresight also tends to be determined by government interests—such as national security [15,24] or technology and innovation in industries [25]—instead of topics that are more relevant to ordinary citizens' lifestyles. Moreover, the stakeholders involved tend to be experts, industry representatives, and policymakers without the involvement of the general public. The public is generally only informed through outreach programmes communicating the foresight outcomes [26]. Thus, the general public and other key actors are often underrepresented in the process despite foresight's enormous potential effect on them [25]. It has also been argued that those setting the agenda on foresight tend to shape the research results and development agenda [27], limiting the discourse on futures to a predetermined framework [28]. In such a setting, foresight serves as an instrument for those actors to reach a consensus rather than being applied to search for alternatives [29].

Addressing the shortcomings in government-led foresight requires that more stakeholders, including the general public, be invited to discuss futures, thereby ensuring the empowerment of citizens [28,30–32]. Citizen participatory foresight is particularly significant in the search for long-term sustainability, which requires systemic structural changes. In this approach, foresight's intrinsic aspect of empowerment to serve the democratic decision-making process is more valuable than its strategic aspect. Inviting marginalised stakeholders or citizens into the foresight process shifts the discourse when setting the agenda, ensures that foresight's proposed policy solutions include 'fundamental changes in the system' [28], and involves these actors to engage and shape the future they want [33]. In this way, foresight is well placed to empower underrepresented actors in the policymaking process [29] using topics that are tangible and familiar to their lifestyles and extending these topics to broader socioeconomic issues surrounding these individuals [13]. Consequently, adopting the concept of 'sustainable lifestyles' in participatory foresight could open the discussion to revisit public policymaking as a method to improve people's quality of life and to empower the public to become agents shaping their own lifestyles through foresight. In this context, this research aims to contribute to participatory foresight in the topic of sustainable lifestyles from two aspects: To analyse the foresighted changes in daily living to understand weaknesses in the current mainstream approach of foresight; and to generate increased understanding on the changing roles of stakeholders in future society. This should be taken into consideration when looking at stakeholder involvement for foresight exercises and when forming collective actions towards shared sustainable lifestyles.

4. Survey Result and Analysis

In terms of policy formation through the participatory foresight process to transition to long-term sustainability, our analysis of the survey results focused on two elements: (1) Determining factors in shaping future lifestyles from four aspects (consumption, infrastructure, work and education, and physical and mental health); and (2) the changing roles of different stakeholders (governments, the private sector, research communities, civil society, local communities, and households and individuals). Among the four aspects of future lifestyles, there are 258 changes reported in total with the following distributions as illustrated in Figure 1: 53 in consumption (32 in food, 21 in manufactured goods), 44 in infrastructure (36 in mobility, 8 in housing), 77 in work and education (47 in work, 30 in education), and 71 in physical and mental health (38 in health, 25 in social connection and relationships, 8 in leisure). There were 13 responses reported as other domains that are excluded from this analysis. The results and analysis in this section supplement the current discourse on sustainable lifestyles and foresight studies, providing directions for policy formation for the long-term transition to sustainable lifestyles. A more detailed analysis of the survey results related to the foresighted changes in society and daily living between now and 2050 has been published as a discussion paper [14], and analysis focused on the technology aspect of foresight has been published in a research article [8].

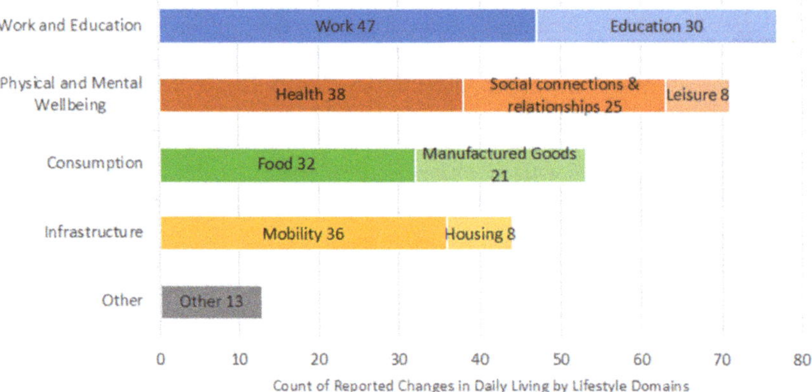

Figure 1. Lifestyles domains in daily living selected by the respondents. Notes: Lifestyle domains in daily living selected by the respondents in the *Global Foresight Survey of Potential Changes in Society by 2050: Perspectives of Research Institutes and NGOs*. Numbers are counts of valid responses. Percentages are the share in total responses. Valid responses only. N = 258.

4.1. Changes in Four Aspects of Lifestyle

4.1.1. Consumption

Thirty-two of the 258 total changes reported in Part 2 of the survey are related to food, with responses focused on three areas: Dietary habits, food security, and location of production (see Table 1). For **dietary habits**, the respondents described changes to the drivers of diet in terms of voluntary and forced change. Voluntary change is due to increased health and environmental concerns and to the preference for convenience in food, whereas forced change is because of food scarcity and the environmental necessity to reduce food's impact on the environment. In terms of **food security**, results covered factors that could see an increase in both food production and food scarcity. As pointed out by the survey respondents, one major element that could increase food production is technological innovation, although the effects of climate change and population growth are likely to outstrip any productivity increase and would result in food scarcity. For **location of production**, there were views that both local production and global production would increase, with localised food production likely to occur in urban areas for self-sufficient food production. There were also views that global-scale food exports and imports will account for a large portion of the economy.

Manufactured goods comprised 21 of the 258 reported changes, with results covering both demand shift and sharing and production patterns (see Table 1). For **demand shift and sharing**, the results show a likely shift from mass consumption to reduced demand for goods due to the increased adoption of the 3Rs (repair, reuse, and recycle), especially in current high-income countries. However, it was also noted that there would be a rise in consumption due to the increasing middle class in emerging economies. **Production patterns** included two major trends: Localised and smaller-scale production by local producers and environmentally friendly production, such as replacing crude oil with algae-based fuel and using biodegradable materials instead of oil-derived plastics.

Table 1. Changes in consumption.

Consumption
Food
Dietary habits: vegetarian (12), health concern (9), insect-based food (6), artificial meat (5), convenient food (5), environmental concern (5), fish/aquaculture (4), white meat (3), restriction/less variation (3), polarisation of dietary habits (3), organic food (2), vegan (2)
Food security: food scarcity (6), reduced food waste (3), use of Information and Communications Technology (ICT) (2), supply chain productivity (2)
Location of production: local production (6), global production (3), urban farming (3)
Manufactured Goods
Demand shift and sharing: repair/reuse/recycle (8), less consumption of goods (6), sharing/servitisation (4), more consumption of goods (3), experience-based consumption (3), more waste (2), less waste (2)
Production pattern: localised/small-scale production (7), environmentally-friendly design/production (4), small-lot production/3D printing (2)

Notes: Consumption changes were mentioned as part of the changes in daily living from the *Global Foresight Survey of Potential Changes in Society by 2050: Perspectives of Research Institutes and NGOs*. Multiple labels were allowed per response, and the frequency of counts is in parentheses. The unit of analysis is reported changes in daily living, and each respondent could report up to two changes. N = 53.

4.1.2. Infrastructure

Mobility received a high level of responses, 36, with the discussion focused on access and demand, mobility technology, and transportation mode (see Table 2). For **access and demand**, there were views on reduced mobility demand due to digital transformation through video conferencing, teleworking, and labour platforms. Respondents also mentioned the drivers of increased mobility demand due to globalisation, conflict, and climate change. Although technological development would result in lower mobility costs, there are also concerns over unequal access to this technology based on affordability. The dominant change in **mobility technology** appears to be in automated vehicles, with electric/hybrid vehicles likely to have a larger share of the automobile market due to increased regulations on vehicles using fossil fuels. Other technology such as high-speed transport will also increase. In **transportation mode**, there were reported changes in considering mobility as a service for sharing, and people use public transport more often which results in fewer cars.

Table 2. Changes in infrastructure.

Infrastructure
Mobility
Access and demand: less mobility demand (7), digital communication (6), unequal accessibility (5), increased accessibility (4), low-cost/efficient safety (3), more mobility demand (3)
Mobility technology: automated vehicle (18), electric/hybrid vehicle (12), high-speed transport (5)
Transportation mode: sharing/mobility as a service (9), public transport (5), less cars (5), traffic congestion (2), urban planning (2)
Housing
Housing supply: housing shortage (4), smaller house (2), green amenities (2)
Technology: renewable energy/lower environmental impacts (2), smart house (2)

Notes: Infrastructure changes were mentioned as part of the changes in daily living from the *Global Foresight Survey of Potential Changes in Society by 2050: Perspectives of Research Institutes and NGOs*. Multiple labels were allowed per response, and frequency of counts is in parentheses. The unit of analysis is reported changes in daily living, and each respondent could report up to two changes. N = 44.

The **housing** domain had only eight responses, and the discussion focused on housing supply and technology (see Table 2). In **housing supply**, a housing shortage due to climate change and rapid urbanisation would result in people shifting to smaller homes; housing would also incorporate green amenities in response to climate change. In **technology**, houses will likely incorporate the latest environmental technology and become 'smart homes', but again, technological innovation in housing will only be available those who can afford it.

4.1.3. Work and Education

Work is the domain that received the most responses—47 of 258—and responses focused on three areas: Labour market, format of work, and meaning of work (see Table 3). For the **labour market**, one key change is the replacement of human labour by AI, robotics, and automation, which will likely cause significant unemployment alongside changes to the labour market involving a demand for irreplaceable highly skilled and new-skilled jobs, as well as service- and human-centred jobs. The labour market itself will become more fluid, and permanent employment is likely to disappear. There were also responses on increased job opportunities from a sharing economy, with the labour market itself becoming more globalised. In terms of **format of work**, communication technology would offer new work formats such as remote work. In addition, the work itself would be more flexible, with more people freelancing or working fewer hours through work-sharing for a better work-life balance. For **meaning of work**, people's perceptions on meaning would also shift due to a changing labour market. This market would also require a new remuneration system, such as universal basic income, so individuals could have more leisure time and seek new means of fulfilment. Inequality is also likely to grow with less social security due to structural unemployment.

Table 3. Changes in work and education.

Work and Education
Work
Labour market: AI/robotics/automation (16), unemployment (12), high-skill/new-skill jobs (6), fluid labour market (5), human-centred jobs (5), less work (4), service jobs (4), digital technology (3), sharing economy (2), globalised labour market (2)
Format of work: telework (7), flexible employment/freelance (6), less working hours/work-sharing (4)
Meaning of work: new remuneration system (5), more leisure time (3), income inequality/low income (3), less social security (2), life-work mixture (2), new means of fulfilment/redefined meaning of work (2)
Education
Purpose of education: continuous education (8), vocational training/practical education (5), new skill requirement (5), restructuring of higher education (4), soft-skill development (3), new purpose of education (3), education for sustainability (2)
Access to and format of education: technology-assisted education (7), more access to education (5), out-of-school learning (3), less opportunity of education (2)

Notes: Work and Education changes were mentioned as part of the changes in daily living from the *Global Foresight Survey of Potential Changes in Society by 2050: Perspectives of Research Institutes and NGOs*. Multiple labels were allowed per response, and frequency of counts is in parentheses. The unit of analysis is reported changes in daily living, and each respondent could report up to two changes. N = 77.

Thirty of the 258 responses focused on **education**, namely on the purpose of education, as well as access to and format of education (see Table 3). Looking at the **purpose of education**, people might continue their education throughout their life to update their skills in a society rapidly changing due to technological disruptions. Furthermore, education's purpose might shift to prevent mental degradation and a sense of irrelevance when mass human labour is replaced by robotics and machine learning. The content of education would also change, with more vocational training, practical education, soft-skill development, and education for sustainability. For **access to and format of education**, technology's role will increase the use of online education, interactive education via computers, and earlier education on computer science. Access to education may well increase through digital means, but there were also views that educational opportunities might be limited to specific groups despite improvements in education technology.

4.1.4. Physical and Mental Health

Eight responses were related to **leisure**, specifically on the **format and purpose of leisure** (see Table 4). Respondents indicated that there would be more experience-based leisure such as holidays and green tourism. Technology development will also provide leisure using virtual reality and artificial leisure, although unequal accessibility would be a concern. There were also views that people could have more leisure time due to the replacement of human labour with robotics and AI.

Table 4. Changes in physical and mental health.

Physical and Mental Health
Leisure
Format and purpose: experience-based leisure/self-realisation (4), unequal accessibility (3), more leisure time (2), virtual reality/artificial leisure (2)
Health
Innovation in healthcare: advancement of healthcare technology (10), automation of healthcare (5), self-health/monitoring (5), precision healthcare (5), preventive healthcare (4), remote healthcare (3) New challenges: higher risk of epidemics/antibiotic resistance (8), health risk from environmental degradation (7), lifestyle/non-communicable diseases (4), more mental health problems (4) Longevity: life-prolonging (11), unhealthy older adulthood (2), improvement of old age life (2), end-of-life decisions (2) Access to healthcare: unequal access to healthcare (9), more access to healthcare (2)
Social Connection and Relationships
Isolation and connection: isolation/individualisation (9), social and economic fragmentation (6), more connectivity (4), household relationships (2) Digitalised world: digital connections (11), more surveillance/manipulation (3)

Notes: Physical and mental health changes were mentioned as part of the changes in daily living from the *Global Foresight Survey of Potential Changes in Society by 2050: Perspectives of Research Institutes and NGOs*. Multiple labels were allowed per response, and frequency of counts is in parentheses. The unit of analysis is reported changes in daily living, and each respondent could report up to two changes. N = 71.

Thirty-eight responses focused on **health**, covering four areas: Innovation in healthcare, new challenges, longevity, and access to healthcare (see Table 4). For **innovation in healthcare**, technological advancement would bring innovation to healthcare, which might involve developing more effective treatments of some diseases such as cancer and Alzheimer's disease. Due to the high cost, healthcare will likely only be accessible to those who could afford it. The application of massive datasets will also be common for automated diagnosis and remote healthcare, and healthcare will be more personalised, with self-monitoring, precision medicine, and preventive care. Some **new challenges** were also raised, such as antibiotic resistance, diseases from environmental degradation, lifestyle diseases, and mental diseases. For **longevity**, advances in healthcare technology will mean people living longer but will also cause longer morbidity periods and unhealthy older age, leading to the need to consider end-of-life decisions involving assisted suicide. There were also views that due to increased health concerns, there will be improvements the quality of life for older adults. In terms of **access to healthcare**, the view is that access to healthcare would remain unequal in spite of technology advancement. There were also responses on the emergence of private entities that might make healthcare more accessible in innovative ways.

Twenty-five responses focused on **social connection and relationships**, specifically on isolation and connection and on the digitalised world (see Table 4). In terms of **isolation and connection**, there were views that people would be more isolated due to the spread of Western lifestyles and, thus, people would focus more on self-improvement and individual lifestyle. Other factors include fewer marriages, more divorces, smaller families, digital communication, and more individualised leisure pursuits. Responses citing more connection are supported by the fact that people would return to their extended families due to ageing demographics, more time for social networking due to less work, and globalisation leading to more tolerance for diversity. There were also responses on changing perceptions of gender due to women's empowerment. For **digitalised world**, social interactions are more likely to take place online, with reduced attention spans and less intimacy. There were also views that online connections might collapse, and people would return to face-to-face communication. Due to increased online communication, there were also concerns about greater surveillance and manipulation by using big data to control people's behaviour.

4.1.5. Determining Factors on Shaping Future Sustainable Lifestyles

Based on the survey results, we derived several determining factors in shaping future lifestyles for each aspect of lifestyle. For consumption, respondents discussed reducing the level of consumption of food (meat) and manufactured goods, but with different rationales depending on the group: For some, it would be due to scarcity of resources, while for others, it would be to fulfil their values on health and the environment. According to research using quantitative analysis to determine the required level of consumption to achieve long-term sustainability [34], these survey results add value to the discussion in the form of various drivers underlying the changes in consumption level.

In terms of infrastructure, technological advancement appears to be an important factor in providing more choices, although respondents noted affordability and equal accessibility as concerns. Thus, technological improvement—and the implication that it could fulfil people's pursuit of a meaningful life—requires additional research [8]. Changes in work and education are disruptive due to technological innovation, meaning they require engaging the general public in a more comprehensive assessment of the desired direction [35]. In terms of physical and mental health, uncertainties remain regarding mental health's effects on people's lifestyles, such as on whether longevity contributes to a more active later life and on whether digitalisation would contribute to more meaningful connections or result in more isolation.

The interpretations on sustainable lifestyles are diverse and non-static, and the respondents of the survey provided a diverse set of possibilities in how the future lifestyles could change. Thus, based on the individuals' interpretation of their desirable future lifestyles, discussions on whether foresight future changes in lifestyles are sustainable or not could be included in the public discussion on the participatory foresight process.

4.2. Stakeholder Roles

4.2.1. The Strength of Different Stakeholders' Roles

Based on the free-text responses, Figure 2 depicts the changing strength of different stakeholders between now and 2050. Government (national governments) is the only stakeholder whose power is perceived by respondents as neutral or weaker. Of those responding on government, 25% think its role will be stronger than today, compared to 29% who provided neutral responses and 46% who think its role will be weakened. A major factor raised by the respondents saying the government's role will be stronger was the belief that government will impose greater control and regulations on society, people, and companies. Those selecting neutral said government's structure will be more decentralised and interactive through direct democracy. Those selecting weaker provided reasons such as reduced public trust in politicians, reduced need for politicians since people could more easily reach consensus using digital platforms, government's inability to handle the challenges caused by climate change and a rapidly changing society, and the independence of central entities from nation-states.

The majority of respondents (74%) believe the role of the **private sector** will be stronger due to increased privatisation of social services, the private sector's potential role in the common good or in society's destruction, and mega-corporations increased influence on government policies. Seventeen per cent of respondents consider the private sector's role to be neutral, mentioning changes in its production process and working models, and 9% believe it will be weaker due to regulations and the redistribution of capital to the government, civil society, and customers.

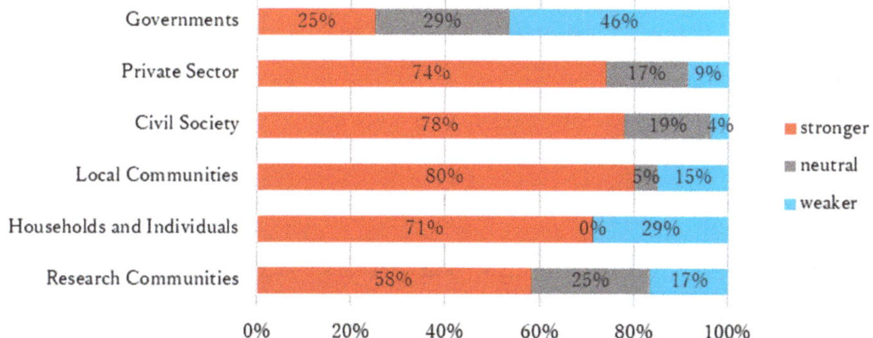

Figure 2. Respondents' projected changes to stakeholders' strength. Notes: the share of the strengths of stakeholders described in the responses of changes in stakeholders' roles in the *Global Foresight Survey of Potential Changes in Society by 2050: Perspectives of Research Institutes and NGOs*. The unit of analysis is reported changes in stakeholders' roles. N = 124.

For **research communities**, 58% of respondents think these communities' role will be stronger due to increasing global complexities. Responses also indicated that different stakeholders are likely to value the knowledge produced by researchers, who would thus have a larger role in the decision-making process. Twenty-five per cent of respondents provided a neutral response, while 17% said researchers' role would be weaker due to a shift from research to consultancy work and due to decreased respect for science due to the overflow of information available online.

Civil society, local communities, and households and individuals represent stakeholders at the grassroots level, although there are some overlaps in the classifications of the three groups. Depending on assumptions, local communities, households, and individuals are part of civil society, which could gain the power to act against other stakeholders that are traditionally included in foresight exercises such as the government and private sector [14]. The topic of sustainable lifestyles covers many tangible aspects of daily living and require local communities at the municipal and community levels to provide infrastructure for necessary collective actions [16]. Households and individuals are required make lifestyles decisions and become vital actors in the participatory foresight process. Thus, the distinctions between the three sets of actors are made in this study.

Seventy-eight per cent of respondents who selected **civil society** indicated that its role will be stronger. Civil society is considered a leader in social change through social movements, self-organisation to build large networks through technological platforms, and greater activity in the policy process and service provision; civil society also acts to counter the power balance with the private sector and fills power vacuums created by a weakened government. Those empowering civil society will act at both the local and global level. Nineteen per cent of responses were neutral, mainly because respondents were unsure whether civil society would be stronger or weaker in the future, and 4% said civil society will be weakened due to the increasing influence of large corporations.

As with civil society, the vast majority of respondents (80%) said **local communities** will have a stronger role due to stronger supportive and inclusive networks developed at the community level for local needs such as local production and consumption. Advances in communication technology will also likely enable people to collectively influence national policies. These changes will probably fill the gaps and bridge the inability of services and governance that only rely on the government. Five per cent of respondents were neutral, and 15% said that local communities would be weaker due to young people becoming more detached from the real world, the loss of a sense of community, and the impoverishment of local communities due to urbanisation.

For **households and individuals**, 71% of respondents said that this group would be stronger due to the improvement of communication technology for information-sharing and that it would be more independent due to the home production of goods and services. As consumers, households and individuals would also demand more sustainable products. Conversely, 29% think these actors would become weaker due to structural unemployment and increased inequality, loss of power to corporations and a wealthy minority, and lack of concern for the real world because of increased attention to the virtual world.

4.2.2. Cross-Analysis of the Strengths and Positiveness in Stakeholder Roles

Considering the importance of stakeholders' role in the participatory foresight process, we also conducted a cross-analysis of the strength and positiveness in their changing roles by respondents. For instance, the response below illustrates that local communities will have a stronger role, and this is considered to be positive to the society's sustainability by the respondent:

"Local communities will be more supportive of equity and inclusion, being more integrated socially, economically, and racially; communities will be supportive of co-production of many needs, and of trading/exchanging goods and services internally; they will with their governments and utilities evolve the energy grid to enable in-home and in-community production and storage; they will be more caring for all in the community, supporting the opportunities for all to contribute, grow in meaning and thrive."

Another example below shows that households are getting weaker, and this is considered to be negative to the society's sustainability by the respondent:

"Individuals are losing ground on several counts—corporate rights are stronger, ability for individuals to fight the system are eroding; taxes favor wealthy and corporations; unions are weakening, as are worker rights."

Figure 3 depicts the positiveness of the foresighted future related to stakeholders' changing roles, and Figure 4 depicts the cross-analysis on strength and positiveness.

In terms of the positiveness of the foresighted future related to stakeholders' changing roles, the majority of those who selected governments, the private sector, and research communities selected neutral, which implies that these actors' role could move in either direction. For civil society and local communities, responses were more positive (48% and 60%, respectively). Responses for households and individuals were well-distributed: 29% positive, 36% neutral, and 36% negative.

The results of the cross-analysis on the strengths and positiveness of the changing roles of each stakeholder (see Figure 4) demonstrate that there are variable implications for different stakeholders. For government, the majority of respondents consider a stronger role to be positive and a weakened role to be negative, meaning government's role should be stronger for a positive future. For the private sector, most respondents selected a stronger role, but its positiveness implication is more 'neutral'; this implies that the private sector's strengthened role could be either positive or negative in the future. A weakened role for the research community was considered negative, meaning these communities should be strengthened for a positive outcome.

Most respondents indicated that civil society should have a stronger role, although responses on positiveness were divided between positive and neutral. Nevertheless, this was a negative outcome for those who did not feel civil society would be stronger. Additionally, although most respondents said local communities' role would be stronger, some gave a neutral response, and positiveness was divided among positive, neutral, and negative; this implies that uncertainty exists with regards to the future direction of this stakeholder group. For households and individuals, although the majority felt this group's role would be stronger, some felt its role would be neutral. Nevertheless, weakened households and individuals were considered negative.

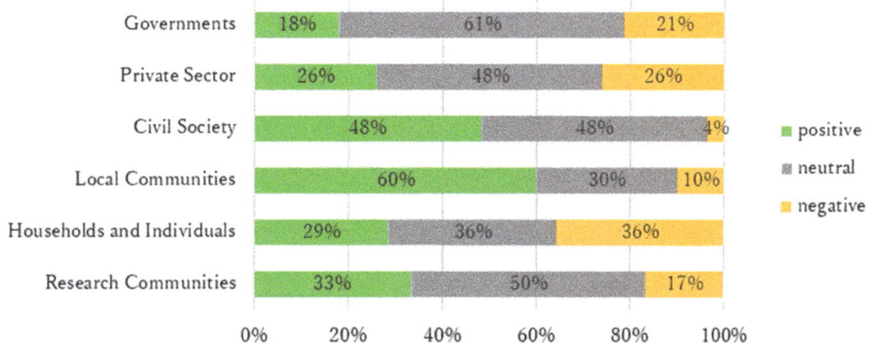

Figure 3. Positiveness of the foresighted future related to stakeholders' changing roles. Notes: the share of the positiveness of future described in the responses of changes in stakeholders' roles in the *Global Foresight Survey of Potential Changes in Society by 2050: Perspectives of Research Institutes and NGOs*. The unit of analysis is reported changes in stakeholders' roles. N = 124.

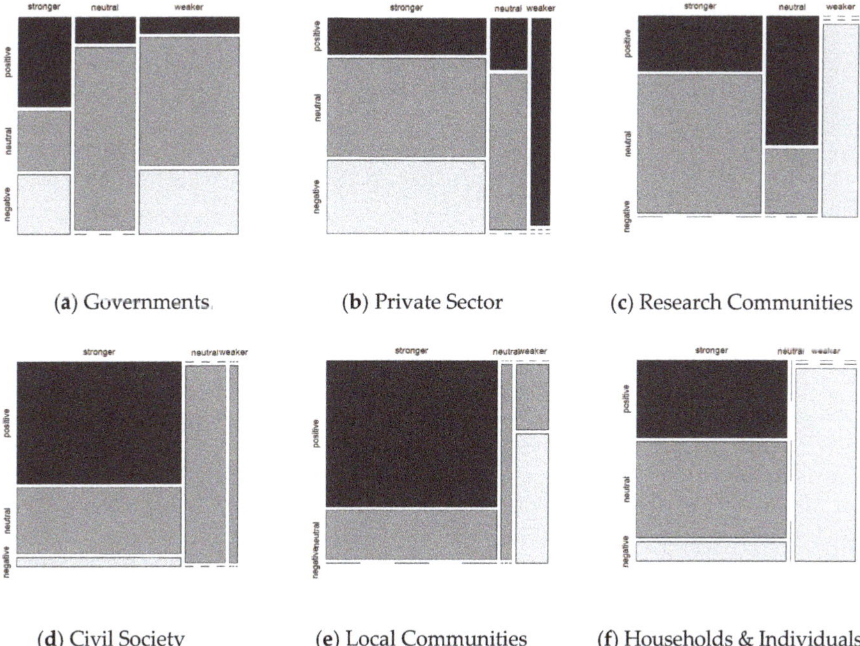

(**a**) Governments　　　　(**b**) Private Sector　　　　(**c**) Research Communities

(**d**) Civil Society　　　　(**e**) Local Communities　　　　(**f**) Households & Individuals

Figure 4. Cross-analysis of the strength and positiveness of stakeholders' changing roles. Notes: mosaic plots of the strengths of the stakeholder and the positiveness of the future described in the responses of changes in stakeholders' roles in the *Global Foresight Survey of Potential Changes in Society by 2050: Perspectives of Research Institutes and NGOs*. Horizontal axis refers to the strengths of stakeholders. Vertical axis refers to the positiveness of future. The size of boxes refers to the number of responses belongs to each combination of strength and positiveness. The unit of analysis is reported changes in stakeholders' roles. N = 124.

4.2.3. Stakeholders' Changing Roles

The results on stakeholders' changing roles between now and 2050 indicate that the government's role could be weaker in the future. More actors at the grassroots level—such as civil society, local communities, and households and individuals—are expected to increase their roles and positively change society. Thus, in foresight design, relying on a government-led process is no longer efficient when other essential stakeholders from the grassroots level are missing. Although not perceived positively when having a stronger role, the private sector was found to be pivotal in shaping futures both positively and negatively. Thus, engaging the private sector in the foresight process could provide a foundation to involve it in discussions on transitioning to a desirable future. For research communities, foresight experts have been key actors in government-led foresight, and the role of research is likely to remain vital, involving different actors and facilitating their future-oriented thinking in the foresight process. Currently, foresight serves to inform policymakers mostly based on expert knowledge, but this must be expanded to gather the general public's collective knowledge and to empower individuals to shape policies for the desired long-term transition.

5. Policy Formation toward Long-Term Sustainable Lifestyles under Participatory Foresight

This analysis of the survey results makes an important contribution to discussions on policy formation towards long-term sustainable lifestyles through participatory foresight in two aspects.

(1)   Changes in the future lifestyles discussed in the survey results revealed that bi-polar changes were foresighted: How to evaluate these changes is a matter of various values and context among people, so we need the participation of households and the local community in the foresight. We could confirm that various changes were expected in daily living, which are beyond technology and macro-level changes. Thus, participation of households and local community is necessary to apply foresight on the lifestyle topic. Based on different sets of interpretations of sustainable lifestyles, foresight exercises could look at the diverse possibilities of change in lifestyles mentioned in the survey and discuss which of these is desirable and which should be avoided. This could be seen as a democratic process to decide the goal of the policymaking in achieving desirable lifestyles. For consumption, although it is easy to simply measure how much consumption levels have increased or decreased, the survey results pointed out that it is the reasoning behind increased or reduced consumption that matters when it comes to understanding whether the choices are forced or voluntary. In infrastructure, technology advancement would likely increase the options. However, there needs to be more scrutiny of these issues in terms of affordability and equal accessibility to advanced choices. In the areas of work and education, the implication of potential disruptive changes brought by robotics and automation is not yet clear. The impact of prolonged life brought by medical advancement and digital connection to mental health should be better investigated depending on different people's aspirations.

(2)   The changing roles of stakeholders in the survey should be considered when designing foresight exercises for stakeholder inclusions and their roles to achieve foresighted futures. In addition to government, private sector, and the experts in the conventional foresight exercises, the survey results show there are more stakeholders at the grassroots level of societies. The potential change in those stakeholder roles should be considered so that the whole of society can transition towards sustainable futures. The survey results show that the role of government will weaken, so long-term transition requires other stakeholders that are conventionally ignored in foresight exercises at the grassroots level. These include civil society, which would come under consideration to lead social movements, as well as local communities that provide the base for local production and consumption, and individuals and households that could be strengthened through collective power.

Foresight should not be limited to the future of technology, as is commonly the case; it should also be used to understand and shape societal expectations and more nuanced areas such as value systems, which tend to shape norms and lifestyle choices. Policy processes designed to benefit from the foresight approach should contribute to the decision-making process by involving various stakeholders and maintaining an openness to a variety of possible futures [10]. Such openness and wider participation would give policy more legitimacy, and thus lead to greater uptake and more effective implementation. However, to be successful, policy processes enabled by foresight would need to be grounded in science-based information, supported by stakeholder education on issues and risks needing to be addressed, and led through a well-facilitated process. Moreover, ongoing efforts to discuss people's lifestyles in foresight exercises are mostly being carried out in European countries [13,31], whereas this survey provides a more global context on this topic.

This paper has provided examples of involving different stakeholders in the participatory foresight process to illustrate factors to consider in the future development of policies on the long-term sustainability of lifestyles. The benefits of foresight in policymaking could have both instrumental value in a more informed decision-making process and also intrinsic value through empowering citizens in a participatory policymaking process [8]. However, a major challenge in future foresight design is how to engage with different stakeholders in the foresight process when the speed of change is increasing due to technological disruption and the interconnectedness of the global system. The current period of uncertainty also provides opportunities to move in a more sustainable direction, with the recent outbreak of Coronavirus Disease 2019 (COVID-19) resulting in destructive changes to society during lockdowns that have impacted people's employment, mobility, leisure, and mental well-being. Many of the possibilities discussed in the expert survey provide determining factors to move forward, such as the discussion on universal basic income and the implications of the rapid digitalisation of people's lives for privacy. Although COVID-19's long-term sustainability implications for society and for people's lives require greater scrutiny and discussion, the destructive changes provide a space to boldly reimagine different futures, especially among those who have the agency in democratic societies. Rather than waiting with anxiety and passiveness, applying a participatory foresight mindset means that inventing a new future—not merely extending business as usual—could be the way forward.

When looking at instrumental values, Slaughter (1995) [36] considers foresight to be 'a process that attempts to broaden the boundaries of perceptions in four ways: By assessing the implications of present actions, decisions, etc. (consequent assessment); by detecting and avoiding problems before they occur (early warning and guidance); by considering the present implications of possible future events (proactive future formulation); and by envisioning aspects of desired futures (normative scenarios).' Instrumental or more strategic aspects [8] of contributions from this research in the field of foresight demonstrate that there have been changes to different aspects of future lifestyles and to stakeholders' roles. The survey results discussed here illustrate that the determining factors extend beyond a domain-based approach to lifestyles, which could facilitate more holistic and systemic views when engaging stakeholders in foresight discussions. One cross-cutting trend is the polarisation of those who will have more options as technology advances and those who will likely be deprived and marginalised due to increasing inequity. Moreover, even for individuals with more options, the implications of a 'better quality' of life are complex and require discussion with the general public.

From an intrinsic perspective [8], there are limitations to engaging 'non-expert' groups such as citizens in this process. Insights into stakeholders' changing role demonstrate that greater engagement with stakeholders from local communities, civil society, and households and individuals could empower them and lead to positive changes for society. Meanwhile, the roles of government, the private sector, and research communities have the potential to move in either a positive or negative direction in the future. Interestingly, the survey results indicate the private sector has the potential to make a transition to be more—or less—sustainable. As the private sector is a key stakeholder, its involvement and cooperation with other stakeholders should be more strongly facilitated in the participatory foresight process.

Due to the potentially increasing role of grassroots-level stakeholders, a foresight process with a more bottom-up approach feeding into policy design could mean a more inclusive policy process. Such opportunities should be used in engaging stakeholders in foresight design. However, when focusing on future lifestyles, it is even more relevant and tangible for citizens to be part of the policy-formation process. In fact, even a bottom-up foresight process is normally facilitated by experts in the foresight field who then engage the general public to discuss their futures [13,31]; thus, more collaboration among experts and grassroots-level stakeholders could empower both sets of stakeholders. Due to the perceived declining strength of government in the survey results, engaging other stakeholders in policymaking would not only ensure a more inclusive process to better reflect the needs of all stakeholders but also delegate some of the government's roles to other stakeholders as a part of their democratic participation.

6. Conclusions

Foresight has been serving the policymaking process by engaging experts and different stakeholders to put expert knowledge into action. However, foresight has clear, untapped potential to empower the general public to become involved. Positioning foresight to address lifestyles opens up discussions that could be relevant to the general public, thereby influencing policymaking in tangible ways. To date, government, the private sector, and research communities have been involved in the discourse on foresight. Now, it is time for more actors at the grassroots level—such as civil society, local communities, and households and individuals—to be considered in the process. In particular when sustainable lifestyles become the focus of the foresight exercise, civil society could be the actor to gain power in society to counter the underlying assumptions that it is government and industry which take the lead in conventional foresight. Local communities have the potential to mobilise collective actions to assert their local needs in providing enabling conditions for sustainable lifestyles. Lastly, households and individuals should pursue their own desirable future lifestyles by engaging in the formation of future policy decisions as part of the democratic participation process.

Foresight and participatory policymaking processes are especially important in areas such as food, social connections, and relationships, which are less driven by expert knowledge and more affected by perceptions and societal norms. Policy on the future of food consumption (and thus, production) is not just an issue of the technical production and distribution of food; it is also about food cultures and traditional practices, which determine food patterns. When communities are allowed to engage in the deliberative process, they gain a better understanding about the links between food choices and diets, health, and environmental factors such as biodiversity effects and food security. Such an understanding means that people are more willing to accept the need for alternatives, as well as to providing informed consent for a more sustainable future direction.

Finally, government and governance require greater foresight for their development. The results of the survey point to national government as the only stakeholder perceived to have a weaker role by 2050. Since perception can easily become reality in democratic governance, this calls for changes to more traditional approaches, and thus, for government to begin embodying the types of sustainable changes envisaged for the future. Such changes involve greater transparency in governance, added clarity on the different roles that stakeholders have and how they participate, as well as a critical distinction between public mandates and private decision-making. A foresighted and orderly approach to governance would include a deliberative process for the evolution of public governance, as well as the design of new institutions that recognise an evolving society and stronger sustainability governance.

Author Contributions: Conceptualization, C.M.; data curation, C.M. and R.K.; formal analysis, C.M. and R.K.; funding acquisition, L.A.; investigation, C.M. and R.K.; methodology, C.M. and R.K.; project administration, L.A.; supervision, L.A.; visualization, R.K.; writing—original draft, C.M.; writing—review and editing, R.K. and L.A. All authors have read and agreed to the published version of the manuscript.

Funding: This study was funded by the Environment Research and Technology Development Fund (S-16-3) of the Environment Restoration and Conservation Agency, Japan and the 10-Year Framework of Programmes on Sustainable Consumption and Production (10YFP).

Acknowledgments: The authors are grateful to the editors and anonymous reviewers for their valuable comments on earlier versions of the manuscript.

Conflicts of Interest: The author declares no conflict of interest.

References

1. United Nations. *Paris Agreement*; United Nations: Paris, France, 2015.
2. IPCC. *Global Warming of 1.5°C: An IPCC Special Report on the Impacts of Global Warming of 1.5°C above Pre-Industrial Levels and Related Global Greenhouse Gas Emission Pathways, in the Context of Strengthening the Global Response to the Threat of Climate Change*; The Intergovernmental Panel on Climate Change: Geneva, Switzerland, 2018.
3. The Government of Japan. *The Long-Term Strategy under the Paris Agreement*; The Government of Japan: Tokyo, Japan, 2019.
4. United Nations. *Transforming Our World: The 2030 Agenda for Sustainale Development, United Nations Department of Economic and Social Affairs*; United Nations: New York, NY, USA, 2015.
5. Dubois, G.; Sovacool, B.; Aall, C.; Nilsson, M.; Barbier, C.; Herrmann, A.; Bruyère, S.; Andersson, C.; Skold, B.; Nadaud, F.; et al. Energy Research & Social Science It starts at home? Climate policies targeting household consumption and behavioral decisions are key to low-carbon futures. *Energy Res. Soc. Sci.* **2019**, *52*, 144–158. [CrossRef]
6. Koide, R.; Lettenmeier, M.; Kojima, S.; Toivio, V.; Amellina, A. Carbon Footprints and Consumer Lifestyles: An Analysis of Lifestyle Factors and Gap Analysis by Consumer Segment in Japan. *Sustainability* **2019**, *11*, 5983. [CrossRef]
7. Lettenmeier, M.; Liedtke, C.; Rohn, H. Eight Tons of Material Footprint—Suggestion for a Resource Cap for Household Consumption in Finland. *Resources* **2014**, *3*, 488–515. [CrossRef]
8. Mao, C.; Koide, R.; Brem, A.; Akenji, L. Technological Forecasting & Social Change Technology foresight for social good: Social implications of technological innovation by 2050 from a Global Expert Survey. *Technol. Forecast. Soc. Chang.* **2020**, *153*, 119914. [CrossRef]
9. Solem, E. Foresight in government: A necessary but rare component. *Futures Can.* **1986**, *8*, 3–7.
10. Nazarko, J.; Radziszewski, P.; Dębkowska, K.; Ejdys, J.; Gudanowska, A.; Halicka, K.; Kilon, J.; Kononiuk, A.; Kowalski, K.J.; Król, J.B.; et al. Foresight Study of Road Pavement Technologies. *Procedia Eng.* **2015**, *122*, 129–136. [CrossRef]
11. Dreyer, I.; Stang, G. Foresight in governments—practices and trends around the world. *Yearb. Eur. Secur.* **2013**, *1368*, 7–32.
12. Costanza, R.; Alperovitz, G.; Daly, H.; Farley, J.; Franco, C.; Jackson, T.; Kubiszewski, I.; Schor, J.; Victor, P. Building a sustainable and desirable economy-in-society-in-nature. In *State of the World 2013: Is Sustainability Still Possible?* Island Press: Washington, DC, USA, 2013; pp. 126–142. [CrossRef]
13. Dumitru, A.; Mira, R.G. *Green Lifestyles, Alternative Models and Upscaling Regional Sustainability-GLAMURS: Description of Work. Glamus, Supporting Green Lifestyles*; University of A Coruna: A Coruna, Spain, 2017.
14. Mao, C.; Koide, R.; Akenji, L. *Society and Lifestyles in 2050: Insights from a Global Survey of Experts*; Institute for Global Environmental Strategies: Hayama, Japan, 2019.
15. Habegger, B. Strategic foresight in public policy: Reviewing the experiences of the UK, Singapore, and the Netherlands. *Futures* **2010**, *42*, 49–58. [CrossRef]
16. Akenji, L.; Chen, H. *A Framework for Shaping Sustainable Lifestyles*; United Nations Environment Programme: Paris, France, 2016.
17. Stiglitz, J.E.; Sen, A.; Fitoussi, J.-P. *Report by the Commission on the Measurement of Economic Performance and Social Progress*; Commission on the Measurement of Economic Performance and Social Progress: Paris, France, 2009.
18. Wiedenhofer, D.; Smetschka, B.; Akenji, L.; Jalas, M.; Haberl, H. Household time use, carbon footprints, and urban form: A review of the potential contributions of everyday living to the 1.5 °C climate target. *Curr. Opin. Environ. Sustain.* **2018**, *30*, 7–17. [CrossRef]
19. Teubler, J.; Buhl, J.; Lettenmeier, M.; Grei, K.; Liedtke, C. A Household's Burden—The Embodied Resource Use of Household Equipment in Germany. *Ecol. Econ.* **2018**, *146*, 96–105. [CrossRef]
20. Tukker, A.; Cohen, M.J.; Hubacek, K.; Mont, O. The Impacts of household consumption and options for change. *J. Ind. Ecol.* **2010**, *14*, 13–30. [CrossRef]

21. Calof, J.; Smith, J. Firesight impacts from around the world: A special issue. *Foresight* **2012**, *14*, 5–14. [CrossRef]
22. Martin, B.R. The origins of the concept of "foresight" in science and technology: An insider's perspective. *Technol. Forecast. Soc. Chang.* **2010**, *77*, 1438–1447. [CrossRef]
23. Cespedes, Q.M.; Martin, D.P. Technology foresight in traditional Bolivian sectors: Innovation traps and temporal unfit between ecosystems and institutions. *Technol. Forecast. Soc. Chang.* **2017**, *119*, 280–293. [CrossRef]
24. Nemeth, B.; Dew, N.; Augier, M. Understanding some pitfalls in the strategic foresight processes: The case of the Hungarian Ministry of Defense. *Futures* **2018**, *101*, 92–102. [CrossRef]
25. Weber, K.M.; Gudowsky, N.; Aichholzer, G. Foresight and technology assessment for the Austrian parliament—Finding new ways of debating the future of industry 4.0. *Futures* **2018**, *109*, 240–251. [CrossRef]
26. Kuosa, T. *Practising strategic foreisght in government: The cases of Finland, Singaper and European Union*; S. Rajaratnam School of International Studies: Singapore, 2011.
27. Grethe, H.; Dembele, A.; Duman, N. *How to Feed the World's Growing Billions: Understanding FAO World Food Projections and Their Implications*; WWF Germany: Berlin, Germany, 2011.
28. Bourgeois, R.; Penunia, E.; Bisht, S.; Boruk, D. Technological Forecasting & Social Change Foresight for all: Co-elaborative scenario building and empowerment. *Technol. Forecast. Soc. Chang.* **2017**, *124*, 178–188. [CrossRef]
29. Ahlqvist, T.; Rhisiart, M. Emerging pathways for critical futures research: Changing contexts and impacts of social theory. *Futures* **2015**, *71*, 91–104. [CrossRef]
30. Amanatidou, E. Foresight process impacts: Beyond any official targets, foresight is bound to serve democracy. *Futures* **2017**, *85*, 1–13. [CrossRef]
31. Jegou, F.; Gouache, C. Envisioning as an enabling tool for social empowerment and sustainable democracy. In *Responsible Living: Concepts, Education and Future Perspectives*; Thoresen, V., Didham, R., Klein, J., Doyle, D., Eds.; Springer: Cham, Switzerland, 2015.
32. Poli, R. The implicit future orientation of the capability approach. *Futures* **2015**, *71*, 105–113. [CrossRef]
33. Konnola, T.; Brummer, V.; Salo, A. Diversity in foresight: Insights from the fostering of innovation ideas. *Technol. Forecast. Soc. Chang.* **2007**, *74*, 608–626. [CrossRef]
34. Institute for Global Environmental Strategies, Aalto University, D-mat Ltd. *1.5-Degree Lifestyles: Targets and Options for Reducing Lifestyle Carbon Footprints*; Technical Report; Institute for Global Environmental Strategies: Hayama, Japan, 2017.
35. Rhisiart, M.; Störmer, E.; Daheim, C. From foresight to impact? The 2030 Future of Work scenarios. *Technol. Forecast. Soc. Chang.* **2017**, *124*, 203–213. [CrossRef]
36. Slaughter, R.A. *The Foresight Principle: Cultural Recovery in the 21st Century*; Adamantine Press: London, UK, 1995.

 *sustainability*

MDPI

Review

Are We Missing the Opportunity of Low-Carbon Lifestyles? International Climate Policy Commitments and Demand-Side Gaps

Janet Salem [1,*], **Manfred Lenzen** [1] **and Yasuhiko Hotta** [2]

[1] Integrated Sustainability Analysis, School of Physics A28, The University of Sydney, Sydney, NSW 2006, Australia; manfred.lenzen@sydney.edu.au
[2] Sustainable Consumption and Production Area, Institute for Global Environmental Strategies, Hayama 240-0115, Kanagawa, Japan; hotta@iges.or.jp
* Correspondence: janet.salem@sydney.edu.au

Abstract: Current commitments in nationally determined contributions (NDCs) are insufficient to remain within the 2-degree climate change limit agreed to in the Paris Agreement. The Intergovernmental Panel on Climate Change (IPCC) states that lifestyle changes are now necessary to stay within the limit. We reviewed a range of NDCs and national climate change strategies to identify inclusion of low-carbon lifestyles. We found that most NDCs and national climate change strategies do not yet include the full range of necessary mitigation measures targeting lifestyle change, particularly those that could reduce indirect emissions. Some exceptional NDCs, such as those of Austria, Slovakia, Portugal and the Netherlands, do include lifestyle changes, such as low-carbon diets, reduced material consumption, and low-carbon mobility. Most countries focus on supply-side measures with long lag times and might miss the window of opportunity to shape low-carbon lifestyle patterns, particularly those at early stages of development trajectories. Systemic barriers exist that should be corrected before new NDCs are released, including changing the accounting and reporting methodology, accounting for extraterritorial emissions, providing guidance on NDC scope to include the menu of options identified by the IPCC, and increasing support for national level studies to design demand-side policies.

Keywords: climate change policies; UNFCCC; demand-side management; behavioral change; consumption-based emissions; low-carbon lifestyles; indirect emissions; carbon footprint

Citation: Salem, J.; Lenzen, M.; Hotta, Y. Are We Missing the Opportunity of Low-Carbon Lifestyles? International Climate Policy Commitments and Demand-Side Gaps. *Sustainability* **2021**, *13*, 12760. https://doi.org/10.3390/su132212760

Academic Editor: Grigorios L. Kyriakopoulos

Received: 29 September 2021
Accepted: 8 November 2021
Published: 18 November 2021

Publisher's Note: MDPI stays neutral with regard to jurisdictional claims in published maps and institutional affiliations.

1. Introduction

Human-induced climate change threatens ecosystems and populations around the world today and increasingly in the future [1]. The majority of countries around the world recognize that only collective action will mitigate climate change. This led to 197 countries coming together in 1992 to adopt a multilateral environmental treaty called the United Nations Framework Convention on Climate Change. Their objective was "stabilization of greenhouse gas concentrations in the atmosphere at a level that would prevent dangerous anthropogenic interference with the climate system. Such a level should be achieved within a time frame sufficient to allow ecosystems to adapt naturally to climate change, to ensure that food production is not threatened and to enable economic development to proceed in a sustainable manner" [2]. The timing of this agreement is relevant given that there was less scientific evidence at the time regarding climate change, and yet member States were driven "to act in the interests of human safety even in the face of scientific uncertainty". It took more than 20 years to agree on the common goal of keeping climate change-related temperature increases to less than 2 degrees, and to pursue efforts to limit global heating to 1.5 degrees, through the 2015 Paris Agreement [3].

The first round of nationally determined contributions was largely complete (186 submissions from 197 members) by 2020. Studies have modeled the implementation measures and goals set out in the nationally determined contributions and the consensus is that they are not sufficient to reach the 2-degree binding, or 1.5-degree aspirational, goal [4]. UNEP's 2020 Emission Gap Report estimates that the global emissions resulting from nationally stated mitigation ambitions currently submitted under the Paris Agreement would lead to global greenhouse gas emissions in 2030 of 53–56 $GtCO_2$-eq per year, aligned with 3 degrees of global heating. Modeled trajectories of global anthropogenic emissions limiting global heating to 1.5 degrees are in the 25 $GtCO_2$-eq per year range [4].

The implications of this mismatch between goals and trajectories are significant. It means more widespread disruption to climate as well as changing ecosystem services that are fundamental to supporting a functional economy and global population of 10 billion [1].

In most cases, analysis indicates that each country's NDC indicates that it would use most of its allowed emissions space for the entire 21st century by 2030 [5], showing that there is a need to look beyond the conventional strategies for climate change mitigation, or that planned measures are not being included in the NDCs. There is consensus that technological solutions alone will not ensure that the 1.5-degree threshold is not crossed [6–9]. Furthermore, the time scales needed to implement these are long term due to the time needed to transition to low-carbon energy and infrastructure, thus they will not generate the immediate reductions needed, and even when they are in place, studies have shown that carbon reductions from technological improvements are far outweighed by carbon increases from economic growth [10].

The Paris Agreement states that *"sustainable lifestyles and sustainable patterns of consumption and production, with developed country parties taking the lead, play an important role in addressing climate change"*. Recent studies estimate that two-thirds of GHG emissions are linked to household consumption—around 6 tCO_2-eq per capita globally, and double that in North America [11]. Since household consumption is driving the majority of emissions, sustainable lifestyles and consumption patterns are a large opportunity for reductions.

This paper will explore whether sustainable lifestyles and sustainable consumption are reflected in nationally determined contributions to climate change mitigation and make the case for shifting the focus in the next round of NDCs from supply-side and territorial emissions to demand-side strategies encompassing low-carbon lifestyles.

We will first make and support our argument that demand-side strategies involving drastic lifestyle changes are required to meet the 1.5-degree goal. We will then review whether, where, and to what extent NDCs embrace measures that consider and support lifestyle changes and explore what barriers might exist that prevent governments from implementing such measures. Finally, we will conclude with recommendations and a future outlook for policy.

2. Methodology

This paper addresses the research question in three steps: (1) reviewing different conceptual frameworks regarding impacts to understand and analyze the role of low-carbon lifestyles (Section 2.1), (2) reviewing the UNFCCC through the lens of a typical policy cycle in order to identify various instruments where low-carbon lifestyles could be included (Section 2.2), and (3) application of these two frameworks to conduct a review of whether and how various policy documents under the UNFCCC address low-carbon lifestyles (Section 2.3).

2.1. Framing Low-Carbon Lifestyles

In order to explore whether sustainable lifestyles and sustainable consumption are included in NDCs and other climate policy documents relevant to the Paris Agreement, we will first summarize different frameworks that have been used to make the case that lifestyles are a vital element of mitigation strategies.

Mitigation pathways can be analyzed in different ways when looking for the role of individual or collective lifestyles in achieving these targets. Various frameworks help to understand the dynamics that contribute to emissions, and hence the leverage points along supply chains where responses to climate change mitigation would be most effective and efficient.

In this section, we will review frameworks before selecting a categorization methodology to assess to what extent sustainable lifestyles and sustainable consumption are included in NDCs and climate change policies.

2.1.1. IPAT

The IPAT equation [12] takes a top-down view to understanding which driving forces have the largest impact on greenhouse gas emissions. Population, wealth, and technology are the three compounding factors:

$$\text{Impact } (kgCO_2\text{-eq}) = \text{Population (cap)} \times \text{Affluence (\$/cap)} \times \text{Technology } (kgCO_2\text{-eq/\$})$$

Population, economic growth, and technology each contribute to greenhouse gas emissions, and conversely, strategies to reduce emissions can—at the macro level—focus on one of the variables.

Technology change, such as renewable energy replacing fossil-fuel based energy, is most often the central and dominating approach of climate change mitigation plans. Many countries, particularly but not exclusively developing countries, set reduced carbon intensity ($kgCO_2$-eq/\$) as their climate mitigation goal, rather than absolute reduction. Studies using this framework have shown that technology change aimed at decarbonizing our economy has resulted in large and steady reductions in absolute growth of carbon emissions [6,13]. It should also be noted here that studies have pointed out that 'technology' measured by production-based carbon intensity can reduce emissions in a country due to outsourcing of carbon-intensive sectors, and hence is different to consumption-based carbon intensity which includes full supply chains [7].

The same studies have shown that economic growth, or affluence measured in \$/cap, overtakes reductions from technology and results in absolute increases in greenhouse gas emissions [6,9,14,15]. For example, Hubacek et al. show that in China, affluence drove increases in emissions by 136%, 85%, and 154% in 1979–1988, 1989–1998, and 1999–2008 respectively. In those same periods, technologies reduced carbon emissions by 44%, 29%, and 30%. Population only increases emissions by between 6–14% in each of those periods [6]. Other studies map out global trends. Lan et al. show that while energy efficiency has reduced energy use by 550 EJ between 1990 and 2010, but this was offset by changes in the production recipe, which added 40 EJ, and eclipsed by increases in energy demand driven by affluence (528 EJ) and final demand composition (56 EJ) [13]. Far reaching changes to the 'affluence' part of the equation are now considered essential to complement technological change if we are to transition to sustainability, as overconsumption by affluent consumers is the overwhelming driver of global environmental impacts [8]. Climate mitigation goals and strategies aiming at the affluence dynamics are less common, particularly in NDCs. This is despite the rise in calls for degrowth or a steady state economy [8,16].

Population does not change as dramatically as technology or affluence, and thus has moderate impacts on climate change, although there are geographic variations. It is generally not featured in climate change policy, likely due to the human rights challenges associated with population control and the economic challenges of an aging population.

In reviewing policy, 'affluence' options are most relevant to sustainable lifestyles, and would steer choices that reduce the overall volume of consumption or shift expenditures to lower impact sectors. Examples of reducing volume include downsizing housing or reducing food waste (as long as this reduces the amount of food purchased, not overeating to avoid food waste). Examples of shifting to lower impact sectors include moving from 100% personal vehicle use, to mixed mobility including car rental, public transport, and non-motorized transport. Technology options would include shifts to more efficient goods and services within the same sector, such as switching to produce from low impact agriculture, or an electric vehicle, or an energy efficient fridge.

Nationally determined contributions mostly focus on the "T" in the IPAT equation, making goods and services low-carbon rather than invoking discernable change for people, and may as a result be missing opportunities to move closer to the absolute reductions set out in the Paris Agreement.

2.1.2. Direct vs. Indirect Emissions (Carbon Footprints; Territorial vs. Imported; Scope 1, 2, 3)

Direct emissions refer to the physical emission of greenhouse gases, such as burning fossil fuels and releasing CO_2. Indirect emissions are greenhouse gas emissions that are a consequence of an activity but occur at an upstream or downstream stage of a supply chain; for example, red meat consumption in households drives methane emissions from cows. A wide range of terminology exists to describe sub-classifications of direct and indirect emissions [17], such as production- vs. consumption-based accounting, carbon footprints, Scope 1, 2, or 3 emissions, and territorial vs. extraterritorial emissions. Direct emissions are associated with territorial, Scope 1, production-based accounting, whereas indirect emissions are associated with Scope 2 and 3, extraterritorial, consumption-based accounting, and carbon footprints.

The GHG Protocol [18] classifies direct emissions as Scope 1. Indirect emissions can be classified into Scope 2 and Scope 3 emissions. Scope 2 emissions include emissions from purchased electricity, steam, heating or cooling, and occur very close in the supply chain from a specified activity. Scope 3 emissions are all other indirect emissions, both upstream and downstream in the supply chain that are not included in Scope 2. In the case of consuming milk, for example, there would be no Scope 1 emissions, Scope 2 emissions would include the share of electricity needed to power the fridge, and Scope 3 emissions would include the share of methane emissions from the cow, greenhouse gas emissions from any heating or cooling purchased during processing, vehicle emissions from transport, all indirect emissions from packaging, and finally emissions such as methane during waste management of the portion of discarded milk and packaging.

Carbon footprints are the sum of direct and indirect emissions, though some accounting frameworks omit downstream emissions.

Another classification distinguishes between territorial and extraterritorial emissions. Territorial emissions are direct emissions within a given territory, whereas extraterritorial emissions are the indirect emissions of domestic activities that occur beyond national borders. When it comes to emissions reporting under the UNFCCC and its climate agreements, only territorial emissions that occur within a country's borders are counted. These include Scope 1, Scope 2, and Scope 3 emissions of consumption only as far as the supply chain is domestic.

Supply driven mitigation strategies align more closely with territorial emissions, whereas demand-side mitigation pathways are more closely associated with non-territorial emissions given the globalized nature of supply chains. In European countries, 25–30% of emissions related to lifestyles occur abroad, due to their highly globalized supply chains [19]. Therefore, including sustainable lifestyles in international or policy commitments is disincentivized as the emission reductions credited to the country are only 70–75% of what they may have achieved in reality. This is a potential area of opportunity in the near term, since reduction strategies for territorial emissions are rapidly reaching their limitations in developed countries with highly globalized supply chains [20,21], and it might be more cost effective to reduce extraterritorial emissions than further territorial emissions. Another indication of the growing importance of indirect emissions is the disproportionate growth of company level Scope 2 and 3 emissions compared to Scope 1. A recent study found that between 1995 and 2015, Scope 1 emissions had grown by 47%, whereas Scope 2 and Scope 3 emissions grew by 78% and 84%, respectively.

Addressing only direct or indirect emissions in climate policy leaves out significant opportunities to implement and achieve climate goals [22]. Once direct domestic emission reductions are achieved, neglecting to include the outsourced, indirect impacts will potentially undermine global mitigation efforts on climate change [8].

2.1.3. Final Demand Categories

At the meso, or sectoral, level, studies have analyzed the contribution of different sectors that contribute to climate change from two perspectives—supply and demand.

The supply side looks at the sectors where emissions directly occur, for instance in the conversion of fossil fuels to CO_2. Energy and transport dominate the sectors contributing to direct emissions, with waste, forestry, and agriculture coming in on the next tier [23]. Nationally determined contributions tend to align well with these identified priorities with most outlining actions addressing energy, transport, and forestry.

A demand perspective, through indirect emission accounting, looks at the consumption categories that drive the volume of output of these sectors and hence the quantity of associated emissions. This is an important approach, because structural decomposition analyzes based on the IPAT equation have shown that changing the carbon intensity of supply while keeping demand constant or increasing will not result in the absolute reductions needed to achieve the Paris Agreement. Methodologies such as input-output analyses and life-cycle assessment enable quantification of the contributions of different demand sectors to overall greenhouse gas emissions. They also enable quantification of demand-side climate mitigation measures [24]. The findings are consistent and "unambiguous", with food, mobility, and housing accounting for 70–80% of life-cycle carbon emissions [23]. On the demand side, buildings feature prominently as drivers of indirect emissions from electricity use and are also frequently included in nationally determined contributions.

2.1.4. Sustainable Consumption and Production, SCP 1.0, 2.0, 3.0

Sustainable consumption and production (SCP) was first adopted as an international policy goal under Agenda 21 in 1992 [25]. Ten years later, the UN adopted the 10 Year Framework Programme of Sustainable Consumption and Production, which included a task force on sustainable lifestyles and education that aimed to advance sustainable lifestyles policy and mainstreaming. In 2015, the UN adopted the Sustainable Development Goals, which included a goal on responsible consumption and production that specifically mentioned "lifestyles in harmony with nature". Worth also mentioning here is the inclusion of material footprints as an indicator for sustainable resource management (target 12.2) and resource efficiency (target 8.4), showing that international policy frameworks have included extraterritorial environmental pressures in other domains.

Over this period, the conceptual framework of sustainable consumption and production broadened from an approach anchored on industry support for cleaner production and education support for sustainable lifestyles, to creating the right enabling frameworks for sustainable consumption, including policy. Hotta et al. [26] developed a classification of the evolution of the SCP policy discourse that is useful for analyzing approaches to sustainable consumption in climate policies.

SCP 1.0 refers to more nascent stages of SCP policy development, which focus largely on cleaner production and pollution prevention. The policy discourse centered on the supply side, direct emissions, with little reference to sustainable consumption or sustainable lifestyles. Although at the international level this perspective was mainstream in the 1970s and 1980s, and later evolved to include life-cycle and demand-side, it is still a dominant approach in countries at early stages of their SCP journey.

SCP 2.0 broadened the perspective of SCP to include the product life cycle in the 1990s. A few factors led to this expansion, such as increasing globalization and fracturing of supply chains, increases in the visibility of environmental impacts, and the development of life-cycle assessment and other supply chain accounting methodologies. Another factor was the maturing of ecological economic theory, which found compatibility between economic competitiveness and environmental sustainability to be possible through resource efficiency, the 'technology' component of the IPAT equation. The introduction of life-cycle assessment made the connection between emissions from the production stage to the consumption phase and started to make the case for sustainable consumption decisions.

SCP 3.0 policy approaches are society-wide, multidisciplinary, and could have significantly higher benefits than SCP 1.0 and 2.0 paradigms. They encompass concepts such as planetary boundaries and sufficiency, which are closely linked to degrowth and dematerialization. Sustainable lifestyles also emerge as a core concept, but are closely linked with creating conducive social context and infrastructure. Sustainable lifestyles policy under this framework is less about appealing to consumers directly through awareness raising campaigns, and more about using policy to change social design. Mao et al. [27] suggest applying this broader, societal context approach to foresight analyses that can support the formulation of sustainable lifestyles policy frameworks.

2.1.5. Sustainable Lifestyles, Sustainable Consumption, Individual Action

This section reviews the literature that frames sustainable consumption or sustainable lifestyles in more granular or nuanced ways. Creutzig et al. [28] distinguish between demand and supply sides, with the demand side including a broad spectrum of "technology choices, consumption, behavior, lifestyles, coupled production–consumption infrastructures and systems, service provision, and associated socio-technical transitions". Moran et al. [29] consider a slightly narrower subset of "consumer options" only including low-carbon choices that are possible for consumers today without requiring government or supply-side actions. The broad categories include: reduce consumption; reduce disposal; change consumption pattern/purchase alternative product; change use behavior; change disposal behavior; purchase more efficient products.

Addressing individual (or household) action directly is shown to be worthwhile. Estimates from Dubois, Sovacool et al. [30] show that households drive 72% of global greenhouse gas emissions. Other studies [31] found that seventeen actions could collectively reduce household (territorial) emissions in the US by 20%, equating to almost 2% of global emissions and more than France's total emissions. Moran, Wood et al. found that a portfolio of household actions achievable today without infrastructure investments can reduce carbon footprints by 25% in Europe [29]. Of the 6 tons CO_2-eq per capita per year that is attributed to households, 1.7 tons CO_2-eq per capita could be reduced from sustainable transport choices such as car-free living, 0.9 tons CO_2-eq per capita from food choices including a plant-based diet, and 1.6 tons CO_2-eq per capita could be reduced in housing including shifting to renewable electricity and renovating [11]. Sector-specific studies also support the shift to demand-side policies, such as food policy which has long addressed consumption patterns in order to achieve health policy goals [32] and in household electrification [33,34].

These findings make individuals key actors in reaching the 1.5-degree goal of the Paris Agreement. However, there is limited understanding and treatment of behavioral change and the relevant policies in mitigation pathways currently submitted by countries to contribute to achieving 1.5-degree ambition of the Paris Agreement. This is surprising as the Paris Agreement itself states that *"sustainable lifestyles and sustainable patterns of consumption and production, with developed country parties taking the lead, play an important role in addressing climate change"*, and the IPCC's 2018 special report dedicated a chapter to behavior change strategies [1].

Numerous studies have found that citizens would accept and moreover expect governments to put in place policies that control consumption choices [35]. However, most demand-side-oriented policies use financial instruments that still largely depend on market forces to steer behavior change, crucially leave low-cost carbon-intensive options on the market, and furthermore generally target low-impact behaviors [30]. They generally neglect the most carbon-intensive consumption patterns (meat and air travel).

2.1.6. Environmental Kuznets Curve

A fundamental principle of the Paris Agreement is 'common but differentiated responsibility' (CBDR) with developed countries taking the lead. This principle refers to the cumulative, or historic, carbon emissions, most often higher for developed countries, which have a higher share of the total greenhouse gas emission concentrations in the atmosphere. Based on this rationale, developing countries, or countries with low cumulative historic greenhouse gas emissions, have lower responsibility and/or economic capacity to mitigate climate change now, even if their annual emissions are high.

The environmental Kuznets curve is a hypothesis that there is an inverted U-shape relationship between economic development and environmental pollution. Countries start with small economies and small pollution levels, and both factors grow until pollution peaks, at which point economic growth continues while pollution reduces. Under this proposition, and in line with CBDR, developing countries may not include ambitious greenhouse gas mitigation goals in their NDCs. This matter is an important factor when analyzing how developing country NDCs tackle sustainable lifestyles and consumption.

Grottera, La Rovere et al. argue that developing countries have greater potential to apply demand-side mitigation strategies [36]. This is partly because developing and middle-income countries like India and China have fast-growing GDP rates, and affluence has been proven to be the driving force behind emissions. This is particularly so because, despite having emerging economy or developing country status, and low per capita GDP rates, they still are home to a large and growing absolute number of affluent consumers. For instance, there are more billionaires in China and India combined than in Europe or the US [37]. Given that affluent consumers drive environmental impacts, it is important to consider affluence in developing country NDCs where there are a large number of affluent consumers.

As countries that will grow the most, and still make decisions and policies that affect consumption patterns that are not yet locked in, much of the mitigation potential lies with developing countries. In normal trajectories, countries may argue to follow an environmental Kuznets curve, developing first then integrating environmentally friendly practices later. However, this has been shown to be ineffectual for global environmental issues such as carbon footprints (more effective for highly local impacts like smog) [38]. As low-income consumers rapidly shift into middle- and upper-class consumption patterns, and countries that are classified as 'low-income' and yet still are home to large numbers of high-income consumers, they will need to address the environment impacts of consumption. Particularly countries such as China and India have the opportunity for "lifestyle leapfrogging" where they skip the carbon-intensive lifestyles of the industrialized countries, but improve their quality of life [39].

This means that developing countries need to integrate climate change into their development agenda. A transition to lower carbon-intensive lifestyles is not easy due to systemic barriers: lack of existing capital, lack of awareness, upfront costs, inertia, and other priorities. Hence a proactive policy is needed at early stages of development trajectories. The dominant development approach of grow first clean up later, along a Kuznets curve, does not occur from a consumption-based perspective [38,40].

2.2. Framing the UNFCCC Policy Process

The policy instrument this paper focuses on, the NDC, generally does not refer to demand-side mitigation to a large extent. In order to identify where barriers to inclusion of demand-side measures might occur, we will present and refer to the policy cycle and how it applies to the Paris Agreement. The Paris Agreement is a policy framework managed by the UNFCCC, which follows a typical policy cycle to shape and implement. UNEP describes a typical environmental policy as follows [41]:

1. Problem framing. This is when information is gathered, analyzed and the nature of the problem is agreed on. In the context of global cooperation on climate change, this is done through the science policy interface called the Intergovernmental Panel on Climate Change, which was established in 1988 "to provide policymakers with regular assessments of the scientific basis of climate change, its impacts and future risks, and options for adaptation and mitigation [1]. IPCC assessment reports help to shape the

PA, and subsequent assessment reports are intended to provide independent scientific evidence to support national action and global cooperation.

2. Policy framing. Once enough knowledge is gathered through the problem framing stage, policy goals are defined, along with guiding principles. The climate change policy goal finally agreed on was staying under 2 degrees warming, or 1.5 degrees ideally, and is described in the Paris Agreement [3]. The UNFCCC also includes guiding principles, the most prominent being "principle of equity and common but differentiated responsibilities and respective capabilities, in the light of different national circumstances."

3. Policy implementation. Once the policy goals are established, policymakers must outline how they will be achieved, through a selection of policy instruments and allocation of budgets and other resources, and how they will be monitored. Under the Paris Agreement, member States are required to submit successively more ambitious nationally determined contributions every five years, outlining how the member States will contribute to the policy goal and enable monitoring [42]. In many cases, these are kept high level in nature, and are complemented by national climate policy documents that provide detail about the strategies that will be employed to meet the targets outlined in the NDCs.

4. Policy monitoring and evaluation. In this step, regular monitoring and reporting support evaluation of the selection of policy implementation mechanisms compared to the stated policy goal. Nationally determined contributions enable monitoring of progress, as well as modeling and forecasting whether we are on track for the 2-degree target and reviewing who is contributing to what extent. In addition, biennial updates provide details about mitigation plans and progress, and hence include significant amounts of data not covered in the NDCs. The measurement framework of carbon accounting is the IPCC's 2006 reporting guidelines [43], which quantify greenhouse gas emissions by country.

2.3. Reviewing Climate Policy Documents to Identify How Sustainable Lifestyles Are Integrated

Reviews of climate policies mostly assess their headline targets on absolute or intensity reductions. There has not yet been a review of whether the call to include sustainable lifestyles has been reflected in climate change policy under the Paris Agreement.

In order to determine whether climate policies are including sustainable lifestyles, we first differentiated between different types of climate policies relevant to international climate change policy development and monitoring. Table 1 summarises which types of policies have been included in the study.

Climate policies exist at the international, national, sectoral, and local level. It was beyond the scope of this paper to determine which levels are the most effective on climate change mitigation. Local and sectoral policies may have a closer alignment between mitigation options and the respective mandates and budgets. However, only headline NDCs are counted in the UNFCCC and third-party international monitoring of climate change policy therefore this was selected as a first step. In some cases, particular for non-Annex 1 countries, the NDCs did not offer significant details, but the biennial update report submissions from non-Annex I parties [44] compiled significant detail on mitigation strategies (India, Indonesia). This might be because they are reports, rather than binding commitments. Particularly for Annex 1 countries, NDCs do not always include details of strategies or policy measures that will be applied in order to reach the targets. Therefore, the second tier of policies to review includes the national climate change policies. This was particularly useful in the case of EU NDCs, which all follow a common template despite each country having vastly different contexts and mitigation plans. In some cases (Malaysia, China), national socioeconomic policies were useful to include since they included commitments on sustainable lifestyles that were missing in the climate change strategies and were considered binding enough to include.

Table 1. Climate change policy instruments under consideration.

Type of Policy Document	What Is It?	Rational for Inclusion or Exclusion
Nationally Determined Contribution (NDC)	Documentation of climate policy goal, together with the actions that will be taken to achieve the goal. To date, 197 NDCs have been submitted.	Included, as it is the formal, universal policy instrument in international climate policy and assessments.
Biennial Update Report (BUR)	An update of climate mitigation, support needed and greenhouse gas accounts by non-Annex-1 parties. To date, 24 BURs have been submitted.	Included, as it is a formal climate policy instrument, and contains relevant details that are lacking in NDCs.
National socioeconomic development strategy	Macro-level economic and/or development strategies that serve as a chapeau for national policy. Examples include China's 5 Year Plans [45], or the Eleventh Malaysia Plan [46].	Partially included where known to include policy relevant to sustainable lifestyles.
Sectoral policy	Detailed policies for sectors, which in the case of energy, transport and buildings, often include specific mention of key measures to achieve greenhouse gas emission reductions.	Not included in this study, as too large in number to be feasible to systematically review.
Subnational policy	Detailed planning strategies for states, provinces or cities, which also often have highly specific mentions of climate change mitigation measures and are much closer to the point of greenhouse gas emissions.	Not included, as too large in number to be feasible to systematically review.

The second step was to determine the country selection of the study. Given that 187 countries had submitted NDCs, it was not possible to review each of them. Three criteria were used to determine the geographic scope. First was alignment of other reviews of NDCs to facilitate cross referencing. The Climate Action Tracker [47], for instance, reviews 7both the headline commitments of countries to GHG reductions as well as actions in five sectors—energy, industry, transport, buildings, and forestry. This omits two sectors key to demand-side management and affluence: food and consumer goods/waste. The UN Emissions Gap report reviews the NDCs of G20 countries, mainly analyzing the headline commitments, and reviewing the national policies most relevant [48]. This report reviews progress against key sectoral climate change goals in energy, industry, transport, buildings, and agriculture, but limited targets related to demand-side management. This aligned closely with the second criterion, which was to capture the bulk of global GDP, given that carbon-intensive lifestyles requiring policy attention are more likely to be in wealthier countries. Our third criterion was to include countries that had relevance to sustainable lifestyles, but were not included in the above two criteria. This included countries with carbon-intensive lifestyles, or countries known for a compelling approach to sustainable lifestyles.

The final scope included the following selection: first, G20 member states, in line with the UN Emissions Gap report (Argentina, Australia, Brazil, Canada, China, India, Indonesia, Japan, Mexico, Republic of Korea, Russia, Saudi Arabia, South Africa, Turkey, USA, EU28 (countries with recent climate change strategies available in English: Germany, Netherlands, Norway, Sweden, Denmark, Estonia, Portugal, France, Slovakia, Austria)). Secondly, additional countries that had high per capita carbon footprints (Monaco, Qatar), and thirdly those with a compelling approach in their climate change mitigation policy (Norway, Bhutan, Seychelles, Sri Lanka, Thailand, Vietnam, Kenya, Israel, Pakistan, Switzerland, Malaysia, New Zealand).

3. Results

This section will specifically review the lifestyles or affluence strategies captured in NDCs, in national policies mentioned in NDCs, or apex national climate change strategies. The aim is to assess whether NDCs and relevant national policies are capturing the full potential range of climate change mitigation options. We will first explain our approach in reviewing NDCs and policy documents, followed by a structured presentation of findings.

3.1. Structuring Findings

For the purpose of this paper, the choice of how to structure sustainable lifestyles elements of NDCs was made with the aim of (1) providing insights into how NDCs and relevant climate change policies refer to low-carbon lifestyles, and (2) supporting the identification of the type of lifestyle changes missing from NDCs that could contribute significantly to the additional climate change reductions needed to remain within the 1.5-degree target.

First a distinction was made between five categories—housing, mobility, food, goods/waste, and leisure—in line with Section 2.1.3. Within each category, mitigation measures that would satisfy the criteria of Creutzig et al. [28] in Section 2.1.6 were included, in the sense that they had to involve a decision by a consumer or an otherwise involuntary change in their life. This extended more broadly than the criteria of Moran et al. [29], because their classification focused on present-day possibilities, whereas NDCs and climate change mitigation strategies are planned to 2050, thereby including options not yet available or requiring initial investment in infrastructure by government. Two further categories were then added, one reflecting SCP 3.0 approaches (infrastructure, social norms, sufficiency, lifestyles) in line with Section 2.1.5, and a category for special references to extraterritorial emissions.

3.2. Results of the Policy Review

The table in the Note Information outlines which countries included sustainable lifestyles in their NDCs or national climate change policies, and attempts to distinguish between territorial emissions, which are within the scope of the Paris Agreement, and non-territorial emissions (or footprints), which are not directly included yet in reporting but referred to in the IPCC reports as necessary to mitigating climate change.

We make four general observations. First, all NDCs reviewed included housing and/or mobility, and therefore did touch on sustainable lifestyles. Most NDCs cover building energy efficiency, many cover public transport. Increasingly, they also cover indirect emissions, such as from food and consumer goods. The circular economy is also emerging as a cross-cutting strategy with many consumer-facing implications. Second, some NDCs or national climate change policies go further to mention reducing consumption, particularly reducing transport consumption through flexible work policies and urban planning, food waste prevention, and switching to share economy systems over personal ownership. Thirdly, several specifically include references to carbon footprinting (Switzerland, France, Japan, Republic of Korea), which was also included in the IPCC AR5 (though omitted from the summary for policymakers). Last, policies released most recently are more likely to include lifestyles, possibly due to the findings of the IPCC 2018 special report. There are exceptions, including China and Japan. In China's case, 85% of its carbon footprint is domestic, therefore the motivation may be to reduce territorial emissions, and remain within carrying capacity of its own environmental systems. In Japan's case, there is a larger footprint abroad, therefore the motivation may be based on common but differentiated responsibility.

More specific observations are listed in Table 2.

Table 2. Summary of findings: four different levels of inclusion of sustainable lifestyles in climate policy.

Level of Inclusion of Sustainable Lifestyles	Findings
Inclusion of lifestyles/carbon footprints directly in the NDC	One NDC included direct reference to demand-side policy goals directly in the text of the NDC. China's 2016 NDC [49] includes a section on "promoting the low-carbon way of life" which calls for a reduction of materials consumption: "moderate consumption, encourage the use of low-carbon products and curb extravagance and waste". Other sections also include behavior or consumption change measures that fit in direct emissions, such as low-carbon buildings, spatial planning, public transport, and pedestrian and bicycle infrastructure.
Inclusion of lifestyles in national climate change policy	In many cases where NDCs do not directly include demand-side or lifestyle change, the respective national strategy does include demand-side strategies. This includes the EU countries, which did not specify in their joint NDC [50] how they would achieve their mitigation goals but were required to submit long term mitigation strategies (Slovak Republic [51], Portugal [52], Austria [53], Denmark [54], Estonia [55], France [56], The Netherlands [57], as well as Switzerland [58], Norway [59], Monaco [60], Malaysia [61]). Food: Austria [53], France [56], Slovak Republic [51], Norway [59], Switzerland [58], the Netherlands [57] included climate-friendly diets, specifically lowering consumption of meat and dairy. Portugal [52] and Denmark [54] referenced shifting diets towards local and organic produce. Several countries also mentioned food waste reductions. Goods/Waste: Austria [53], Denmark [54], Estonia [55], the EU [50], France [56], Japan [62], Malaysia [61], Monaco [60], Sweden [63], Portugal [52], Seychelles [64], Slovak Republic [51], Switzerland [58], included references to shifting consumption habits towards share, reuse, rental, repair, and extended product lifespans. Cross-cutting: Several countries made a specific reference to carbon footprints and emissions outside of national boundaries, and linked this to product labeling or calculators, including Sweden [63], Republic of Korea [65], Switzerland [58], France [56], Japan [62], EU [50], Denmark [54], Austria [53].
Reference to sustainable lifestyles without referring to specific measures	In several cases, inclusion of sustainable lifestyles occurred in a headline or macro level manner, without significant, specific or quantified details. Germany [66] did reference food waste reductions, but did not quantify them or refer to dietary change. New Zealand [67], Singapore [68], Sri Lanka [69], Thailand [70], and Pakistan [71] each reference lifestyle changes, but either in a broad way, or in a way that would not significantly reduce emissions (e.g., reducing packaging waste).
No inclusion of lifestyles in NDC or national climate policy	Many NDCs and the climate change policies reviewed did not include reference to lifestyles or footprint/indirect emission reductions, including those of Brazil [72], Qatar [73], Australia [74], India [75], Indonesia [76], Israel [77], Kenya [78], South Africa [79], USA [80].

4. Discussion

The global community now has access to a broad range of studies confirming that lifestyle change across all consumption domains will be needed to keep climate change within 1.5 degrees of warming. In this paper we confirmed that the majority of NDCs do not significantly include lifestyle change, in particular the large emitters (USA, Australia, Singapore, India, Russia). There are signs that this is changing, and some more recent NDCs and climate change strategies do include demand-side measures including those related to indirect emissions (food, goods and services). Given the limited time left to change course, climate change stakeholders must address barriers to addressing the full range of mitigation measures recommended by the IPCC, including significant demand-side measures. Table 3 provides an overview of potential barriers that may explain the trends in Section 3.

Table 3. Potential barriers to inclusion of demand-side mitigation measures in climate policy.

Nature of the Barrier	Explanation
Agreed scope of NDCs	The Paris Agreement refers to NDCs in Article 4. This negotiated text refrains from laying out a mandatory scope for NDCs; therefore it is up to each country to develop its own scope and format. However, there are some keywords in the text that may serve as barriers to inclusion of lifestyles related emissions. For instance, Article 4/2 states that "parties shall pursue domestic mitigation measures", which can be interpreted as territorial emissions, and hence may disincentivize action on indirect emissions that partly reduce emissions abroad rather than domestically. An exception to the focus on the territorial emissions rule is the case of offsets, whereby a country may take credit for emission reductions abroad, but not emission increases.
Leaving room for improvement	Article 4/3 states that "successive nationally determined contribution will represent a progression beyond the party's then current nationally determined contribution". Considering that the Paris Agreement calls for increasing levels of ambition every five years in successive NDCs, some countries may wish to reserve the full portfolio of mitigation measures for future NDCs. They may also withhold early ambition in order to negotiate deals in the future if they are developing countries not required to achieve absolute reductions.
GHG accounting does not include extraterritorial emissions	Extraterritorial emissions are not included in the IPCC accounting framework. Article 4/13 states that member States must "ensure the avoidance of double counting." Therefore although carbon-intensive lifestyle choices, with inherent extraterritorial emission footprints, are not accounted for, since the supply side emissions are accounted for in another NDC [81]. A significant proportion of GHG footprints occur abroad (23–30%) [19] and therefore reductions will be accounted for in other country NDCs. If reductions in extraterritorial emissions associated with domestic consumption are not measured, and not reported in the Member State updates to the UNFCCC, there is no incentive to reduce them in mitigation strategies, despite the IPCC stating that demand-side strategies are critical to meeting Paris Agreement goals.
Lag time between NDCs and the IPCC special report, unclear science-policy link	There is a lag time between the IPCC report clear messages on lifestyles, and the time it takes to formulate new NDCs is at least 2 years. Most NDCs pre-date the call from the IPCC's special report in 2018 for inclusion of sustainable lifestyles in climate mitigation plans. Some evidence of this is that more recent climate change policies (notably from Europe and Japan) arising approximately two years after the IPCC special report have included sustainable lifestyles. A related issue is that although the Paris Agreement specifically called for the IPCC special report, it did not specify how the findings would be applied in climate change mitigation strategies. This is a missing link in the policy cycle, between problem framing and policy implementation, as outlined in Section 1.
Perspective of negotiators vs. the perspectives of practitioners	The Paris Agreement and NDCs are the responsibility of negotiators skilled in strategic foreign policy. This skill set may be more biased towards more conservative levels of ambition, particularly given the nascent nature of the Paris Agreement and the requirement for continual increases in ambition. However, the skill set needed in designing national mitigation strategies would need to be more practical, ambitious, and risk tolerant in order to achieve the magnitude of change required. Practitioners and experts are thus key stakeholders in the NDC drafting process.
Prescription vs. consensus	There are no templates or internationally agreed guidelines for NDC development that outline a menu of mitigation options, aligned with IPCC recommendations, to support those tasked with NDC design. Although officially recognized (and costly) scientific assessments such as the IPCC special report lay out policy relevant *findings* regarding mitigation options, these remain separate from the policy *guidance* on mitigation options, for instance through templates or manuals. One reason behind this is to avoid prescriptive policy messaging that may jeopardize the consensus that is critically needed as a minimum to maintain the Paris Agreement. NDCs are nationally, not internationally, determined, as the name indicates, so all member States are able to arrive at their mitigation strategies independently of any international recommendation.

Table 3. *Cont.*

Nature of the Barrier	Explanation
Developing countries are not required to commit to absolute reductions	The Paris Agreement states that developed countries should take the lead with absolute reductions, whereas developing countries should "continue enhancing their mitigation efforts" (Article 4/4). Not all developed countries do commit to absolute reductions (e.g., Singapore has committed to a reduction in carbon intensity), and none of the developing countries did so. This is another disincentive to reach for mitigation measures that maximize reductions that are not required. Countries with low per capita or cumulative emissions may also wish to avoid politically and officially accepting responsibility for mitigation through consumption under the principle of common but differentiated responsibility.
Lack of awareness or appetite for demand-side climate change mitigation	There are several factors that could be at play. First, the fear of public backlash to policies that affect lifestyles. Second is the perception that they do not have responsibility for extraterritorial emissions, particularly in the case of import-dependent countries (such as Singapore) or countries with low per capita emissions (such as developing countries with significant and rapidly growing affluent communities, like India, Thailand, and Indonesia). Third is the mismatch between the mandate of the national agency setting aspirational mitigation targets (often a ministry of environment or climate change), and the know-how of the sectoral agencies tasked to provide the mitigation strategies (ministries of industry, transport or agriculture) which are less accustomed to dealing with demand-side strategies that involve significant understanding of behavior change. There are exceptions, however, as many developing countries are addressing the impacts of consumption at early stages of their development trajectory (Sri Lanka, Bhutan, China), and developed countries that are integrating deep behavior change strategies in their national mitigation plans (Slovakia, Austria, Portugal, the Netherlands).
Open question on whether the demand-side should be prioritized	While demand is the key driver of environmental impacts, it is not the point of actual emissions. Consumers do not directly control or easily find information about impacts behind supply chains, even if they do have the ultimate power over the consumption decision [19]. Both actors, producers and consumers, have responsibility and opportunity to mitigate climate change, since one approach of the other will not be sufficient in isolation. Different macroeconomic theories can be used to support this [82].
Reduced control over implementation effectiveness from reduced consumption	Indirect emission reductions are not easy to guarantee, since demand reduction or change may not eliminate the upstream emissions associated with the consumption activity. Reduced demand for a carbon-intensive product may have unexpected impacts such as a reduction of price that increases demand elsewhere.
Lack of methodological frameworks to support policy action on behavior change	Many countries, particularly in Europe, have research institutions that can support the development of evidence-based, demand-side policies, as well as quantification of carbon footprints. In countries that do not yet have this expertise, the lack of scientific basis is a barrier to demand-side policy commitments. The outlook is positive here, as the body of literature on sustainable lifestyles and other demand-side solutions is "growing exponentially", though slower than the growth in climate-related studies [83].

5. Conclusions: Recommendations

The policy cycle framework outlined in Section 1 shows how environmental policies, including those on climate action, should be developed and monitored based on scientific evidence. Climate policies, such as NDCs, should apply the full range of scientifically identified climate change mitigation strategies in order to reduce emissions sufficiently and efficiently, and to more accurately monitor the combined commitments. However, there are a range of barriers when it comes to demand-side policies. In order to overcome them, researchers and policymakers need to collaborate far more to increase the uptake of methodological frameworks that can quickly and comprehensively support countries in selecting the right policy goals and instruments. Below are some recommendations on potential solutions to address the barriers identified in Section 4.

The accounting and reporting methodology plays a fundamental role, and is currently not conducive to demand-side or extraterritorial emission reductions. Despite this, indirect emissions, or footprints, are increasingly referred to in climate change mitigation strategies (Korea, France, Austria, Japan) and can be powerful strategies to bridge the gap to climate change goals. However, the pressure to deliver on domestic emission reductions, and concerns about double counting extraterritorial emissions remain unresolved. Therefore, member States should agree on a globally accepted accounting methodology that enables them to report on indirect extraterritorial emission reductions while addressing double counting concerns.

Second, a move away from national borders would help optimize climate mitigation measures and inclusion of demand-side measures. The globalized nature of our supply chains has stifled the optimization of climate change action. Instead of aiming for the largest mitigation opportunities, countries are focused on domestic mitigation, with the exception of the EU. The EU has shown that submitting one common regional NDC can encourage inclusion of supply chain emissions or footprints, since a large percentage of EU countries' extraterritorial emissions are still within the EU [29] and hence contribute to the common emission reduction target. Climate policymakers should consider the transboundary impacts and mitigation opportunities related to domestic demand within their NDCs. There is some precedent for this. The international nature of emissions is already acknowledged in the Paris Agreement through offsets, where countries can be credited for reducing extraterritorial emissions through offset programs. Article 14 of the Paris Agreement opens doors for this under the terms of "collective progress" and "enhancing international cooperation for climate action".

Third, there could be more official guidance on NDC scope, particularly on linking the NDCs to the IPCC findings. The inconsistencies between NDCs make it difficult to monitor progress against mitigation pathways and compare countries. More guidance could be provided on how to arrange demand-side climate mitigation actions according to existing or additional sectoral categorization. This could encourage countries to reflect existing measures already in national policy in the NDC, and also provide a nudge to include demand-side mitigation strategies. Climate mitigation is generally categorized according to energy, transport, industry, agriculture, Land Use, Land-Use Change and Forestry (LULUCF), and waste; demand-side measures could be added as a cross-cutting sector or as subsectors in the existing sectors.

Related to the above, a strengthening of the science–policy interface could be achieved through dedicated training for NDC developers on how to include demand-side and extraterritorial mitigation measures recommended by the IPCC. The decision to adopt the Paris Agreement (1/CP.2) included a paragraph (Article II/21) under the article covering NDCs that the IPCC "provide a special report in 2018 on the impacts of global warming of 1.5 °C above pre-industrial levels and related global greenhouse gas emission pathways"). There is no mention of how this report's findings should be used in the design of nationally determined contributions going forward. Training for NDC developers can help them address domestic political and strategic concerns, while also employing the full spectrum of climate change mitigation options identified through international scientific assessment.

Lastly, countries need nationally tailored support to establish the evidence base for, and design of, demand-side policies. There is an urgent need for more country level assessments of options for demand-side climate change mitigation to even out the asymmetry in availability of such assessments for countries. While all countries should reduce supply side and territorial emissions, there may be cost effective, fast options available to them that are either indirect or occurring abroad in upstream supply chains or both. National studies should provide a quantification of the demand-side mitigation options, and also seek solutions to harmonize demand-side options with national political and economic contexts. This is particularly urgent in developing countries undergoing rapid growth and hence holding significant future cumulative responsibility for GHG emissions under business-as-usual projections.

6. Future Outlook

From an academic point of view, findings from different multi-regional input–output assessments are converging [84], institutional and governance requirements are clear [85], implications of consumption-based accounting are understood [86], and case studies from policy practice exist [30]. In short, the scientific domain of carbon footprinting and sustainable lifestyles policy has reached a level of maturity and agreement to be ripe for application in climate policy.

Member States have already agreed to including a different footprint metric, the material footprint, in another international agreement—the Sustainable Development Goals indicator framework. It serves as an indicator for SDG 8.4 on resource efficient growth, and SDG 12.2 on sustainable resource management. The metadata include a methodology, based on the multi-regional input output framework [87], under the caretaker organization, the UN Environment Programme. Its inclusion in the SDGs is not binding, since all member States can select their own indicators, and the goals themselves and reporting processes are also voluntary. However, the endorsement of this methodology and survival in a multilateral agreement process dependent on consensus give some hope that carbon footprints can also be integrated into the methodological and reporting framework under the Paris Agreement. Linkages between climate policy and the Sustainable Development Goals may also create synergies that enable demand-side mitigation strategies [88].

Climate policy and research has made promising progress in the spirit of the Paris Agreement's statement that *"sustainable lifestyles and sustainable patterns of consumption and production, with developed country parties taking the lead, play an important role in addressing climate change"*. Six years into the Paris Agreement implementation, it is urgent that all countries apply the best available knowledge on the full range of climate mitigation options to the NDCs. Sustainable lifestyles are considered essential to achieving targets, therefore the barriers to including them in NDCs and national climate policies need to be further investigated so that the solutions can be shaped and implemented well before the global carbon budget is depleted.

Author Contributions: Conceptualization, J.S.; data curation, J.S.; formal analysis, J.S.; investigation, J.S.; methodology, J.S. and M.L.; supervision, M.L.; validation, M.L.; writing—original draft, J.S.; writing—review and editing, M.L. and Y.H. All authors have read and agreed to the published version of the manuscript.

Funding: This publication was supported by the Environment Research and Technology Development Fund (S-16-3: JPMEERF16S11630) of the Environmental Restoration and Conservation Agency, Japan.

Institutional Review Board Statement: Not applicable.

Informed Consent Statement: Not applicable.

Data Availability Statement: Not applicable.

Conflicts of Interest: The authors declare no conflict of interest.

References

1. IPCC. *Global Warming of 1.5 °C: An IPCC Special Report on the Impacts of Global Warming of 1.5 °C above Pre-industrial Levels and Related Global Greenhouse Gas Emission Pathways, in the Context of Strengthening the Global Response to the Threat of Climate Change, Sustainable Development, and Efforts to Eradicate Poverty*; Intergovernmental Panel on Climate Change: Geneva, Switzerland, 2018.
2. UN. *United Nations Framework Convention on Climate Change*; United Nations: New York, NY, USA, 1992.
3. UN. *Paris Agreement*; United Nations: New York, NY, USA, 2015.
4. UNEP. *Emissions Gap Report 2020*; UN Environment Programme: Nairobi, Kenya, 2020.
5. Pan, X.; Elzen, M.D.; Höhne, N.; Teng, F.; Wang, L. Exploring fair and ambitious mitigation contributions under the Paris Agreement goals. *Environ. Sci. Policy* **2017**, *74*, 49–56. [CrossRef]
6. Hubacek, K.; Feng, K.; Chen, B. Changing Lifestyles towards a Low Carbon Economy: An IPAT Analysis for China. *Energies* **2012**, *5*, 22–31. [CrossRef]
7. Xiao, H.; Sun, K.-J.; Bi, H.-M.; Xue, J.-J. Changes in carbon intensity globally and in countries: Attribution and decomposition analysis. *Appl. Energy* **2019**, *235*, 1492–1504. [CrossRef]
8. Wiedmann, T.; Lenzen, M.; Keyßer, L.T.; Steinberger, J.K. Scientists' warning on affluence. *Nat. Commun.* **2020**, *11*, 3107. [CrossRef] [PubMed]
9. Le, T.-H. Drivers of Greenhouse Gas Emissions in ASEAN + 6 Countries: A New Look. *Environ. Dev. Sustain.* **2021**, 1–20. [CrossRef]

10. Malik, A.; Lan, J.; Lenzen, M. Trends in global greenhouse gas emissions from 1990 to 2010. *Environ. Sci. Technol.* **2016**, *50*, 4722–4730. [CrossRef] [PubMed]
11. Ivanova, D.; Barrett, J.; Wiedenhofer, D.; Macura, B.; Callaghan, M.; Creutzig, F. Quantifying the potential for climate change mitigation of consumption options. *Environ. Res. Lett.* **2020**, *15*, 093001. [CrossRef]
12. Ehrlich, P.R.; Holdren, J.P. Impact of Population Growth. *Science* **1971**, *171*, 1212–1217. [CrossRef]
13. Lan, J.; Malik, A.; Lenzen, M.; McBain, D.; Kanemoto, K. A structural decomposition analysis of global energy footprints. *Appl. Energy* **2016**, *163*, 436–451. [CrossRef]
14. Lenzen, M. Structural analyses of energy use and carbon emissions—An overview. *Econ. Syst. Res.* **2016**, *28*, 119–132. [CrossRef]
15. Wiedenhofer, D.; Smetschka, B.; Akenji, L.; Jalas, M.; Haberl, H. Household time use, carbon footprints, and urban form: A review of the potential contributions of everyday living to the 1.5 °C climate target. *Curr. Opin. Environ. Sustain.* **2018**, *30*, 7–17. [CrossRef]
16. Vandeventer, J.S.; Cattaneo, C.; Zografos, C. A Degrowth Transition: Pathways for the Degrowth Niche to Replace the Capitalist-Growth Regime. *Ecol. Econ.* **2019**, *156*, 272–286. [CrossRef]
17. Hertwich, E.G.; Wood, R. The growing importance of scope 3 greenhouse gas emissions from industry. *Environ. Res. Lett.* **2018**, *13*, 104013. [CrossRef]
18. Bhatia, P.; Cummis, C.; Brown, A.; Rich, D.; Draucker, L.; Lahd, H. *Corporate Value Chain (Scope 3) Accounting and Reporting Standard*; Greenhouse Gas Protocol: Washington, DC, USA, 2011.
19. Wiedmann, T.; Lenzen, M. Environmental and social footprints of international trade. *Nat. Geosci.* **2018**, *11*, 314–321. [CrossRef]
20. Malik, A.; Lan, J. The role of outsourcing in driving global carbon emissions. *Econ. Syst. Res.* **2016**, *28*, 168–182. [CrossRef]
21. Hoekstra, R.; Michel, B.; Suh, S. The emission cost of international sourcing: Using structural decomposition analysis to calculate the contribution of international sourcing to CO2-emission growth. *Econ. Syst. Res.* **2016**, *28*, 151–167. [CrossRef]
22. Steininger, K.W.; Munoz, P.; Karstensen, J.; Peters, G.P.; Strohmaier, R.; Velázquez, E. Austria's consumption-based greenhouse gas emissions: Identifying sectoral sources and destinations. *Glob. Environ. Chang.* **2018**, *48*, 226–242. [CrossRef]
23. Tukker, A.; Cohen, M.; Hubacek, K.; Mont, O. The Impacts of Household Consumption and Options for Change. *J. Ind. Ecol.* **2010**, *14*, 13–30. [CrossRef]
24. Hertwich, E.G. The life cycle environmental impacts of consumption. *Econ. Syst. Res.* **2011**, *23*, 27–47. [CrossRef]
25. Nations, U. Agenda 21: The United Nations Programme of Action from Rio. In Proceedings of the United Nations Conference on Environment & Development, Rio de Janeiro, Brazil, 3–14 June 1992.
26. Hotta, Y.; Tasaki, T.; Koide, R. Expansion of Policy Domain of Sustainable Consumption and Production (SCP): Challenges and Opportunities for Policy Design. *Sustainability* **2021**, *13*, 6763. [CrossRef]
27. Mao, C.; Koide, R.; Akenji, L. Applying Foresight to Policy Design for a Long-Term Transition to Sustainable Lifestyles. *Sustainability* **2020**, *12*, 6200. [CrossRef]
28. Creutzig, F.; Roy, J.; Lamb, W.F.; Azevedo, I.M.L.; de Bruin, W.B.; Dalkmann, H.; Edelenbosch, O.Y.; Geels, F.W.; Grubler, A.; Hepburn, C.; et al. Towards Demand-Side Solutions for Mitigating Climate Change. *Nat. Clim. Chang.* **2018**, *8*, 260–263. [CrossRef]
29. Moran, D.; Wood, R.; Hertwich, E.; Mattson, K.; Rodriguez, J.F.D.; Schanes, K.; Barrett, J. Quantifying the potential for consumer-oriented policy to reduce European and foreign carbon emissions. *Clim. Policy* **2020**, *20*, S28–S38. [CrossRef]
30. Dubois, G.; Sovacool, B.; Aall, C.; Nilsson, M.; Barbier, C.; Herrmann, A.; Bruyère, S.; Andersson, C.; Skold, B.; Nadaud, F.; et al. It starts at home? Climate policies targeting household consumption and behavioral decisions are key to low-carbon futures. *Energy Res. Soc. Sci.* **2019**, *52*, 144–158. [CrossRef]
31. Dietz, T.; Gardner, G.T.; Gilligan, J.; Stern, P.C.; Vandenbergh, M.P. Household actions can provide a behavioral wedge to rapidly reduce US carbon emissions. *Proc. Natl. Acad. Sci. USA* **2009**, *106*, 18452–18456. [CrossRef] [PubMed]
32. Temme, E.H.; Vellinga, R.E.; de Ruiter, H.; Kugelberg, S.; van de Kamp, M.; Milford, A.; Alessandrini, R.; Bartolini, F.; Sanz-Cobena, A.; Leip, A. Demand-Side Food Policies for Public and Planetary Health. *Sustainability* **2020**, *12*, 5924. [CrossRef]
33. Almohaimeed, S.; Suryanarayanan, S.; O'Neill, P. Reducing Carbon Dioxide Emissions from Electricity Sector Using Demand Side Management. Available online: https://www.tandfonline.com/doi/abs/10.1080/15567036.2021.1922548 (accessed on 1 November 2021).
34. Sakamoto, S.; Nagai, Y.; Sugiyama, M.; Fujimori, S.; Kato, E.; Komiyama, R.; Matsuo, Y.; Oshiro, K.; Herran, D.S. Demand-side decarbonization and electrification: EMF 35 JMIP study. *Sustain. Sci.* **2021**, *16*, 395–410. [CrossRef]
35. Sköld, B.; Baltruszewicz, M.; Aall, C.; Andersson, C.; Herrmann, A.; Amelung, D.; Barbier, C.; Nilsson, M.; Bruyère, S.; Sauerborn, R. Household Preferences to Reduce Their Greenhouse Gas Footprint: A Comparative Study from Four European Cities. *Sustainability* **2018**, *10*, 4044. [CrossRef]
36. Grottera, C.; La Rovere, E.L.; Wills, W.; Pereira, A.O., Jr. The role of lifestyle changes in low-emissions development strategies: An economy-wide assessment for Brazil. *Clim. Policy* **2020**, *20*, 217–233. [CrossRef]
37. Ankel, S. The 15 Top Countries for Billionaires, Ranked by How Many Live There. Business Insider 2020. Available online: https://www.businessinsider.in/slideshows/miscellaneous/the-15-top-countries-for-billionaires-ranked-by-how-many-live-there/slidelist/74266779.cms (accessed on 1 November 2021).
38. Haberl, H.e.a. A systematic review of the evidence on decoupling of GDP, resource use and GHG emissions, part II: Synthesizing the insights. *Environ. Res. Lett.* **2020**, *15*, 065003. [CrossRef]

39. Schroeder, P.; Anantharaman, M. "Lifestyle Leapfrogging" in Emerging Economies: Enabling Systemic Shifts to Sustainable Consumption. *J. Consum. Policy* **2016**, *40*, 3–23. [CrossRef]
40. Parrique, T.; Barth, J.; Briens, F.; Kerschner, C.; Kraus-Polk, A.; Kuokkanen, A.; Spangenberg, J.H. *Decoupling Debunked: Evidence and Arguments against Green Growth as a Sole Strategy for Sustainability*; European Environmental Bureau: Bruxelles, Belgium, 2019.
41. UNEP. *Sustainable Consumption and Production—A Handbook for Policymakers*; United Nations Environment Programme: Nairobi, Kenya, 2015.
42. UN. *Report of the Conference of the Parties on Its Twenty-First Session, Held in Paris from 30 November to 13 December 2015*; United Nations: New York, NY, USA, 2016.
43. IPCC. *2006 IPCC Guidelines for National Greenhouse Gas Inventories*; IPCC: Geneva, Switzerland, 2006.
44. UNFCCC. *Biennial Update Reports*; UNFCCC: Bonn, Germany, 2020.
45. NDRC. *The 13th Five-Year Plan for Economic and Social Development of the People's Republic of China*; National Development and Reform Commission, NDRC: Beijing, China, 2016.
46. Malaysia Economic Planning Unit. *Eleventh Malaysia Plan 2016–2020: Anchoring Growth on People*; Malaysia Economic Planning Unit: Putrajaya, Malaysia, 2015.
47. CAAN Institute. The Climate Action Tracker. Available online: https://climateactiontracker.org/about/ (accessed on 1 November 2021).
48. UNEP. *Emissions Gap Report 2019*; United Nations Environment Programme: Nairobi, Kenya, 2019.
49. NDRC. *Enhanced Actions on Climate Change: China's intended Nationally Determined Contributions*; National Development and Reform Commission, NDRC: Beijing, China, 2015.
50. Latvian Presidency of the Council of the European Union. Intended Nationally Determined Contribution of the EU and Its Member States. Available online: https://www4.unfccc.int/sites/ndcstaging/PublishedDocuments/European%20Union%20First/LV-03-06-EU%20INDC(Archived).pdf (accessed on 1 November 2021).
51. Ministry of Environment of the Slovak Republic. *Low-Carbon Development Strategy of the Slovak Republic until 2030 with a View to 2050*; Ministry of Environment of the Slovak Republic: Bratislava, Slovakia, 2020.
52. Republica Portuguesa. *Roadmap for Carbon Neutrality 2050 (rnc2050) Long-Term Strategy for Carbon Neutrality of the Portuguese Economy by 2050*; Ambiente e Transicao Energetica: Lisbon, Portugal, 2019.
53. Federal Ministry of Agriculture, Regions and Tourism. *Langfriststrategie—Österreich*; Federal Ministry of Agriculture, Regions and Tourism: Vienna, Austria, 2019.
54. Danish Ministry of Climate Energy and Utilities. *Denmark's Integrated National Energy and Climate Plan*; Danish Ministry of Climate Energy and Utilities: Copenhagen, Denmark, 2019.
55. Government of the Republic of Estonia. *Resolution of the Riigikogu General Principles of Climate Policy until 2050*; Federal Ministry for the Environment, Nature Conservation, Building and Nuclear Safety (BMUB): Tallinn, Estonia, 2017.
56. Ministry of Ecological Transition. *La Stratégie Nationale Bas-Carbone: La Transition Écologique et Solidaire Vers la Neutralité Carbone*; Ministry of Ecological Transition: Paris, France, 2020.
57. Ministry of Economic Affairs and Climate Policy. *Long Term Strategy on Climate Mitigation*; Ministry of Economic Affairs and Climate Policy: The Hague, The Netherlands, 2019.
58. Federal Office for the Environment. *Switzerland's Climate Policy—Implementation of the Paris Agreement*; Federal Office for the Environment (FOEN): Bern, Switzerland, 2018.
59. Norwegian Ministry of Climate and Environment. *Norway's Climate Strategy for 2030: A transformational approach within a European cooperation framework*; Norwegian Ministry of Climate and Environment: Oslo, Norway, 2017.
60. UNFCCC. *Principality of Monaco National Contribution*; UNFCCC: Bonn, Germany, 2015.
61. Ministry of Environment and Water. *Malaysia Third Biennial Update Report to the UNFCCC*; Ministry of Environment and Water: Putrajaya, Malaysia, 2020.
62. The Government of Japan. *The Long-Term Strategy under the Paris Agreement*; The Government of Japan: Tokyo, Japan, 2019.
63. Regeringskansliet. *En samlad politik för klimatet—klimatpolitisk handlingsplanwork*; Regeringskansliet: Stockholm, Växel, 2019.
64. Minister for Agriculture, Climate Change and Environment. *Seychelles' Updated Nationally Determined Contribution*; Minister for Agriculture, Climate Change and Environment: Mahé, Seychelles, 2021.
65. Ministry of Environment. *The 2030 Greenhouse Gas Reduction Roadmap and the Allocation Plan for 2018–2020 Emissions*; Ministry of Environment: Sejong-si, Korea, 2018.
66. Federal Ministry for the Environment, Building and Nuclear Safety (BMUB). *Climate Action Plan 2050—Principles and Goals of the German Government's Climate Policy*; Nature Conservation, Building and Nuclear Safety (BMUB): Berlin, Germany, 2016.
67. UNFCCC. *Submission under the Paris Agreement: Communication and Update of New Zealand's Nationally Determined Contribution*; UNFCCC: Bonn, Germany, 2020.
68. NCCS. *Singapore's Climate Action Plan: Take Action Today, For a Carbon-Efficient Singapore*; National Climate Change Secretariat: Singapore, Singapore, 2016.
69. Ministry of Mahaweli Development and Environment. *Nationally Determined Contributions*; Ministry of Mahaweli Development and Environment: Colombo, Sri Lanka, 2016.
70. UNFCCC. *Thailand's third Biennial Update Report (BUR)*; UNFCCC: Bonn, Germany, 2020.
71. UNFCCC. *Pakistan's Intended Nationally Determined*; Contribution (Pak-Indc); UNFCCC: Bonn, Germany, 2015.

72. UNFCCC. *Brazil's Third Biennial Update Report to the United Nations Framework Convention on Climate Change*; UNFCCC: Bonn, Germany, 2019.
73. Ministry of Environment. *Intended Nationally Determined Contributions (INDCs) Report*; Ministry of Environment: Doha, Qatar, 2015.
74. Australian Government. *Australia's Intended Nationally Determined Contribution to a New Climate Change Agreement*; Australian Government: Canberra ACT, Australia, 2015.
75. Ministry of Environment, Forest and Climate Change. *India Third Biennial Update Report to the United Nations Framework Convention on Climate Change*; Ministry of Environment, Forest and Climate Change, Government of India: New Delhi, India, 2021.
76. Minister of Environment and Forestry. *Indonesia Second Biennial Update Report under the United Nations Framework Convention on Climate Change*; Minister of Environment and Forestry: Jakarta, Indonesia, 2018.
77. Israel Ministry of Environment Protection. *Israel's First Biennial Update Report—Submitted to the United Nations Framework Convention on Climate Change*; Israel Ministry of Environment Protection: Jerusalem, Israel, 2015.
78. UNFCCC. *Kenya's Intended Nationally Determined Contribution (INDC)*; UNFCCC: Bonn, Germany, 2015.
79. UNFCCC. *South Africa's 3rd Biennial Update Report to the United Nations Framework Convention on Climate Change*; UNFCCC: Bonn, Germany, 2019.
80. UNFCCC. *The United States of America Nationally Determined Contribution—Reducing Greenhouse Gases in the United States: A 2030 Emissions Target*; UNFCCC: Bonn, Germany, 2021.
81. Lenzen, M. Double-Counting in Life Cycle Calculations. *J. Ind. Ecol.* **2008**, *12*, 583–599. [CrossRef]
82. Lange, S. Macroeconomics without growth: Sustainable economies in neoclassical, Keynesian and Marxian theories. In *Wirtschaftswissenschaftliche Nachhaltigkeitsforschung*; Metropolis-Verlag: Marburg, Germany, 2018.
83. Creutzig, F.; Callaghan, M.W.; Ramakrishnan, A.; Javaid, A.; Niamir, L.; Minx, J.C.; Müller-Hansen, F.; Sovacool, B.K.; Afroz, Z.; Andor, M.; et al. Reviewing the scope and thematic focus of 100,000 publications on energy consumption, services and social aspects of climate change: A big data approach to demand-side mitigation. *Environ. Res. Lett.* **2020**, *16*, 033001. [CrossRef]
84. Moran, D.; Wood, R. Convergence between the eora, wiod, exiobase, and openeu's consumption-based carbon accounts. *Econ. Syst. Res.* **2014**, *26*, 245–261. [CrossRef]
85. Wiedmann, T.; Wilting, H.C.; Lenzen, M.; Lutter, S.; Palm, V. Quo vadis MRIO? Methodological, data and institutional requirements for Multi-Region Input-Output analysis. *Ecol. Econ.* **2011**, *70*, 1937–1945. [CrossRef]
86. Peters, G.P. From production-based to consumption-based national emission inventories. *Ecol. Econ.* **2008**, *65*, 13–23. [CrossRef]
87. Leontief, W.W.; Strout, A.A. Multiregional input-output analysis. In *Structural Interdependence and Economic Development*; Barna, T., Ed.; Macmillan: London, UK, 1963; pp. 119–149.
88. Roy, J.; Some, S.; Das, N.; Pathak, M. Demand side climate change mitigation actions and SDGs: Literature review with systematic evidence search. *Environ. Res. Lett.* **2021**, *16*, 043003. [CrossRef]

 *sustainability*

Article

Measurement of the Importance of 11 Sustainable Development Criteria: How Do the Important Criteria Differ among Four Asian Countries and Shift as the Economy Develops?

Tomohiro Tasaki *, Ryo Tajima and Yasuko Kameyama

National Institute for Environmental Studies, 16-2 Onogawa, Tsukuba 305-8506, Ibaraki, Japan; tajima.ryo@nies.go.jp (R.T.); ykame@nies.go.jp (Y.K.)
* Correspondence: tasaki.tomohiro@nies.go.jp

Abstract: Understanding the criteria underlying development in a country is crucial to formulating developmental plans. However, it is not always clear which criteria are more important than others in different countries and at different times. The relationship between developmental criteria and the stage of economic development is also unclear in many countries. Therefore, we devised an indirect stated preference approach for the measurement of the importance of developmental criteria and employed it in four Asian countries—Japan, South Korea, Thailand, and Vietnam—to measure the importance of sustainable development (SD) criteria perceived by the general public. Specifically, we evaluated the importance of 58 national goals linked to 1 of 11 SD criteria. Security, efficiency, accessibility, capability, and environmental capacity were perceived as relatively important by respondents in all four countries. The respondents perceived that the currently important criteria would be important in the future as well. The order of the importance in each country differed. For example, environmental capacity was ranked lower, and inclusiveness was ranked higher as the gross domestic product of a country increased. Thai and Vietnamese respondents had similar perceptions and, overall, tended to have higher levels of importance than South Korean and Japanese respondents, who also had similar perceptions of importance.

Keywords: sustainability criteria; national target; country development stage; indirect stated preference; sustainable development goals (SDGs)

Citation: Tasaki, T.; Tajima, R.; Kameyama, Y. Measurement of the Importance of 11 Sustainable Development Criteria: How Do the Important Criteria Differ among Four Asian Countries and Shift as the Economy Develops? *Sustainability* **2021**, *13*, 9719. https://doi.org/10.3390/su13179719

Academic Editor: Alan Randall

Received: 21 July 2021
Accepted: 26 August 2021
Published: 30 August 2021

Publisher's Note: MDPI stays neutral with regard to jurisdictional claims in published maps and institutional affiliations.

1. Introduction

The United Nations General Assembly adopted 17 Sustainable Development Goals (SDGs) with 169 targets in 2015 [1]. The SDGs are a universal agenda taking various aspects in development into account and applying them to both developing and developed countries in the post-2015 period. Each government is supposed to set its own national targets contributing to the achievement of SDGs on the global level. However, how to determine these national targets is left up to each country to decide, and supporting methodologies are not necessarily sufficient even though there is some movement to develop SDG indicators that monitor countries' progress toward sustainable development [2–4]. For example, Hák et al. [5] pointed out that there is still little agreement or consensus on criteria for evaluating indicators, such as correctness of underlying assumptions and concepts, relevance of various phenomena for sustainable development, and data quality. Fukuda-Parr and McNeill [6] asserted that the SDGs are vehicles—or instruments—that convey norms and that the criteria for SDG indicator selection should be based more on norms and less on data availability. Allen et al. [7] reviewed 80 models that have the potential to support national development planning within the context of the SDGs; however, the selection of a model based on the specific circumstances or needs of a country was not discussed.

Having criteria underlying the development of each country is crucial for countries to formulate the direction of their development. The ideas of social development and human development have been discussed since the 1960s to avoid the negative consequences of economy-centered development. For example, the UN mentioned "qualitative and structural changes in the society must go hand in hand with rapid economic growth" in 1970 [8]. The United Nations Conference on Environment and Development (UNCED), known as the Earth Summit, which was held in 1992, agreed on the principles that human beings are at the center of concern for sustainable development (Principle 1) and environmental protection shall constitute an integral part of the development process and cannot be considered in isolation from it (Principle 4) [9]. These principles urged countries to change the direction of their development. Furthermore, the SDGs, adopted in 2015, encompass concrete criteria for development. For instance, SDG 7 ("ensure access to affordable, reliable, sustainable and modern energy for all") encompasses the developmental criteria of accessibility, security, inclusiveness, and environmental capacity. Accessibility as a national minimum is no longer an important criterion for developed countries, but the use of renewable energy (i.e., the criterion of environmental sustainability) has become more important, as stated by SDG target 7.2. For SDG 12 ("ensure sustainable consumption and production patterns"), SDG target 12.1 mentions implementing "the 10-Year Framework of Programmes taking into account the development and capabilities of developing countries.", and SDG target 12.2 is to "achieve the sustainable management and efficient use of natural resources". Capability and efficiency are thus included in the criteria for sustainable development.

Understanding such criteria is very important, especially when a country enters into another stage of development and fails to introduce new criteria into its public policy. For a hypothetical example, energy systems criteria could develop as shown in Figure 1, from accessibility in the 1st phase to efficiency in the 2nd phase and further and to advanced criteria in subsequent phases of development. Understanding the importance of such criteria is also critical to properly reflect citizens' opinions of national policy. So far, Rostow [10] delineated five stages of economic development, and Hotta et al. [11] asserted the evolution and three versions of sustainable consumption and production policies. Meadowcraft and Fiorino [12] illustrated a conceptual innovation process of environmental policies toward sustainability; for example, it shifted from pollution to sustainable development and climate change as well as from the polluter pay principle to decoupling over the last few decades. These examples indicate that development criteria could and should change according to the phases of development. Even so, identifying which criteria are the most important remains unclear. Interestingly, Khoshnava et al. [13] analyzed 23 criteria related to SDGs and the green economy to identify the most effective ones, and Su et al. [14] analyzed 22 criteria of sustainable supply chain management. However, these criteria were not the *criteria* this study refers to; rather, they were policy or management *goals*.

We therefore aimed to measure the importance of criteria for the sustainable development (hereinafter, referred to as "SD criteria") of countries. We also attempted to compare the importance levels among four Asian countries at different stages of economic development to gain insights on the evolution of SD criteria with the following research questions: (1) Which SD criteria change their importance as the economy develops and how? (2) What SD criteria retain their importance regardless of economic development? (3) Do non-economic factors have influences on the perception of the importance of SD criteria? For research question 1, we conducted a cross-sectional survey of four countries at different levels of economic development (1a) and surveyed the future importance of SD criteria as well (1b). The intended difference between 1a and 1b is that 1a addresses the perceptions of respondents at different levels of economic development while the 1b addresses the perception of respondents at a certain level of economic development for different times periods.

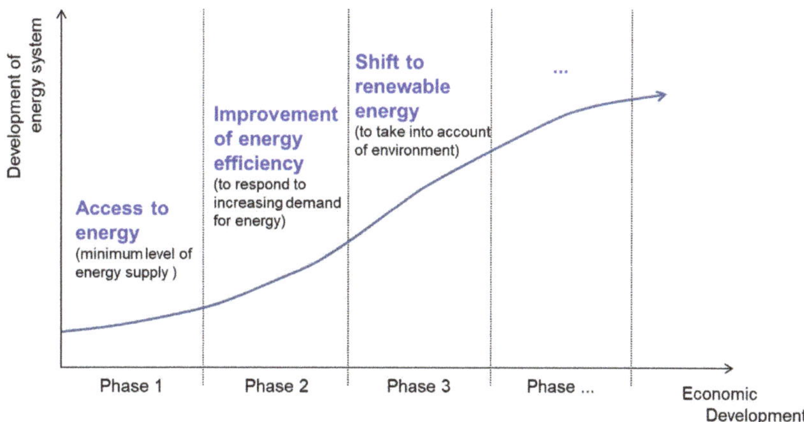

Figure 1. Hypothetical shift of the most important criteria of developing energy systems in each phase of economic development of a country.

2. Materials and Methods

2.1. Sustainable Development Criteria

To determine the SD criteria to be analyzed in this study, we reviewed the literature in the field of sustainable development [15–34], documents about principles and criteria used by a variety of certification programs [35–51], and the 169 SDG targets. We found that the following 11 SD criteria were embedded in these references, at the least: accessibility, capability, convenience, efficiency, environmental capacity, inclusiveness, resilience and stability ("resilience" in short), security, self-sufficiency, social justice, and variety of choice ("variety", in short). We therefore used these 11 SD criteria in our analysis. The working definitions of the criteria used in this survey are given in Table 1.

Table 1. Working definitions of the 11 sustainable development criteria.

Criterion	Definition
Accessibility	The quality of being able to attain and use that which provides for various human needs.
Capability	The extent of human ability to achieve sustainable development.
Convenience	The quality of being able to easily or suitably fulfill needs.
Efficiency	Ratio of output to a given input.
Environmental capacity	The property of the natural environment to sustain and accommodate human activities such as the exploitation of natural resources and the emission of environmental pollutants.
Inclusiveness	The quality of not excluding any race, gender, religion, culture, etc.; understanding the perspectives and contributions of all people; and striving to incorporate diverse needs into society.
Resilience and Stability	The capacity of a system to absorb and/or adapt to disturbances, and even change the system itself in some cases, so that the system maintains its basic function and structure.
Security	The quality of being free from danger or threat.
Self-sufficiency	The state of needing no external support to satisfy human needs, such as food and energy.
Social Justice	The state where basic human rights are not violated, and benefits and costs are equitably allocated.
Variety of Choice	The extent of abundance of options and goods such that people can choose among them.

2.2. Indirect Stated Preference Approach

We devised an indirect stated preference approach for the measurement of the importance of SD criteria because it would be difficult for ordinary people to give direct answers about the importance of the 11 criteria (i.e., use a direct stated preference approach). Instead, we prepared 58 national goals covering 6 domains that directly and exclusively link to one of the 11 SD criteria. An example for energy is shown in Figure 2. The 58 goals in this study were created by the authors by combining the six domains and the eleven SD criteria (see Table A1 in Appendix A for all of the national goals used in this study). The six domains used in this study were energy, economy, health, ecosystem, education, and food. They were chosen because of their importance as national sustainable development indicators [28].

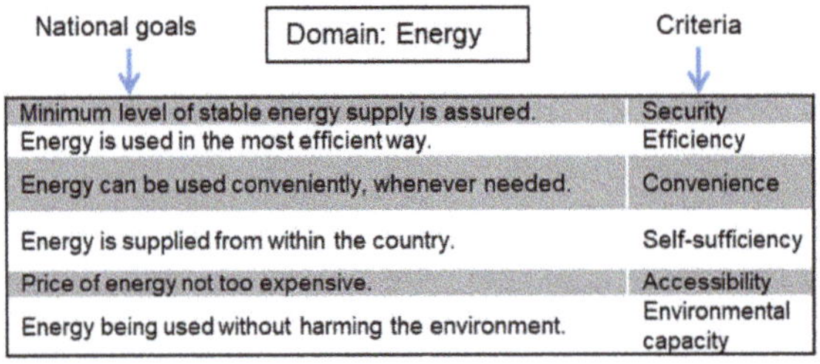

National goals	Domain: Energy	Criteria
Minimum level of stable energy supply is assured.		Security
Energy is used in the most efficient way.		Efficiency
Energy can be used conveniently, whenever needed.		Convenience
Energy is supplied from within the country.		Self-sufficiency
Price of energy not too expensive.		Accessibility
Energy being used without harming the environment.		Environmental capacity

Figure 2. Examples of pairs of national goals and criteria used in the indirect stated preference approach of this study.

We asked the respondents to rate the importance of each of the 58 goals with a 10-point Likert scale (from very important to not important at all). We also asked them to rate the importance of the goal in the future relative to that of the present (hereinafter, referred to as "relative future importance") with a 3-point Likert scale (becomes more important (+1), importance will not change (0), becomes less important (−1); the statement used in the survey was "*How do you think the importance of the goals will change in the future? Please answer assuming a period up to 20 years from now*"). We calculated the average importance (*I*) of the national goals linked to the same criterion (*C*) of respondents *j*, $I_{C,j}$, by using Equation (1):

$$I_{C,j} = \frac{\sum \left(i_{g \in C,j}\right)}{n_{g \in C}} \tag{1}$$

A country may have an urgent and severe problem in a certain domain, which could result in that domain scoring higher than the others in that country and also higher than its importance in other countries. To counterbalance this effect, we calculated the standardized importance of the criterion, $S_{C,j}$, by using Equation (2), which standardizes the importance of each goal with the average importance of goals in the same domain (*d*) for each respondent, given by Equation (3):

$$S_{C,j} = \frac{\sum \left(i_{g \in C,j} / I_{d,j}\right)}{n_{g \in C}} \tag{2}$$

$$I_{d,j} = \frac{\sum \left(i_{g \in d,j}\right)}{n_{g \in d}} \tag{3}$$

Here, $i_{g \in C,j}$ is the importance of national goal *g* with criterion *C* as reported by the respondents *j*, $n_{g \in C}$ is the number of goals with the same criterion *C*, and $n_{g \in d}$ is the number of goals in domain *d*.

The importance and standardized importance of SD criterion *C* in each country were then calculated by Equations (4) and (5), respectively:

$$I_C = \frac{\sum_j I_{C,j}}{n_j} \tag{4}$$

and

$$S_C = \frac{\sum_j S_{C,j}}{n_j} \tag{5}$$

Here, n_j is the number of respondents in each country.

We calculated the relative future importance and the standardized relative future importance of each SD criterion C for each country in the say way.

2.3. Survey and Analysis

An online questionnaire survey was conducted from 2013 to 2015 in four Asian countries: Japan, South Korea, Thailand, and Vietnam, which are at different levels of economic development (see GDP per capita in Table 2). The respondents were the monitors of two survey companies, Cross Marketing in Japan and Cross Marketing Asia, who were 20 years of age or older. There were 500 respondents for each country, except for Japan, which had 1408. Quota sampling was applied for each country, with eight equal quotas for the combinations of the two sexes and the ages of the participants who were in their 20s, 30s, 40s, and over 50 (See Table A2 in Appendix B for the profiles of the respondents). Questions were prepared in Japanese and in English and were then translated from English to Korean, Thai, and Vietnamese. After the survey, we calculated the current/future importance of the above-mentioned 11 SD criteria perceived by members of the general public of the four Asian countries.

Table 2. Standardized importance ranks of the 11 SD criteria in Japan, South Korea, Thailand, and Vietnam. Criteria ranked in top five for at least one country are presented. Per capita gross domestic product (GDP, PPP based) in 2014 is also shown.

Rank	Japan	South Korea	Thailand	Vietnam
1	Security	Security	Security	Security
2	Efficiency	Efficiency	Env. capacity	Env. capacity
3	Resilience	Env. capacity	Efficiency	Efficiency
4	Inclusiveness	Resilience	Capability	Capability
5	Capability	Accessibility	Accessibility	Accessibility
6	Accessibility	Inclusiveness	Inclusiveness	Resilience
7				
8		Capability		Inclusiveness
9	Env. capacity		Resilience	
GDP/capita	$37,390	$35,277	$14,354	$5,635

3. Results and Discussion

3.1. Current and Future Importance of SD Criteria

The calculated current and relative future importance values of the 11 SD criteria as perceived by members of the general public are presented in Figure 3.

The current and future results were positively correlated ($r^2 = 0.766$), meaning that the respondents in all four Asian countries perceived that the more important a criterion was at present, the more important it will become in the future (20 years). No criterion was perceived to become less important (i.e., all of the future values are positive), but differences in the degree of change in terms of future importance changed the rank between the present and the future. This means that the future importance rank of the criteria located relatively far to the right in Figure 3 can become more important than those located to the left, even if the ones on the left are higher. For example, compare security and self-sufficiency in Japan with inclusiveness and accessibility, respectively.

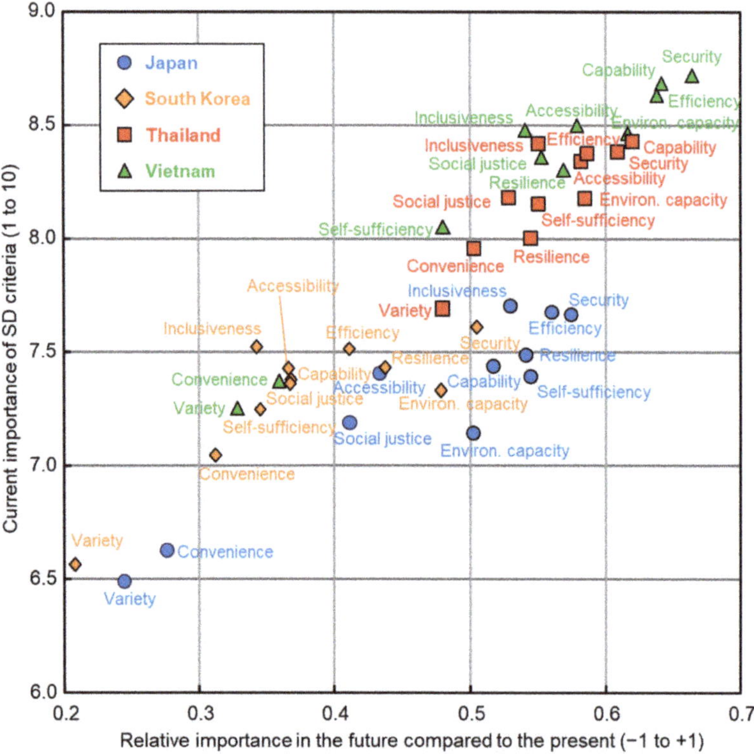

Figure 3. Current and relative future importance values for the 11 SD criteria in Japan, South Korea, Thailand, and Vietnam.

Overall, Thai and Vietnamese respondents tended to evaluate the importance of the SD criteria higher than the South Korean and Japanese respondents. Previous studies (e.g., [52–54]) have argued that these kinds of differences may be rooted in the different response styles of people in these countries. That is, respondents in some countries tend to select middle answers, whereas others choose extreme answers. The former style is called the middle response style, and the latter is called the extreme response style. According to a literature review by Harzing [52], Japanese and Korean respondents tend to have a middle response style. This will be discussed in more detail in Section 3.4.

We identified criteria that could be considered to have the same level of importance between pairs of the four countries by using a *t*-test (see Table A3 in in Appendix C). Most of the importance values were significantly different, but the current importance of the six criteria and the future importance of one criterion between Japan and South Korea were not significantly different. In addition, the current importance of four criteria and the future importance of five criteria between Thailand and Vietnam were also not significantly different. In other words, respondents in Japan/South Korea and those in Thailand/Vietnam had relatively similar perceptions on the importance of SD criteria.

3.2. Standardized Importance of SD Criteria

The results of standardized importance are presented in Figure 4. The relationship between the current and relative future importance values was stronger ($r^2 = 0.917$) than it was in the unstandardized results shown in Figure 3. This indicates that measurement by standardized importance is less influenced by the countries' specific circumstances in terms of domains and respondent styles. People may think of the importance of national goals based on the importance of a domain of concern first and then differentiate the importance of each goal based on criteria (i.e., the perception of domains is more influential). This type of two-phased consideration could be employed by people intentionally or unintentionally. To determine whether the domain or criteria is more similar, we applied cluster analysis to the current importance of the SD criteria for each country. The results (Figure A1 in Appendix D) showed that many clusters included goals in the same domains but did not include many goals within the same criteria. More study is needed on this topic.

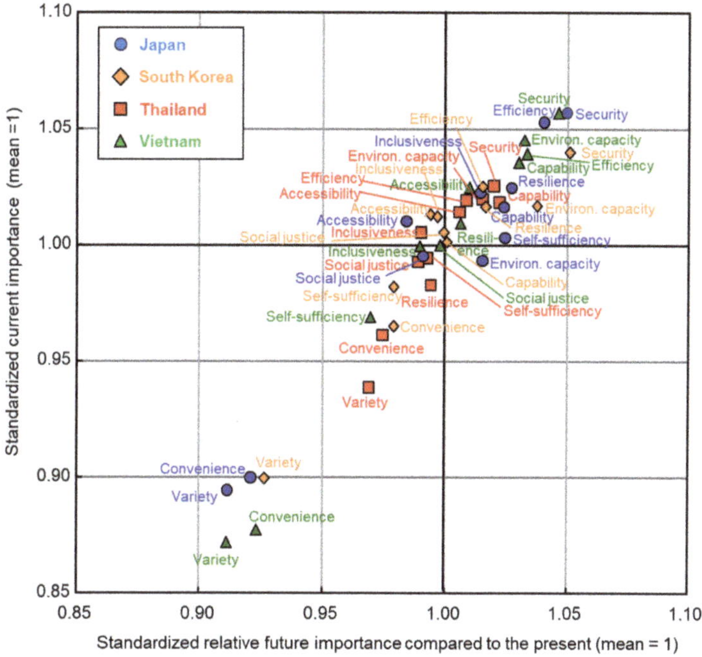

Figure 4. Standardized current and relative future importance values of 11 SD criteria in Japan, South Korea, Thailand, and Vietnam.

We split Figure 4 into four quadrants by drawing a line through the value of one on each axis. Security, efficiency, and environmental capacity are the prominent SD criteria in the first (upper right) quadrant. These criteria are currently important relative to the others and the respondents anticipated that they would become even more important in the future. In contrast, convenience and variety were prominent in the third (lower left) quadrant, meaning that they are both currently less important and the respondents determined that they would become relatively less important in the future.

These are important findings in light of the objectives of this study. However, the question arises: do market prices properly reflect the importance placed on these SD criteria? In general, the results in Figure 4 do not appear to consistently align with actual market prices. Rare products and convenient products tend to be valued higher, but these criteria were located in the third (lower importance) quadrant. Security and efficiency are valued in the market but perhaps not to the extent that Figure 4 shows. In addition, environmental capacity is often externalized by market mechanisms. Thus, the relative importance of the SD criteria found in this study may differ from that inferred from current market prices. As is well known, market mechanisms place prices on products and services based on exchange values. In contrast, our methodology measures the inherent values of the SD criteria. The relationship between the inherent importance and market pricing of the SD criteria is an interesting topic for future study.

Only a few SD criteria were located in the second (upper left) and fourth (lower right) quadrants. Accessibility and inclusiveness were perceived to be relatively important at the present but less so in the future in some cases; for example, Japanese and South Korean respondents evaluated accessibility in this manner. These countries are developed and have higher levels of accessibility to a variety of infrastructure and public services, which can be taken for granted. Thus, it is not surprising that accessibility was located in the second quadrant. The Japanese respondents evaluated environmental capacity as relatively less important at the present but that it would become more important in the future. People's attention in Japan has shifted from local environmental pollution, which can draw strong attention, to global environmental issues, which can be harder to grasp on a personal level and may cause respondents to rank them as being of relatively lower importance. Worsening global warming has been found to draw the most attention among various environmental issues in Japan [55,56], which may also explain this result.

3.3. Differences in the Ranks of Importance of SD Criteria

Here, we focus on the order of the standardized importance of the 11 criteria of each country and compare the ranks among countries for research questions 1 and 2. Harzing [52] concluded that ranking is generally a superior method for working with scores obtained from Likert scales and also thought that ranking can better avoid the issue of different response styles. Table 2 shows a summary of the ranking results. Among the 11 criteria, security, efficiency, accessibility, capability, and environmental capacity were commonly perceived as relatively important by respondents from all four countries; however, the ranks differed by country.

For example, environmental capacity was ranked lower, and inclusiveness was ranked higher as the per capita gross domestic product (GDP) (converted based on purchasing power parity [PPP]) increased. Environmental capacity and capability were seen as more important in Thailand and Vietnam, whereas resilience was more important in Japan and South Korea. Severe environmental pollution, such as $PM_{2.5}$ air pollution (particulate matter < 2.5 μm in diameter) in Thailand and Vietnam [57], could influence the respondents' evaluations. Accessibility was ranked higher than resilience in Thailand and Vietnam but lower in Japan and South Korea. This probably relates to insufficient basic infrastructure and public services in Thailand and Vietnam, whereas the infrastructure issues have shifted from initial provision to maintenance in the other two countries. The rank of capability was higher than that of inclusiveness in Thailand and Vietnam but was lower in Japan and South Korea. This may imply that the Japanese and South Korean respondents believe that individual efforts are no longer sufficient and that society should care for vulnerable people.

3.4. Influences of Non-Economic Factors

Not only economic factors but also non-economic factors might affect the importance of certain SD criteria for a country. Several studies have paid much attention to the cultures of different countries, and we hereby discuss the possibility of influences of such factors on the perception of the importance of the SD criteria (research question 3).

The World Value survey led by Inglehart and Welzel [58] and the survey by Hofstede et al. [59] are famous examples because they covered many countries. The latest results of the World Value Survey [60] present a new version of the so-called Inglehart–Welzel cultural map, which has two major axes of cross-cultural variation—traditional values versus secular-rational values (the vertical axis) and survival values versus self-expression values (the horizontal axis). This new map shows that Japan and South Korea are located in secular areas (in the vertical axis), whereas Thailand and Vietnam are located in between secular and traditional. All four countries are located near the center of the horizontal axis, indicating moderate self-expression values. The difference between the Japan/South Korea pair and the Thailand/Vietnam pair may be attributed to differences in secular-rational values, or they may just reflect the degree of economic development as shown by the per capita GDP differences in Table 2. The latest data from the Hofstede group's survey [61] are summarized in Table 3. The two abovementioned pairs apparently differ in uncertainty avoidance and long-term orientation. These two cultural tendencies could result in high ranks for resilience in Japan and South Korea. In contrast, Thailand and Vietnam had high ranks for capability, which can be interpreted that, at least in the short term, they place more importance on the capability to solve current issues.

Table 3. Hofstede's six indices of national culture and their values in 2015 for Japan, South Korea, Thailand, and Vietnam.

	Index Value (0–100)					
	Power Distance	Individualism	Masculinity	Uncertainty Avoidance	Long-Term Orientation	Indul-Gence
Japan	54	46	95	92	88	42
South Korea	60	18	39	85	100	29
Thailand	64	20	34	64	32	45
Vietnam	70	20	40	30	57	35
Avg. (A)	62	26	52	68	69	38
World average (B)	59	45	49	67	45	45
Difference, (A)–(B)	3	−19	3	0	24	−7

[1] Data retrieved from Geerthofstede.com [61]; the averages and differences were calculated by the authors.

Harzing [52] conducted a regression analysis between response styles and Hofstede's cultural values and found that people with a high power distance (a tendency to accept an unequal distribution of power) and individualism tended not to have a middle response style ($p < 0.01$). Power distance explains the results of our survey on the importance of national goals—Thai and Vietnamese respondents tended to rate the importance of national goals higher—however, those with a high level of individualism do not. Other factors such as the perceived seriousness of the issues and/or a strong motivation for improvement in each country's context could play an influential role in the responses.

4. Conclusions

We measured the importance of 11 SD criteria as perceived by the general public in Japan, South Korea, Thailand, and Vietnam. The 11 SD criteria were accessibility, capability, convenience, efficiency, environmental capacity, diversity and choice, inclusiveness, resilience and stability, security, self-sufficiency, and social justice. We used an indirect stated preference approach and employed 58 questions in 6 domains.

The main findings and the answers to the three research questions are as follows:

- Among the 11 SD criteria, security, efficiency, and accessibility were commonly perceived as relatively important in the four Asian countries. Security and efficiency retain their importance regardless of economic development (research question 2);
- The respondents in each country, i.e., in a certain development phase, perceived that the currently important criteria would also be important in the future. This suggests that SD criteria are considered to apply in a similar manner regardless of time unless the phase of development changes (research question 1b);
- Japan and South Korea had relatively similar perceptions on the importance of the SD criteria, as did Thailand and Vietnam. The Thai and Vietnamese respondents tended to have higher importance values than the South Korean and Japanese respondents overall; this difference could be partly attributed to differences in the power distance values (acceptance of an unequal power distribution) between these countries (research question 3). Additional analysis is necessary to identify important factors related to this phenomenon;
- We inferred that people may first think of the importance of national goals based on the importance of a domain of concern and then differentiate the importance of each

goal based on the SD criteria. Perception of the importance of the domains may be more influential than that of the criteria;

- The order (rank) of importance of the 11 SD criteria differed by country to a certain extent, which may be related to the economic development of the countries. For example, environmental capacity was ranked lower, and was ranked inclusiveness higher in the countries with a higher per capita GDP (research question 1a).

The main academic contributions of this study perceived by the authors are the development of the method for measuring the importance of SD criteria and the results of attempting the measurement. As many studies at their initial stages have, this study has some limitations. First, this study focused on six domains, but there are others. Second, there were only four target countries. Expanding the scope and number of countries remains as a future research task. The same survey applying to a country at a different time also remains as a future task. A third limitation is that we did not identify what the explanatory variables of the predictors of the importance of SD criteria of general public are. To do so, in-depth analyses of the results are needed. The fourth limitation is the number of SD criteria. Establishing a more complete set of SD criteria and the questions that should be used to elucidate relevant responses also remains as a future research task. Finally, although we devised and employed an indirect stated preference approach in this study, the development of different approaches to measure the importance of SD criteria and to compare the results among the different methodologies should allow us to produce more reliable results in the future.

Author Contributions: Conceptualization, T.T., R.T. and Y.K.; methodology, formal analysis, investigation, visualization, supervision, writing—original draft preparation, T.T.; writing—review and editing, R.T. and Y.K.; funding acquisition, Y.K. and T.T. All authors have read and agreed to the published version of the manuscript.

Funding: This research was conducted based on the research funded by the Environment Research and Technology Development Fund (S-11) of the Environmental Restoration and Conservation Agency, Japan.

Conflicts of Interest: The authors declare no conflict of interest.

Appendix A

Table A1. Fifty-eight statements about national goals in six domains and their corresponding sustainable development (SD) criteria.

	National Goals	Corresponding SD Criteria
	A stable supply of required minimum energy for daily living, etc., is secured.	Accessibility
	Energy is used efficiently without any waste.	Efficiency
	Energy can be used freely whenever people want to.	Convenience
	Energy sources that are managed to reduce accidents are used.	Security
Energy	Renewable energy (e.g., natural energy and biomass energy) is used within sustainable limits.	Environmental capacity
	People are self-sufficient in supplying energy within my country, local communities, and/or households.	Self sufficiency
	The price for using energy is low.	Accessibility
	Energy is used in a way to avoid causing environmental problems such as global warming and air pollution.	Environmental capacity
	Energy that can be supplied consistently even in an emergency is used.	Resilience & Stability
	People are allowed to choose an energy source out of various options.	Variety of choice

Table A1. *Cont.*

	National Goals	Corresponding SD Criteria
Economy	Daily necessities are priced low.	Accessibility
	The economy is highly productive.	Efficiency
	The distribution of wealth is fair.	Social Justice
	Stable employment opportunities are secured for people.	Security
	Various products and services are available, and people can choose according to their own preference.	Variety of choice
	There is a balance between the real economy and the financial economy (nominal economy).	Resilience & Stability
	The economy is booming.	Capability
	Economic activities are not too dependent on other countries.	Self sufficiency
	My country is striving for a green economy (economy that is in balance with the environment).	Environmental capacity
	My country is not exacerbating social issues in other countries (e.g., not doing business with operators that are infringing the rights of local residents and workers in other countries, etc.).	Social Justice
Health	Medical institutions are available not far from home.	Accessibility
	Public finance will not collapse as a result of the government providing healthcare security to the people.	Efficiency
	People can choose a better medical service by paying an additional fee.	Variety of choice
	Education or information for maintaining health is commonly available.	Capability
	Everyone can receive medical services equally regardless of being rich or poor.	Social Justice
	Services for maintaining health will become common so that the number of people who need medical care declines.	Resilience & Stability
	The quality of the environment such as air and water is maintained to prevent health problems.	Security
	Sports facilities are enhanced in order to promote health.	Accessibility
	People are becoming responsible for their own health and their make best effort to manage their health.	Self sufficiency
Ecosystem	There is green (nature) in an easily accessible area close to home.	Accessibility
	Public finance will not collapse as a result of the government stepping up their effort in nature conservation.	Efficiency
	Activities such as bass fishing involved in personal hobbies will continue to be available in the future.	Variety of choice
	Education or information to help protect the ecosystem is commonly available.	Capability
	People who contribute to nature conservation do not incur a loss (e.g., the government to buyout the forests that have been conserved, etc.).	Social Justice
	Untouched natural areas remain in my country. Measures such as restricted access are implemented as needed.	Resilience & Stability
	The lives of animals such as monkeys, wild boar, and deer are respected even if they devastate the land and people do not easily resort to extermination.	Social Justice
	The genes of endangered species are preserved so that genetic information is not lost even if they become extinct.	Security
	My country provides funds to the international community in order to protect ecosystems overseas.	Environmental capacity
	Capturing species such as bluefin tuna and eel that have been observed to be declining in number is prohibited until population recovery has been confirmed.	Environmental capacity

Table A1. Cont.

	National Goals	Corresponding SD Criteria
Education	There is an elementary school within walking distance for children.	Accessibility
	People can go to college and graduate school regardless of the level of their economic resources.	Inclusiveness
	Gifted and motivated individuals can receive more advanced education.	Efficiency
	There are no truants due to bullying, etc.	Social Justice
	People can access high-quality classes and lectures without being restricted by time or place.	Convenience
	Opportunities for learning and self-improvement are guaranteed over a lifetime.	Accessibility
	The level of basic academic skills in my country is high compared to other countries.	Efficiency
	People have skills such as speaking English to be internationally successful.	Capability
	People are capable of understanding complex issues and applying knowledge and skills that are useful in resolving those issues.	Capability
	People understand and try to accept others with differences by demonstrating compassion for others, etc.	Inclusiveness
Food	Groceries are inexpensive, accounting for a small percentage of the total household expenditure.	Accessibility
	Food contamination with toxic and hazardous substances is prevented and safe/secure food products such as chemical-free vegetables are available.	Security
	People can maintain good nutrition regardless of gender, age, income, etc.	Accessibility
	There is little food waste such as leftovers and expired food, and the environmental impact of food production and disposal is minimal.	Efficiency
	Food items are also available to poor people.	Social Justice
	My country no longer relies on other countries for food supply.	Self sufficiency
	There are opportunities to enjoy a variety of foods, from high-end foodstuffs and fine dining to B-class gourmet food in my country and overseas.	Variety of choice
	Convenient food products and services that do not require the effort of cooking are available.	Convenience
	People can eat what they want whenever they want regardless of the season.	Convenience

Appendix B

Table A2. Profile of the respondents in the four-country survey.

(a) Japanese respondents.

Income (JPY million)	%	Age	%	Sex	%	Area	%
10.00 and over	3.6	20s	25	Male	50	Urban	88.8
7.00–9.99	16.4	30s	25	Female	50	Rural	11.2
5.00–6.99	23.0	40s	25			Other	0
3.00–4.99	31.8	50s+	25				
2.00–2.99	13.5						
1.99 and below	11.7						
n	1150		1408		1408		1408

Table A2. *Cont.*

(b) South Korean respondents.

Income (KRW million)	%	Age	%	Sex	%	Area	%
7.50 and over	9.8	20s	25	Male	50	Urban	85.0
5.00–7.49	23.0	30s	25	Female	50	Rural	15.0
4.00–4.99	12.6	40s	25			Other	0
3.00–3.99	19.2	50s+	25				
2.00–2.99	22.6						
1.99 and below	12.8						
n	500		500		500		500

(c) Thai respondents.

Income (TBH thousand)	%	Age	%	Sex	%	Area	%
70.00 and over	22.8	20s	25	Male	50	Urban	83.0
50.00–69.99	21.4	30s	25	Female	50	Rural	14.4
40.00–49.99	20.2	40s	25			Other	2.6
18.00–39.99	26.8	50s+	25				
7.50–17.99	8.8						
7.49 and below	0.0						
n	500		500		500		500

(d) Vietnamese respondents.

Income (VND million)	%	Age	%	Sex	%	Area	%
30.00 and over	11.8	20s	25	Male	50	Urban	67.4
15.00–29.99	32.6	30s	25	Female	50	Rural	18.6
7.50–14.99	34.2	40s	25			Other	14.0
4.50–7.49	15.6	50s+	25				
3.00–4.49	5.8						
2.99 and below	0.0						
n	500		500		500		500

Appendix C

Table A3. Responses that were not significantly different between pairs of the four Asian countries (*t*-test, $p < 0.05$). Criteria in the same cells were not statistically different between the listed pair of countries. "None" indicates that no criteria were not significantly different.

.	Current Importance	Relative Future Importance
Japan–South Korea	Accessibility Capability Resilience and Stability Security Self-sufficiency Variety of choice	Convenience
Thailand–Vietnam	Accessibility Inclusiveness Self-sufficiency Social justice	Accessibility Capability Environ. Capacity Inclusiveness Resilience and Stability Social justice

Table A3. *Cont.*

.	Current Importance	Relative Future Importance
Japan–Thailand	None	Efficiency Inclusiveness Resilience and Stability Security Self-sufficiency
Japan–Vietnam	None	Inclusiveness Resilience and Stability
South Korea–Thailand	None	None
South Korea–Vietnam	None	None

Appendix D

Each respondent indicated the importance of the 58 national goals (combinations of six domains and 11 SD criteria) in the survey using a 10-point scale. We applied cluster analysis to the responses of the importance to check which domains or SD criteria tend to fall in the same cluster, i.e., which domains or criteria are relatively similar. As shown in Figure A1, many clusters tended to include the same domains rather than the same criteria. The total number of the items (combinations) with the same domains in the same clusters was 22, and the number of the items with the same SD criteria was 12 for Japan. These numbers were 26 and 25 for South Korea, 29 and 14 for Thailand, and 25 and 16 for Vietnam. For the total for all four countries, these numbers were 102 and 67.

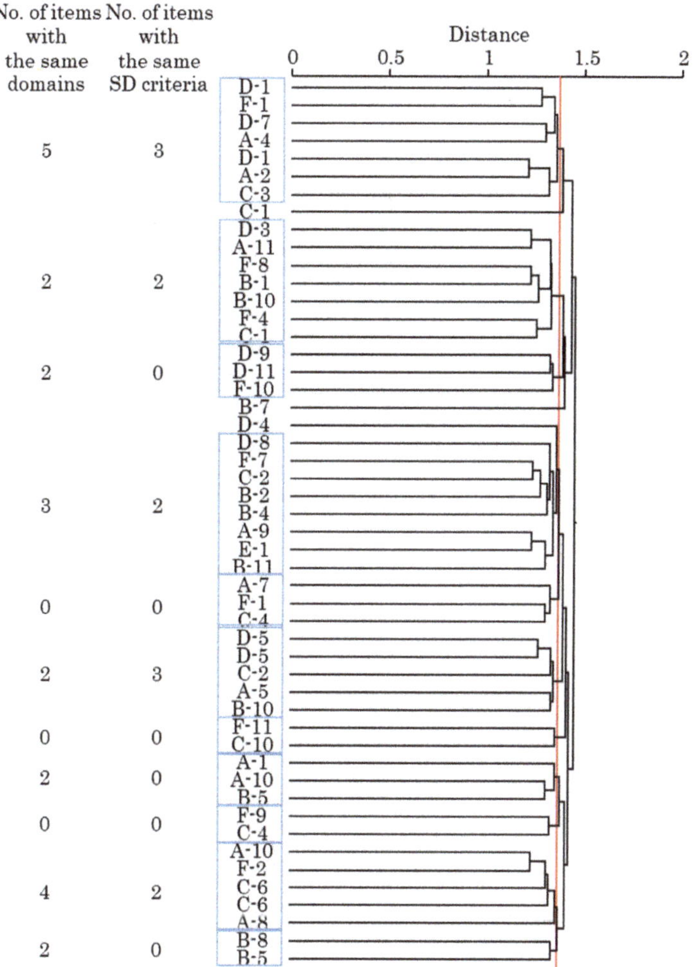

Figure A1. An example result of a cluster analysis for the standardized current importance of the 58 combinations of domains and SD criteria (Japanese respondents only). The items on the left of the dendrogram indicate the combinations. Boxes around the combinations indicate the clusters determined at the red line. The six domains are A: economy, B: ecosystem, C: education, D: energy, E: Food, and F: health. The 11 SD criteria are 1: accessibility, 2: capability, 3: convenience, 4 efficiency, 5 environmental capacity, 6: inclusiveness, 7: resilience & stability, 8: security, 9: self-sufficiency, 10: social justice, and 11: variety of choice.

References

1. United Nations. *Transforming Our World: The 2030 Agenda for Sustainable Development*; United Nations: New York, NY, USA, 2015.
2. United Nations Economic and Social Council. *Report of the Inter-Agency and Expert Group on Sustainable Development Goal Indicators*; United Nations Economic and Social Council: New York, NY, USA, 2016.
3. United Nations. *The Sustainable Development Report 2020*; United Nations Publications: New York, NY, USA, 2020.
4. Sachs, J.; Schmidt-Traub, G.; Kroll, C.; Lafortune, G.; Fuller, G.; Woelm, F. *The Sustainable Development Goals and COVID-19. Sustainable Development Report 2020*; Cambridge University Press: Cambridge, UK, 2020.
5. Hák, T.; Janoušková, S.; Moldan, B. Sustainable Development Goals: A need for relevant indicators. *Ecol. Indic.* **2016**, *60*, 565–573. [CrossRef]

6. Fukuda-Parr, S.; McNeill, D. Knowledge and Politics in Setting and Measuring the SDGs: Introduction to Special Issue. *Glob. Policy* **2019**, *10*, 5–15. [CrossRef]

7. Allen, C.; Metternicht, G.; Wiedmann, T. National pathways to the Sustainable Development Goals (SDGs): A comparative review of scenario modelling tools. *Environ. Sci. Policy* **2016**, *66*, 199–207. [CrossRef]

8. United Nations. *International Strategy for the Second UN Development Decade*; United Nations: New York, NY, USA, 1970.

9. United Nations. *Rio Declaration on Environment and Development*; United Nations: New York, NY, USA, 1992.

10. Rostow, W.W. *The Stages of Economic Growth*, 3rd ed.; Cambridge University Press: Cambridge, UK, 1990; pp. 4–16.

11. Hotta, Y.; Tasaki, T.; Koide, R. Expansion of Policy Domain of Sustainable Consumption and Production (SCP): Challenges and Opportunities for Policy Design. *Sustainability* **2021**, *13*, 6763. [CrossRef]

12. Meadowcraft, J.; Fiorino, D.J. *Conceptual Innovation of Environmental Policy*; MIT Press: Cambridge, UK, 2017; pp. 7–11, 21–51.

13. Khoshnava, S.M.; Rostami, R.; Zin, R.M.; Štreimikienė, D.; Yousefpour, A.; Strielkowski, W.; Mardan, A. Aligning the Criteria of Green Economy (GE) and Sustainable Development Goals (SDGs) to Implement Sustainable Development. *Sustainability* **2019**, *11*, 4615. [CrossRef]

14. Su, C.-M.; Horng, D.-J.; Tseng, M.-L.; Chiu, A.S.F.; Wu, K.-J.; Chen, H.-P. Improving sustainable supply chain management using a novel hierarchical grey-DEMATEL approach. *J. Clean. Prod.* **2016**, *134*, 469–481. [CrossRef]

15. Arrow, K.J.; Dasgupta, P.; Goulder, L.H.; Mumford, K.J.; Oleson, K. Sustainability and the measurement of wealth. *Environ. Dev. Econ.* **2012**, *17*, 317–353. [CrossRef]

16. Barbier, E. The Concept of Sustainable Economic Development. *Environ. Conserv.* **1987**, *14*, 101–110. [CrossRef]

17. Bettencourt, L.M.A.; Kaur, J. Evolution and structure of sustainability science. *Proc. Natl. Acad. Sci. USA* **2011**, *108*, 19540–19545. [CrossRef]

18. Daly, H.E. Toward Some Operational Principles of Sustainable Development. *Ecol. Econ.* **1990**, *2*, 1–6. [CrossRef]

19. Ekins, P. A Four-Capital Model of Wealth Creation. In *Real-Life Economics: Understanding Wealth Creation*; Ekins, P., Max-Neef, M., Eds.; Routledge: London, UK, 1992; pp. 147–155.

20. Folke, C. Resilience: The emergence of a perspective for social-ecological systems analyses. *Glob. Environ. Chang.* **2006**, *16*, 253–267. [CrossRef]

21. Griggs, D.; Stafford-Smith, M.; Gaffney, O.; Rockström, J.; Öhman, M.C.; Shyamsundar, P.; Steffen, W.; Glaser, G.; Kanie, N.; Noble, I. Sustainable development goals for people and planet. *Nature* **2013**, *495*, 305–307. [CrossRef]

22. Hoff, H. *Understanding the Nexus. Background Paper for the Bonn2011 Conference: The Water, Energy and Food Security Nexus*; Stockholm Environment Institute: Stockholm, Sweden, 2011.

23. Holling, C.S. Resilience and Stability of Ecological Systems. *Annu. Rev. Ecol. Syst.* **1973**, *4*, 1–23. [CrossRef]

24. Kates, R.W.; Parris, T.M. Long-term trends and a sustainability transition. *Proc. Natl. Acad. Sci. USA* **2003**, *100*, 8062–8067. [CrossRef] [PubMed]

25. Kates, R.W. What kind of a science is sustainability science? *Proc. Natl. Acad. Sci. USA* **2011**, *108*, 19449–19450. [CrossRef]

26. Meadows, D. *Indicators and Information Systems for Sustainable Development*; The Sustainability Institute: Hartland, WI, USA, 1998.

27. Rockström, J.; Steffen, W.; Noone, K.; Persson, Å.; Chapin, F.S., III; Lambin, E.F.; Lenton, T.M.; Scheffer, M.; Folke, C.; Schellnhuber, H.J.; et al. A safe operating space for humanity. *Nature* **2009**, *461*, 472–475. [CrossRef]

28. Tasaki, T.; Kameyama, Y.; Hashimoto, S.; Moriguchi, Y.; Harasawa, H. A survey of national sustainable development indicators. *Int. J. Sustain. Dev.* **2010**, *13*, 337–361. [CrossRef]

29. United Nations Development Programme. *Human Development Report 2011—Sustainability and Equity: A Better Future for All*; United Nations Development Programme: New York, NY, USA, 2011.

30. Weisz, H.; Suh, S.; Graedel, T.E. Industrial Ecology: The role of manufactured capital in sustainability. *Proc. Natl. Acad. Sci. USA* **2015**, *112*, 6260–6264. [CrossRef] [PubMed]

31. Wiek, A.; Withycombe, L.; Redman, C.L. Key competencies in sustainability: A reference framework for academic program development. *Sustain. Sci.* **2011**, *6*, 203–218. [CrossRef]

32. International Finance Corporation. *International Finance Corporation's Policy on Environmental and Social Sustainability*; International Finance Corporation: Washington, DC, USA, 2012.

33. International Integrated Reporting Council. The International IR Framework. 2013. Available online: https://www.iasplus.com/en/resources/sustainability/iirc (accessed on 26 August 2021).

34. United Nations Global Compact. *Guide to Corporate Sustainability*; United Nations Global Compact: New York, NY, USA, 2014.

35. Forest Stewardship Council. *FSC®International Standard: FSC Principles and Criteria Forest Stewardship*; FSC-STD-01-001 V5-2 EN; Forest Stewardship Council: Bonn, Germany, 2015.

36. Marine Stewardship Council. *MSC Fishery Standard: Principles and Criteria for Sustainable Fishing*; Version 1.1; Marine Stewardship Council: London, UK, 2010.

37. Aquaculture Stewardship Council. *ASC Salmon Standard*; Version 1.0; Aquaculture Stewardship Council: Utrecht, The Netherlands, 2012.

38. Roundtable on Sustainable Palm Oil. *Principles and Criteria for Sustainable Palm Oil Production*; Roundtable on Sustainable Palm Oil: Zurich, Switzerland, 2007.

39. Round Table on Responsible Soy. *RTRS Standard for Responsible Soy Production*; Version 2.0; Round Table on Responsible Soy: Zurich, Switzerland, 2013.

40. Sustainable Agriculture Network. *Sustainable Agriculture Standard*; Version 2; Sustainable Agriculture Network: San José, Costa Rica, 2010.
41. FairWild Foundation. *FairWild Standard*; Version 2.0; FairWild Foundation: Weinfelden, Switzerland, 2010.
42. International Council on Mining and Metals. 10 Sustainable Development Principles. Available online: http://www.icmm.com/our-work/sustainable-development-framework/10-principles (accessed on 15 July 2016).
43. Australian Packaging Covenant. *Sustainable Packaging Guidelines*; Australian Packaging Covenant: Sydney, Australia, 2010.
44. Roundtable on Sustainable Biofuels. RSB Principles & Criteria for Sustainable Biofuel Production; RSB-STD-01-001, Version 2.0. 2010. Available online: https://rsb.org/wp-content/uploads/2017/02/11-03-08-RSB-PCs-Version2.pdf (accessed on 27 August 2021).
45. U.S. Green Building Council. *LEED 2009 for New Construction and Major Renovations Rating System*; Updated November 2011; U.S. Green Building Council: Washington, DC, USA, 2011.
46. Global Sustainable Tourism Council. *Global Sustainable Tourism Criteria*; Global Sustainable Tourism Council: Washington, DC, USA, 2008.
47. Sustainable Apparel Coalition. Available online: http://www.apparelcoalition.org/3.html (accessed on 13 July 2012).
48. Fairtrade International. *Fairtrade Standard for Small Producer Organizations*; Version 1.3; Fairtrade International: Bonn, Germany, 2011.
49. United Nations Environment Programme. *Sustainable Events Guide: Give Your Large Event a Small Footprint*; United Nations Environment Programme: Nairobi, Kenya, 2012.
50. The Association for the Advancement of Sustainability in Higher Education. *STARS (Sustainability Tracking, Assessment & Rating System) Technical Manual*; Version 2.0; The Association for the Advancement of Sustainability in Higher Education: Denver, CO, USA, 2014.
51. Tasaki, T.; Kameyama, Y.; Oshima, M.; Motoki, H. Sustainability Criteria Used in 25 Initiatives and Certification Programs: Towards Setting of Criteria for Sustainable Procurement and Conceptualizing of Sustainability. *Environ. Sci.* **2016**, *29*, 305–314. (In Japanese) [CrossRef]
52. Harzing, A.-W. Response styles in cross-national survey research: A 26-country Study. *Int. J. Cross Cult. Manag.* **2006**, *6*, 243–266. [CrossRef]
53. Harzing, A.-W.; Baldueza, J.; Barner-Rasmussen, W.; Barzantny, C.; Canabal, A.; Davila, A.; Espejo, A.; Ferreira, R.; Giroud, A.; Koester, K.; et al. Rating versus ranking: What is the best way to reduce response and language bias in cross-national research? *Int. Bus. Rev.* **2009**, *18*, 417–432. [CrossRef]
54. Lee, J.W.; Jones, P.S.; Mineyama, Y.; Zhang, X.E. Cultural Differences in Responses to a Likert Scale. *Res. Nurs. Health* **2002**, *25*, 295–306. [CrossRef]
55. Ministry of the Environment, Japan. *Report on the Survey of Environmentally Friendly Lifestyles*; Ministry of the Environment: Tokyo, Japan, 2014. (In Japanese)
56. The Asahi Glass Foundation. Results of the 23rd Annual "Questionnaire on Environmental Problems and the Survival of Humankind". 2014. Available online: https://www.af-info.or.jp/en/ed_clock/assets/pdf/result/2014jresult_fulltext.pdf (accessed on 7 July 2021).
57. Asia Development Bank. *Key Indicators for Asia and the Pacific 2016*; Asia Development Bank: Manila, Philippines, 2016.
58. Inglehart, R.F.; Welzel, C. *Modernization, Cultural Change, and Democracy: The Human Development Sequence*; Cambridge University Press: Cambridge, UK, 2005.
59. Hofstede, G.; Hofstede, G.J.; Minkov, M. *Cultures and Organizations: Software of the Mind, Intercultural Cooperation and Its Importance for Survival*, 3rd ed.; McGraw-Hill: New York, NY, USA, 2010.
60. The World Value Survey Association. Inglehart-Welzel Cultural Map, Wave 7 Provisional Version. Available online: https://www.worldvaluessurvey.org/WVSContents.jsp (accessed on 28 March 2021).
61. Geerthofstede.com. Dimension Data Matrix. Available online: https://geerthofstede.com/research-and-vsm/dimension-data-matrix/ (accessed on 28 March 2021).

Article

Impact of Gaps in the Educational Levels between Married Partners on Health and a Sustainable Lifestyle: Evidence from 32 Countries

Xiangdan Piao [1],*, Xinxin Ma [2],*, Chi Zhang [1],* and Shunsuke Managi [1],*

[1] Urban Institute & Department of Civil Engineering, School of Engineering, Kyushu University, Fukuoka 819-0395, Japan

[2] Faculty of Social Sciences, University of Toyama, Toyama 930-8555, Japan

* Correspondence: piao.xiangdan@doc.kyushu-u.ac.jp (X.P.); xxma@eco.u-toyama.ac.jp (X.M.); zhang.chi.169@s.kyushu-u.ac.jp (C.Z.); managi@doc.kyushu-u.ac.jp (S.M.)

Received: 8 April 2020; Accepted: 2 June 2020; Published: 5 June 2020

Abstract: Using original cross-sectional internet survey data from 32 countries covering six continents, we investigated the impact of education gaps between married partners on their health status and sustainable lifestyles using the instrumental variable method. A self-rated health status index, mental health index, and an objective health status index were utilized to assess the health statuses of individuals, and six unique indices were used to investigate the sustainable lifestyles. According to the main findings, work-family conflicts may be severe for both wives and husbands with high education levels, and the hypothesis regarding the positive effect of income was not supported. Two major conclusions were derived. First, in general, as opposed to couples with equal education levels, the probability of reporting a worse health status was higher, and the activities related to sustainable development such as improving environmental sustainability were less for couples with education gaps. Second, a comparison of the effects of education gaps on the health status of couples in various groups reveals that highly educated groups, women, and people in Asian or middle-income countries had a higher negative effect on their health status.

Keywords: intrahousehold education gap; marriage; health status; instrumental variable; level of education; self-rated health; sustainable lifestyle

---

1. Introduction

The United Nations published the sustainable development goals (SDGs), which include responsible consumption and production (goal 12), good health and well-being (goal 3), quality education (goal 4), gender equality (goal 5), and reduced inequalities (goal 10), as well as the 2030 agenda for sustainable development in 2015 [1]. From the SDG perspective, this study investigates the impact of the gaps in education levels between married couples on health and a sustainable lifestyle [2]. Moreover, it employs an international comparison on the issue based on original international survey data collected from 32 countries across six continents.

The main contributions of this study can be considered as follows. First, regarding the issue of the correlations between education and health, since exploring the determinants of individual health is an important issue for policymakers seeking to improve the health status of the national population, many researchers have conducted empirical studies on this issue. Regarding self-rated health (SRH) and mental health, it has been found that socioeconomic factors such as income and health behaviors (e.g., avoiding alcohol consumption and smoking) are associated with health outcomes. An individual's level of education is the most controversial index in human development. It is a primary factor in the labor market, and it may also be an important factor in determining the health status of individuals. As such, many studies have investigated the relationship between education and health outcomes. Previous studies have shown that education has a positive effect on health [3,4]. Regarding mental health, recent studies have indicated that there is a lower probability of developing a mental health disorder if the education level of an individual is high [5–9]. Moreover, Fletcher [9] argued that there is a negative relationship between education and mental health for women. Although many previous studies have investigated the relationship between education and health, some issues have yet to be discussed. For example, based on the collective model proposed by Chiappori [10], intrahousehold bargaining power may affect intrahousehold economic resource allocation, which may, in turn, affect household members' health outcomes [11–19]. Additionally, the education gap between wives and husbands has been utilized as an index of intrahousehold bargaining power [20,21].

Based on the abovementioned previous research, it is assumed that an education gap between wives and husbands may influence the health outcomes of intrahousehold members. From an economic perspective, there are four reasons why an intrahousehold education gap (IHEG) could affect health. First, an individual with a high level of education may be more likely to find a better job and have a higher income than an individual with a low level of education. Therefore, he or she may be able to accumulate more wealth than other household members and invest more money to improve his or her health status [11–18]. As a result, a couple's education gap might positively influence their health status (the positive effect of income hypothesis). Second, based on gender role consciousness, even highly educated wives who earn more income tend to undertake more housework than their husbands [21–23]. The work-family conflict may worsen the health status of wives (the negative effect of the gender role consciousness hypothesis). Third, highly educated individuals may have higher professional abilities and skills [24–27]. Because skill gaps exist between wives and husbands, a highly educated individual may not get help from the less-educated partner in overcoming difficulties in life and work, which may worsen their mental health (the negative effect of skill gap hypothesis). Fourth, the stress of work hours can have negative effects on health. In reality, a highly educated individual may have longer working hours than less-educated partner. Long working hours may negatively affect the health status of individuals (the negative effect of long working hours hypothesis) [28–31]. Because of the positive and negative effects mentioned above, and because gender-roles consciousness regarding family responsibility differs by country, an empirical study based on cross-country data should be undertaken to investigate the relationship between couples' education gaps and their health in order to determine the impact of education on health. However, although Groot and Van Den Brink [20] and Ma and Piao [21] reported that an education gap between wives and husbands affects the happiness of the individuals concerned, empirical studies on this issue involving international comparisons are scarce. This study bridges the gap in the literature.

Second, since human activities started to significantly modify the global environment, investigation on sustainable production and consumption has been expanding, prompted by the willingness to reduce the impact of the throwaway culture. There are several examples of recent studies carried out to bring the human behavior close to a sustainable lifestyle and try to build a systematical resource management and an efficient recycling system, reducing the electric waste and improving the electric management [2,32–42]. Regarding the issue of a sustainable lifestyle, which is related to the SDGs, it is assumed that the intrahousehold education gap may influence a sustainable lifestyle by reshaping patterns of individuals' behaviors. According to Akenji and Chen [2], the sustainable lifestyle is defined as "a cluster of habits and patterns of behavior embedded in a society and facilitated by institutions, norms, and infrastructures that frame individual choice, in order to minimize the use of natural resources and generation of wastes, while supporting fairness and prosperity for all". Therefore, to achieve the SDGs, the patterns of reducing consumption of natural resources and generation of wastes have attracted the attention of scholars (e.g., [2,32–37]). Shove [32] provided a theoretical framework that combines end users who are comfortable with their service and sustainable technology consumption, which might play a role in reshaping patterns of resource consumption. Peattie and Collins [37] highlight the important role of sustainable consumption in releasing environmental resource consumption pressure. They argue that this is expected to sustain the environment. Moreover, Zakaria et al. [34] argued that consumers' consumption choices are associated with a sustainable lifestyle. However, a sustainable lifestyle affects sustainable consumption practice. Therefore, it is necessary to conduct an empirical study to investigate the correlations between the gaps in education levels between married couples, their health, and sustainable lifestyles through consumption choices. Tilman and Clark [42] highlight the crucial relationship between environmental sustainability and public health through food lifecycle analysis.

This study provides empirical evidence about the relationship between IHEGs and the health status of individuals (SRH, mental health, and objective health status) using cross-country household survey data covering 32 countries on six continents (Asia, Europe, North America, South America, Africa, and Australia). To the best of our knowledge, this is the first study on the issue, which may provide new evidence regarding the link between education and health as well as a sustainable lifestyle.

This paper is structured as follows. Section 2 describes the analytic methods, including an introduction to the data and models. Section 3 discusses the analysis results, and Section 4 introduces the quantitative analysis results. The last section summarizes the main conclusions.

2. Methodology

First, the health function was estimated in order to investigate the relationship between IHEGs and health. The instrumental variable method was used to address the endogeneity problem and investigate the causal relationship between couples' education gaps and health. The ordinary least squares (OLS) model is presented in Equation (1).

$$H_{iC} = \alpha + \theta IHEG_{iC} + X'_{iC}\beta + D'_C\delta + \varepsilon_{iC} \tag{1}$$

Here, iC refers to an individual i in country C; H is an individual's health status (SRH, the mental health index, and objective health); $IHEG$ denotes an intrahousehold education gap, which is the couple's education gap; D represents the country dummy variables; α is a constant; θ, β, and δ are the estimated coefficients; and ε is an error term. When θ is a negative value and is statistically significant, it indicates that a high IHEG may worsen an individual's health.

An endogeneity problem is possible in the OLS model, i.e., when an individual with poor health prefers to marry a highly educated partner for financial benefits, and the main independent variables of interest, $IHEG$, is correlated with the error term. Thus, to address this endogeneity problem, the instrumental variable (IV) method was utilized in this study [43,44]. The first-stage and second-stage estimation equations are expressed as Equations (2) and (3), respectively.

$$IHEG_{iC} = b_0 + Z'_{iC}b_1 + X'_{iC}b_2 + D'_C b_3 + u_{iC} \tag{2}$$

$$H_{iC} = \alpha + \theta \widehat{IHEG}_{iC} + X'_{iC}\beta + D'_C\delta + \varepsilon_{iC} \tag{3}$$

In Equations (2) and (3), b_0 is a constant; b_1, b_2, and b_3 represent the estimated coefficients; u is an error term; Z is the set of instrumental variables (e.g., parent's highest education level); and $\widehat{IHEG}$ is an imputed value based on the results of the first-stage regression shown by Equation (2). The weak instrument test and the Sargan test were used to test for the endogenous problem and to judge the statistical validity of the instruments [45].

Second, to investigate the probability channels in order to explain the impact of IHEGs on health, multiple regression models were used, as shown in Equation (4):

$$Y_{iC} = \alpha + \theta IHEG_{iC} + X'_{iC}\beta + D'_C\delta + \varepsilon_{iC} \tag{4}$$

In Equation (2), Y represents income satisfaction, weekly working days, overcoming difficulties, satisfaction with health or medical care, and attending environmental activities as a volunteer.

3. Data

Hence, to investigate the impact of IHEGs on individual health, an internet survey via a website was conducted by a third-party company in Japan in 2015 and 2017. The third party company has provided lots of reliable website survey services in recent decades, and the company also has an extensive panel that allows the conducted sample to match the population distribution by regional area, age, and gender constitution. The original survey was conducted from 2015 to 2017, and data regarding demographics, household income, education level, SRH, mental health, and objective health were collected by matching country-level population age and gender. While web-based surveys randomly select respondents, compared to interview-based surveys, web-based surveys tend to select well-educated respondents because non-internet users are excluded. To address this problem, we carefully analyzed the respondents with high and low education levels separately. The detailed description regarding this original survey is included in [46].

The original cross-country survey data were comprised of 32 developing and developed countries on six continents: Asia, Europe, North America, South America, Africa, and Australia. Thirty-two countries were assessed using a web-based survey, and five countries were assessed using an interview-based survey that was web-based. The information on these specific country-level observations is available upon reasonable request. The countries surveyed in each continent are as follows:

1. Asia (Japan, Thailand, Malaysia, Indonesia, Singapore, Vietnam, the Philippines, India, and China);
2. Europe (Russia, Germany, the United Kingdom, France, Spain, Italy, Sweden, the Netherlands, Greece, Turkey, Hungary, Poland, Czechoslovakia, and Romania);
3. North America (Mexico, Canada, and the United States);
4. South America (Venezuela, Chile, Brazil, and Colombia);
5. Australia (Australia);
6. Africa (South Africa).

The dependent variables were three indices. (1) An individual's SRH was a scale variable from one to five. We coded the health status numbers as "5 = very good, 4 = good, 3 = neither, 2 = poor, and 1 = very poor". (2) The mental health index was a mental health score, which was calculated based on the 12 survey items related to mental health. All answer options for the 12 survey items were from 1 to 4, which indicated mental health status from the worst status to the best. The 12 survey items included concentrating ability, sleeping quality, feelings of stress, behavioral control, depression, feelings of confidence, and positive effects. The 12-item general health questionnaire in the survey was a general short version of the World Health Organization's 60-item questionnaire. (3) Objective health was a dummy variable that was equal to 1 if an individual did not experience an illness or surgery in the past half-year (healthy = 1, unhealthy = 0). Objective unhealthy includes physical illness and mental illness.

The main independent variable was IHEG. Two indices were utilized in this study: (1) IHEG value (IHEG1) and (2) IHEG dummy variable (IHEG2). IHEG1 was a continuous variable which denotes the difference between the educational attainment level of an individual minus their partner's education level. IHEG2 was a IHEG dummy variable (equal or unequal). The scale value of educational attainment level was as follows: never attended school = 1; dropping out of primary school = 2; primary school = 3; junior high school = 4; senior high school = 5; vocational school = 6; college = 7; university = 8; graduate school (master's degree) = 9; and graduate school (doctor degree) = 10. Education gap was a scale variable that ranged from −9 to 9 that was calculated by the scale number of the respondent's education minus that of his/her partner's education. For example, a value of −9 indicates that an uneducated individual married a partner who had a doctorate degree, and a value of 9 is the opposite. The education gap based on years of schooling was also conducted to improve the robustness of the results (the results using years of schooling are available upon request). As mentioned above, since there are both negative and positive effects on the relationship between the education gap and health, the results of the effect of the education gap on health are not clear. When the negative effect was greater than the positive effect, the coefficient of the education gap variable (IHEG1 or IHEG2) was negative, and the value was statistically significant and vice versa. These results are reported in the following section.

Thus, to address the endogenous problem, instrumental variables were utilized, which are as follows. (1) Parents' highest educational attainment, which is a scale variable from 1 to 10 (a parent's highest education level was evaluated as follows: never attended school = 1; dropping out of primary school = 2; primary school = 3; junior high school = 4; senior high school = 5; vocational school = 6; college = 7; university = 8; graduate school (master) = 9; and graduate school (doctorate) = 10.) (2) The dummy variable of a parent's advanced educational attainment (also known as tertiary education (International Standard Classification of Education (ISCED) levels 5 to 8—tertiary education included both commonly accepted academic education and advanced vocational or professional education) defined by UNESCO or higher education referred to as World Bank, mentioned by the World Bank) [47,48].

It is possible that a parent's educational attainment level does not directly influence an adult child's health level but affects the child's educational attainment. The overidentification test was used to test the validity of these instruments. First, the range in evaluated educational attainment was 1 to 10, from the lowest level (uneducated) to the highest level (individuals having doctorate degrees). The mandatory number of years of education differs by country and area, and this type of data was largely missing in the survey; therefore, measurement error may have occurred. Thus, the range score of evaluated education from 1 to 10 was a better variable than the years of schooling variable. Second, nine dummy variables of a parent's highest educational attainment level were utilized as instrumental variables: (1) dropping out of primary school dummy, (2) primary school, (3) junior high school, (4) senior high school, (5) vocational school, (6) college, (7) university, (8) graduate school (master), and (9) graduate school (doctorate). We will discuss the validity and violation of the instrument variables in the following section.

The other explanatory dummy variables were as follows: female dummy variable; work status dummy variables (e.g., individual unemployed, full-time employee, part-time employee, company owner, government employee, professional worker such as physician and professor, self-employed, student, and housewife or househusband); education level dummy variables (e.g., senior high school or lower, vocational school, college or university, and graduate school); housing status dummy variables (e.g., rent and house owner); age dummies (e.g., less than 30 years old, 31–39, 40–49, 50–59, and 60 years or older); number of children dummy variables (e.g., no child, one child, two children, and three or more children); five household income dummy variables (e g , first quintile to the fifth quintile); and country dummy variables were used to control country-level heterogeneity. The original data comprised 32 countries, including developing and developed countries on six continents (Asia, Europe, North America, South America, Africa, and Australia).

The following variables were utilized as dependent variables to investigate the potential channel of the impact of education gaps on health and a sustainable lifestyle. The dummy variables were income satisfaction, weekly working days, overcoming difficulties (based on the question "Have you recently felt that you could not overcome your difficulties?"), the overcome difficulties variable is constructed as "4 = not at all, 3 = no more than usual, 2 = rather more than usual, and 1 = much more than usual"), and satisfaction with health or medical care. Regarding a sustainable lifestyle, six dummy variables are constructed as sustainable lifestyle indices based on Akenji and Chen [2]. Attend environmental activities as volunteers (yes = 1, no = 0), donate to environmental activities (income) (yes =1, no = 0), donate to environmental activities (goods) (yes = 1, no = 0), purchase energy-saving household products (yes = 1, no = 0), energy-saving activities (yes = 1, no = 0), sorting or reducing rubbish (yes = 1, no = 0).

In the robustness check, the variables (1) satisfaction with health/medical care, (2) do not smoke dummy variable, (3) alcohol consumption dummy variable (drink alcohol every day; 4–5 times per week; 2–3 times per week; once per week; less than above; and do not drink alcohol) were selected.

The statistical descriptions of the variables utilized in this study are summarized in Table 1.

Table 1. Descriptive statistics.

Variables	Mean	S.D.	Obs.
Self-rated health (1–5)	3.88	0.84	53,365
Mental health (1–4)	3.06	0.50	53,365
Objective health (0.1)	0.83	0.38	53,365
Intra-household education difference ((−9)–9) (IHEG1 =individual's education–partner's education)	0.28	1.55	53,365
Having education gap (IHEG2) (0.1)	0.51	0.50	53,365
Intra-country household income			
Income first quintile	0.27	0.44	53,365
Income second quintile	0.24	0.42	53,365
Income third quintile	0.14	0.35	53,365
Income fourth quintile	0.19	0.39	53,365
Income fifth quintile	0.17	0.38	53,365
Educational attainment			
Senior high school or lower	0.22	0.41	53,365
Vocational school	0.09	0.29	53,365
College or university	0.56	0.50	53,365
Masters or more	0.13	0.33	53,365
Aged less than 30 years	0.14	0.35	53,365
Aged 31–39 years	0.22	0.42	53,365
Aged 40–49 years	0.25	0.43	53,365
Aged 50–59 years	0.23	0.42	53,365
Aged 60 years or more	0.16	0.37	53,365
Occupational status (ref. Unemployed)			
Full-time employee	0.52	0.50	53,365
Part-time employee	0.07	0.26	53,365
Company owner	0.03	0.16	53,365
Government employee	0.04	0.19	53,365
Professional	0.04	0.20	53,365
Self-employed	0.07	0.26	53,365
Student	0.01	0.11	53,365
Housewife/Househusband	0.09	0.29	53,365
Other	0.07	0.25	53,365
Female dummy	0.47	0.50	53,365
No child	0.17	0.37	53,365
One child	0.40	0.49	53,365
Two children	0.30	0.46	53,365
Three or more children	0.13	0.34	53,365
Rent	0.21	0.41	53,365
Owner	0.77	0.42	53,365
Other	0.02	0.15	53,365
Instrument set 1			
Parents' highest education attainment (1–10)	5.62	2.18	53,365
Parents have advanced education (0.1)	0.47	0.50	53,365
Instrument set 2			
never attended school	0.04	0.19	53,365
dropped out of primary school	0.04	0.20	53,365
primary school	0.09	0.28	53,365
junior high school	0.13	0.33	53,365
senior high school	0.24	0.42	53,365
vocational school	0.11	0.31	53,365
college	0.08	0.27	53,365
university	0.20	0.40	53,365
graduate school (master)	0.05	0.23	53,365
graduate school (doctorate)	0.02	0.15	53,365
Income satisfaction	3.22	0.89	51,384
Weekly working days	5.05	1.03	38,600
Difficulties overcome	2.98	0.81	53,365

Table 1. *Cont.*

Variables	Mean	S.D.	Obs.
Satisfaction of health/medical care	3.41	1.22	53,365
Volunteer attendance at environmental activities	0.26	0.44	18,223
Donation to environmental activities (income)	0.19	0.39	53,365
Donation to environmental activities (goods)	0.17	0.38	53,365
Purchase energy saving household products	0.52	0.50	53,365
Energy saving activities	0.64	0.48	53,365
Sorting/reducing rubbish	0.63	0.48	53,365
Do not smoke	0.68	0.47	18,223
Frequency of drinking alcohol			
Drink alcohol every day	0.22	0.42	18,223
4–5 times per week	0.18	0.38	18,223
2–3 times per week	0.26	0.44	18,223
Once per week	0.06	0.24	18,223
Less than above	0.19	0.39	18,223
Do not drink alcohol	0.09	0.28	18,223

Note: Calculated based on the original international survey from 2015 to 2017. IHEG1: Difference between individual's own educational attainment level and the partner's level. IHEG2: having an education gap dummy variable (1 = having a gap, 0 = no gap).

4. Results

4.1. Impact of a Couple's Education Gap on Health: OLS and Instrumental Variable Two-Stage Least-Squares (IV-2SLS) Estimations

The first stage was based on the OLS model. It estimated various influence factors that included the effect of a parent's highest education level on the IHEG. The results are shown in Table 2. The dependent variables in Model 1 and Model 2 were labeled as "IHEG1," and in Model 3 and Model 4, they were labeled as "IHEG2." The main independent variables were a parent's highest level of education, which was described as instrumental variables in the two-stage least squares (2SLS) model. Stock et al. [45] suggested that if the F-statistic values were greater than 10 for one of the endogenous variables based on the 2SLS estimation, then the selected instrument variable was not weak. From Model 1 to Model 4, the F-statistic values for the joint significance on the coefficients of the instruments were 704.550, 159.136, 230.902, and 60.043, respectively, which were all larger than 10. Therefore, it is clear that the selected instruments in this study are not weak.

The estimation results are summarized as follows. (1) A parent's highest level of education was negatively correlated with both IHEG1 and IHEG2 in Model 1 and Model 3 (−0.092 in Model 1, −0.025 in Model 3), and the results were significant at the 1% level. (When using the number of years of schooling of parents as the educational attainment index of parents, the results were similar. However, the mandatory minimum number of years of schooling differs by country and area. Moreover, the number of years of schooling for individuals who dropped out was unable to be counted, which resulted in measurement errors. Therefore, we determined that the evaluated education from 1 to 10 was more appropriate than transforming the results into years of schooling.) The coefficient of parents with an advanced education was −0.106 in Model 1, and it was statistically significant at the 1% level. The results indicate that the IHEG was smaller for individuals with highly educated parents.

Table 3 presents the estimation results by using various health indices (SRH, the mental health index, and objective health) and various methods of analysis (the OLS and IV-2SLS methods). The dependent variable for Model 1, Model 2, and Model 6 was SRH. The dependent variable for Model 3, Model 4, and Model 7 was the mental health index, and the dependent variable for Model 5 and Model 8 was objective health. The results of the OLS model are shown in Models 1 and 3. The results of the IV-2SLS method are shown in Models 2 and 8. The results of the overidentification tests in Model 2 and from Model 4 to Model 7 were statistically insignificant at the 5% level. These findings indicate that the instruments are statistically exogenous in these models and that the instrument variable methods should be utilized to address the endogeneity problem. This means that there was bias in the results based on the OLS model. Therefore, we report mainly the results based on the IV-2SLS method in the following section. We also compare the results to those based on the OLS model. The main results are as follows.

First, to compare the results obtained by the OLS and IV models, although the coefficients in both the OLS (Model 1) and IV-2SLS (Model 3) were statistically significant at the 1% level, the coefficient of IHEG1 was −0.020 for the OLS model and −0.046 for the IV-2SLS model. Similar results were observed in Models 3 and 4; the coefficient of IHEG1 was −0.003 for the OLS model (Model 3) and −0.029 for the IV-2SLS model (Model 4). The magnitudes of the coefficients in the IV-2SLS model were greater than those in the OLS model, which suggests that the impact of IHEG1 on health might be underestimated by the OLS model. Consistent results were obtained by additionally controlling health insurance satisfaction, alcohol consumption, and smoking.

Second, regarding the impact of IHEGs on health, there were two outcomes. (1) The coefficients of IHEG1 were negative values (−0.046 in Model 2; −0.029 in Model 4; and −0.015 in Model 5; Table 3), and they were statistically significant at the 1% and 5% levels. These findings suggest that health status (SRH, mental health, and objective health) was worse for individuals with a higher level of education than for their partner. Because a negative effect of IHEG1 on health was found for both husbands and wives, we investigate the above effects by gender in the following section. (2) The coefficients of IHEG2 (couples with education gaps) were negative values (−0.205 in Model 6; −0.137 in Model 7; and −0.062 in Model 8—when using the educational attainment dummy variables, it was also found that the IHEG negatively affected objective health status, and the result was statistically significant at the 5% level—and they were statistically significant at the 1% and 10% levels. These findings indicate that, compared with the health statuses of couples with equal levels of education, health status (SRH, mental health, or objective health) was worse for couples with intra-education gaps.

4.2. Estimations by Various Groups

To consider the heterogeneity in various groups, we also made estimations based on education, gender, age, income, and country. As a kind of human capital, a high level of education is associated with more working skills and higher incomes. A couple with a large education gap may also have great skill and knowledge gaps, resulting in communication difficulties. Moreover, the probability of obtaining help from his or her partner may be lower for an individual with a higher level of education because he or she is more likely to do work that requires specific skills and knowledge. To consider the heterogeneity due to individual education level, we made estimations using two groups: (1) a high education level group that completed vocational school or higher (also referred to as tertiary education (ISCED levels 5 to 8) by The United Nations Educational, Scientific and Cultural Organization (UNESCO) (tertiary education included both commonly accepted academic education and advanced vocational or professional education) or higher education by the World Bank) [47,48]; (2) a low education level group that completed senior high school or lower (also known as primary or secondary education (ISCED levels 0 to 4)) by UNESCO). The estimations were also employed by gender (women and men), age (younger than 40 years and older than 40 years), continents (Asia, Europe, and North America, and South America and Australia), and by income (high-income countries and middle-income countries) groups. The results for the high- and low-education groups are summarized separately in Table 4 column (a) and column (b) The value of the IHEG (IHEG1) was used as the education gap index (IHEG) in the estimations. The dependent variables were SRH, mental health, and objective health. The independent variables were similar with those in Table 3, but only the results of the IHEG are summarized in Table 4. The main results are as follows.

Table 2. First-stage estimates of the effects of parental education attainment and intrahousehold couples' educational gap—OLS.

	Model 1		Model 2		Model 3		Model 4	
	IHEG1 Coeff.	S.E.	IHEG1 Coeff.	S.E.	IHEG2 Coeff.	S.E.	IHEG2 Coeff.	S.E.
Parents' highest educational attainment	−0.092 ***	(0.005)			−0.025 ***	(0.002)		
Parents have advanced education attainment	−0.106 ***	(0.022)			0.007	(0.008)		
Parents' highest education attainment (ref. never attended school)								
dropped out of primary school			−0.174 ***	(0.042)			0.065 ***	(0.015)
primary school			−0.196 ***	(0.036)			0.007	(0.013)
junior high school			−0.260 ***	(0.035)			−0.006	(0.012)
senior high school			−0.367 ***	(0.033)			−0.057 ***	(0.012)
vocational school			−0.557 ***	(0.036)			−0.048 ***	(0.013)
college			−0.667 ***	(0.037)			−0.097 ***	(0.013)
university			−0.738 ***	(0.034)			−0.130 ***	(0.012)
graduate school (master)			−0.939 ***	(0.041)			−0.121 ***	(0.015)
graduate school (doctorate)			−0.974 ***	(0.052)			−0.121 ***	(0.019)
Other controls.	Yes		Yes		Yes		Yes	
Country dummy	Yes		Yes		Yes		Yes	
Number of observations	53,365		53,365		53,365		53,365	
Number of countries	32		32		32		32	
F-statistic for instruments	704.550		159.136		230.902		60.043	
R-squared	0.229		0.229		0.044		0.045	

Notes: Standard errors in parentheses. *** $p < 0.01$, ** $p < 0.05$, * $p < 0.1$. IHEG1: Individual's education−partner's education; IHEG2: having an education gap dummy variable (1 = having a gap, 0 = no gap). Other controls are income, education, age, occupation, female dummy, number of children, house status. Sources: Calculated based on original international survey from 2015 to 2017.

Table 3. The effects of intrahousehold couples' educational gap on health—OLS and IV-2SLS estimation results.

	Self-Rated Health		Mental Health		Objective Health	Self-Rated Health	Mental Health	Objective Health
	Model 1 OLS Coeff.	Model 2 IV-2SLS Coeff.	Model 3 OLS Coeff.	Model 4 IV-2SLS Coeff.	Model 5 IV-2SLS Coeff.	Model 6 IV-2SLS Coeff.	Model 7 IV-2SLS Coeff.	Model 8 IV-2SLS Coeff.
IHEG1 (value)	−0.020 ***	−0.046 ***	−0.003 **	−0.029 ***	−0.015 **			
	(0.002)	(0.016)	(0.001)	(0.009)	(0.007)			
IHEG2(Having an education gap)						−0.205 ***	−0.137 ***	−0.062 *
						(0.076)	(0.045)	(0.036)
Household Income (ref. Income first quintile)								
Income second quintile	0.099 ***	0.093 ***	0.067 ***	0.061 ***	−0.003	0.103 ***	0.067 ***	0.000
	(0.010)	(0.011)	(0.006)	(0.006)	(0.005)	(0.010)	(0.006)	(0.005)
Income third quintile	0.095 ***	0.085 ***	0.078 ***	0.068 ***	−0.005	0.094 ***	0.074 ***	−0.002
	(0.012)	(0.013)	(0.007)	(0.008)	(0.006)	(0.012)	(0.007)	(0.006)
Income fourth quintile	0.139 ***	0.125 ***	0.099 ***	0.085 ***	−0.000	0.135 ***	0.090 ***	0.003
	(0.011)	(0.014)	(0.007)	(0.008)	(0.006)	(0.012)	(0.007)	(0.006)
Income fifth quintile	0.206 ***	0.187 ***	0.123 ***	0.104 ***	0.008	0.201 ***	0.112 ***	0.012 *
	(0.012)	(0.016)	(0.007)	(0.010)	(0.008)	(0.014)	(0.008)	(0.006)
Educational attainment (ref. Senior high school or lower)								
Vocational school	−0.024 *	−0.005	−0.006	0.013	−0.019 **	−0.020	0.004	−0.024 ***
	(0.014)	(0.018)	(0.008)	(0.011)	(0.008)	(0.016)	(0.009)	(0.007)
College or university	0.066 ***	0.106 ***	0.026 ***	0.066 ***	0.014	0.037 ***	0.022 ***	−0.008 *
	(0.010)	(0.026)	(0.006)	(0.015)	(0.012)	(0.010)	(0.006)	(0.004)
Graduate school	0.142 ***	0.199 ***	0.025 ***	0.082 ***	0.026	0.129 ***	0.038 ***	0.004
	(0.014)	(0.036)	(0.008)	(0.022)	(0.017)	(0.017)	(0.010)	(0.008)
Other controls.	Yes	Yes	Yes	Yes	Yes	Yes	Yes	Yes
Country dummy	Yes	Yes	Yes	Yes	Yes	Yes	Yes	Yes
Number of observations	53,365	53,365	53,365	53,365	53,365	53,365	53,365	53,365
Number of countries	32	32	32	32	32	32	32	32
Overidentification test (Sargan test)		chi2(1) = 1.701 (p = 0.19)		chi2(1) = 0.217 (p = 0.64)	chi2(1) = 3.549 (p = 0.06)	chi2(1) = 3.109 (p = 0.08)	chi2(1) = 0.952 p = 0.33)	
R-squared	0.106	0.104	0.102	0.097	0.022	0.094	0.095	0.020

Notes: (1) Standard errors in parentheses. *** $p < 0.01$, ** $p < 0.05$, * $p < 0.1$. (2) Instrument variables: Parents' educational attainment and parents have advanced education from Model 1 to Model 7. Model 8 instrument variable: parents' educational attainment. (3) IHEG1 (value) = individual's education–partner's education; IHEG2: having an education gap dummy variable (1 = having a gap, 0 = no gap). (4) Other controls include age, occupation, female dummy, number of children, house status. Source: Calculated based on the original international survey from 2015 to 2017.

First, in general, the coefficients of the IHEG were negative values (−0.084 in (a) for SRH; −0.042 in (a) for mental health; and −0.024 in (a) for objective health), and they were all statistically significant at the 1% level. This finding indicates that for the group with a high education level that completed vocational school or higher, an education gap between a respondent and his or her partner may lower the respondent's SRH, mental health, and objective health status. Individuals with education levels that are higher than those of other family members may have more household financial responsibilities, which may result in long working hours and more feelings of stress. As a result, their mental health and physical health status may be poor.

Second, the effects of the IHEG on health differed in the different groups. (1) For the high-education group (a), the coefficients of the IHEG were negative for women (−0.106 in Model 1(a); −0.034 in Model 3(a) for mental health; and −0.057 in Model 5(a)), and they were all statistically significant at the 1% level. These results suggest that when a wife has a higher level of education than her husband, her SRH, mental health, and objective health may be worse. The coefficients for men were −0.063, −0.043, and −0.001 and were statistically significant at the 1% level for SRH and mental health. An education gap between wives and husbands also negatively affects the husband's health. Comparing the groups of husbands and wives, the negative effect of having a higher education than one's partner was greater for women regarding SRH and objective health. Accordingly, on average, wives have more housework. Therefore, work-family conflicts might be severe for a wife when she has an education level that is greater than that of her partner. However, for the low-education group, most coefficients were not statistically significant for either wives or husbands. Compared with the high-education group, the negative effect of an IHEG seemed to be smaller for the low-education group. This finding indicates that work-family conflicts may be severe for both wives and husbands with high education levels.

(2) For the high-education group, although a negative effect of the IHEG on health was found in both the younger group and the older group, the effect was greater for the younger group based on SRH and mental health than for their counterparts. However, the effect was greater for the older group based on objective health. For the low-education group, most coefficients were not statistically significant for either the younger or older groups.

(3) Comparing the results in various areas of the world, for the high-education group, the coefficients of the IHEG were negative for Asian countries (−0.108 in Model 1(a); −0.057 in Model 3(a); and −0.045 in Model 5(a)), and they were all statistically significant at the 1% level. This finding suggests that, for the high-education group, an IHEG may worsen the SRH, mental health, and objective health of individuals in Asian countries. The coefficients for Europe and North America were −0.082, −0.033, and 0.003, and they were statistically significant at the 1% and 5% levels in Model 1 and Model 2. Comparing Asian countries with European and North American countries, the negative effect of an IHEG on health was greater for individuals in Asian countries. However, for the low-education group, most coefficients were not statistically significant for the Asian, European, and North American countries.

(4) Considering the results of lower and upper middle-income countries and high-income countries internationally, the negative effect of an IHEG on health was greater for lower and upper middle-income countries than for high-income countries regarding mental health and objective health for the lower and upper middle-income countries. This finding may be because high-income countries can provide universal public health insurance and advanced medical or health care service.

Third, for the low-education group, there may be a positive relationship between an IHEG and health. For example, the coefficients of the IHEG were positive values (0.053 in Model 2(b) for the total; 0.020 in Model 6(b)), and they were statistically significant at the 1% and 10% level. This finding suggests that, for the low-education group, reducing the education gap may improve the health status of an individual, particularly regarding the mental health condition of women (0.076 and statistically significant at the 1% level in Model 4 (b) for women).

Table 4. Summaries of the IV-2SLS results by subsamples.

Model	Self-Rated Health		Mental Health		Objective Health	
	Model 1 (a) High Education Coeff. (S.E.)	Model 2 (b) Low Education Coeff. (S.E.)	Model 3 (a) High Education Coeff. (S.E.)	Model 4 (b) Low Education Coeff. (S.E.)	Model 5 (a) High Education Coeff. (S.E.)	Model 6 (b) Low Education Coeff. (S.E.)
Total	−0.084 *** (0.016)	0.039 (0.027)	−0.042 *** (0.009)	0.053 *** (0.016)	−0.024 *** (0.008)	0.020 * (0.012)
Selected subsamples: By gender groups						
women	−0.106 *** (0.028)	0.040 (0.043)	−0.034 ** (0.017)	0.076 *** (0.026)	−0.057 *** (0.014)	0.022 (0.019)
men	−0.063 *** (0.020)	0.039 (0.036)	−0.043 *** (0.012)	0.024 (0.021)	−0.001 (0.009)	0.017 (0.016)
By age groups						
age ≤ 40	−0.051 ** (0.024)	0.056 (0.037)	−0.052 *** (0.017)	0.025 (0.025)	−0.027 ** (0.012)	0.034 ** (0.016)
age > 40	−0.096 *** (0.021)	0.002 (0.040)	−0.035 *** (0.011)	0.056 ** (0.022)	−0.019 ** (0.010)	0.006 (0.018)
By continent groups						
Asia	−0.108 *** (0.022)	−0.061 (0.042)	−0.057 *** (0.013)	0.056 ** (0.024)	−0.045 *** (0.011)	−0.022 (0.018)
Europe and North America	−0.082 *** (0.027)	0.043 (0.038)	−0.033 ** (0.016)	0.040 * (0.023)	0.003 (0.012)	0.042 ** (0.017)
South America and Australia	0.024 (0.031)	0.086 (0.056)	0.023 (0.022)	0.027 (0.035)	−0.009 (0.016)	0.034 (0.025)
By inter-country income level groups						
lower and upper middle-income countries	−0.070 *** (0.020)	0.083 ** (0.039)	−0.043 *** (0.013)	0.103 *** (0.025)	−0.045 *** (0.011)	−0.022 (0.019)
high-income countries	−0.086 *** (0.025)	0.001 (0.038)	−0.033 ** (0.014)	0.022 (0.021)	0.007 (0.010)	0.048 *** (0.016)

Notes: Standard errors in parentheses. *** $p < 0.01$, ** $p < 0.05$, * $p < 0.1$. The control variables are the same with those in Table 4 except for the respondent's educational attainment. 9 kinds of educational attainment dummy variables are utilized. IHEG1 (value): Individual's education–partner's education. Self-rated health scale is from 5 (very healthy) to 1 (very poor); Mental health index score is from 4 (very healthy) to 1 (very poor). High education: vocational school or higher; Low education: senior high school or lower. Sources: Calculated based on original international survey from 2015 to 2017.

4.3. The Impact of IHEGs, Sustainable Lifestyle, and Health

Next, we investigated the potential mechanism of the negative effect of IHEGs on health and sustainability lifestyle. We estimated the effects of IHEG1 (value) on (1) income satisfaction, (2) weekly working days, (3) overcoming difficulties (the impact of IHEGs on feelings of stress was also estimated, and the results were consistent with those for overcoming difficulties in the analyses. These results are available upon request), (4) satisfaction with health or medical care, (5) attending environmental activities as a volunteer, (we also explored the effect of IHEGs on the frequency of drinking alcohol and smoking behavior. The results indicated that the impact of a couple's education gap on healthy behavior was not statistically significant. An intrahousehold education gap may not worsen health behaviors) (6) donation to environmental activities (income), (7) donation to environmental activities (goods), (8) purchase energy-saving household products, (9) energy-saving activities, and (10) sorting or reducing rubbish.

As the mechanism may differ by education level, we made estimations for both the (a) high-education group (vocational school or higher)—this designation is known as tertiary education (ISCED levels 5 to 8) by UNESCO or higher education by the World Bank—and (b) the low-education group (senior high school education or lower). This designation is also known as primary/secondary education (ISCED 0 to 4) by UNESCO. The results are summarized in Table 5.

For the high-education group, all coefficients of the IHEG in the five models were statistically significant at the 5% or 1% level. Based on the results, four channels regarding the effect of IHEGs on health were determined. First, an individual with a high level of education is more likely to find a better job and have a higher income in the labor market than an individual with a low level of education. Therefore, he or she can accumulate more wealth and invest more to improve his or her health status and those of other household members (positive effect of income hypothesis). A high IHEG may decrease an individual's income satisfaction (−0.036) and health or medical care satisfaction (−0.016). These results do not support the positive effect of the income hypothesis. Therefore, the negative effects may be greater than the positive effects.

Second, regarding household responsibilities, a highly educated individual may have longer working hours than their less-educated partner. It was found that long working hours may negatively affect the health status of individuals (negative effect of longer working hours hypothesis). The coefficient of the IHEG was 0.017 for weekly working days, and −0.008 for attending environmental activities as a volunteer. These findings indicate that a highly educated individual with a higher IHEG may have to work longer and that the probability of participation in social activities is lower, which may worsen their health status (For the impacts of long working hours on mental health, please see [28–31]; for the impacts of volunteer activity on health, please see ref. [49–51]). These results support the negative effect of the longer working hours hypothesis. Third, the coefficient of the IHEG for overcoming difficulties was −0.008, which shows that the higher the IHEG, the lower the probability of overcoming difficulties. A couple's education gap may decrease the amount of help provided by a partner for individuals with high education levels. As a result, he or she has to address these problems alone, which may increase loneliness and stress when the individual faces difficulties in life and work. The results support the negative effect of the skill gap hypothesis.

Table 5. The potential mechanism by two different educational attainment groups.

Model	Intrahousehold Education Gap	
	Coeff.	(S.E.)
Model 1 Income satisfaction		
(a) High education	−0.036 ***	(0.003)
(b) Low education	−0.005	(0.005)
Model 2 Satisfaction with health/medical care		
(a) High education	−0.016 **	(0.007)
(b) Low education	0.003	(0.010)
Model 3 Weekly working days		
(a) High education	0.017 ***	(0.004)
(b) Low education	0.012	(0.009)
Model 4 Volunteer attendance at environmental activities		
(a) High education	−0.008 ***	(0.002)
(b) Low education	−0.002	(0.002)
Model 5 Difficulties overcome		
(a) High education	−0.008 **	(0.003)
(b) Low education	0.004	(0.005)
Model 6 Donation to environmental activities (income)		
(a) High education	−0.008 ***	(0.001)
(b) Low education	0.001	(0.002)
Model 7 Donation to environmental activities (goods)		
(a) High education	−0.004 ***	(0.001)
(b) Low education	0.000	(0.002)
Model 8 Energy saving household products		
(a) High education	0.001	(0.002)
(b) Low education	−0.008 ***	(0.003)
Model 9 Energy saving actions		
(a) High education	0.001	(0.002)
(b) Low education	−0.006 **	(0.003)
Model 10 Sorting and reducing rubbish		
(a) High education	0.001	(0.002)
(b) Low education	−0.009 ***	(0.003)

Notes: (1) Standard errors in parentheses. *** $p < 0.01$, ** $p < 0.05$, * $p < 0.1$. (2) The control variables are similar with those in Table 4, except for the respondent's educational attainment. (3) The OLS regression is utilized in Model 1 to Model 10. (4) High education: vocational school or higher; Low education: senior high school or lower. (5) The independent variable is intrahousehold education (value) Source: Calculated based on the original international survey from 2015 to 2017.

Third, regarding a sustainable lifestyle, the results suggest that the intrahousehold education gap may decrease the sustainable lifestyle activities of improving environmental sustainability. For example, the coefficient for the intrahousehold education gap regarding volunteer attendance at environmental activities is −0.008 for the high-education group. Statistically, it is significant at 1%. It suggests that individuals who completed vocational school education or higher (high education) and experience an education gap in marriage are less likely to volunteer for environmental activities. On the contrary, for the low-education group, the coefficient is positive and statistically insignificant; it indicates that for individuals who completed high school or lower and experience a low-education gap in marriage are more likely to volunteer for environmental activities. Similar trends are found in income donation or goods donation to environmental activities (models 6 and 7). The results show that the negative influence of the education gap in marriage on the environmental activities may influence environmental sustainability. Regarding household-consumption-related environmental activities, there is a similar negative relationship with a sustainable lifestyle. For example, the coefficient of the education gap for energy-saving household products is a negative value for the low-education group, and it is statistically significant at the 1% level. It suggests that individuals who completed high school or lower and experience an education gap have a lower probability of purchasing energy-saving household products. However, no significant influence is found in the high-education group. Moreover, the results are almost similar for energy-saving actions and sorting or reducing rubbish (models 9 and 10). These results suggest that the intrahousehold education gap may worsen the sustainability lifestyle (i.e., reducing the activities of improving environmental sustainability).

Besides the workload and income effects on the linkages between health and education gaps, our results show the negative role played by the intrahousehold education gap on reshaping the household's choice to harmonize environment through energy consumption, recycling, separate collection and reduction of rubbish, volunteer attendance, and charity. As highlighted by Tilman and Clark [42], the crucial relationship between sustainability lifestyle and public health is through environmental sustainability and food lifecycle. These results demonstrate the importance of the linkages between health and education gaps and a sustainable lifestyle and suggest that education equality in marriage plays a crucial role in enhancing consumption and production sustainability.

4.4. Robustness Check to Consider Intergeneration Influences in the Relation between Education and Health

The exclusion restriction in this study was that the impact of a parent's educational attainment level on an adult child's health must be indirect (such as a child's education level or a couple's educational difference) and not via direct channels [52–59]. As such, the two sets of selected instrumental variables using a parent's educational attainment for the endogenous variables of a couple's education difference should satisfy this exclusion restriction condition. It has been argued that increasing the educational attainment level of parents improves the educational level of children [53,58]. When individuals attain a high education level (e.g., complete a doctorate degree in graduate school), well-educated individuals are more likely to have higher education levels than their partners. Therefore, it is acceptable that parents that are more educated may potentially affect their adult child's choice of an educated partner versus a less-educated partner, but a parent's education level is not directly associated with an adult child's health. As far as we know, evidence of a direct correlation of health status and parent educational attainment has rarely been shown.

The indirect effects of parental education on health may also result from reshaping an adult child's unhealthy behavior (such as smoking). Individuals with relatively better education levels are thought to exhibit healthier behaviors; for example, these individuals are more likely to have a ealthy weight, consume a healthy diet, exhibit healthier behaviors, have a reduced likelihood of disaster, and have an enlarged social network [52–59]. When parents are well educated, they have extensive knowledge on the harmful effects of smoking and consuming alcohol in large amounts and have a more efficient way of selecting health insurance. These kinds of knowledge might influence the health behaviors or choices of their adult children, and as a consequence, their adult children may be healthy (indirect channel). To consider this possible indirect channel, we conducted a further robustness check by controlling additional control variables (Panel 2). The variables were (1) satisfaction with health/medical care, (2) nonsmoker, (3) alcohol consumption (drink alcohol every day; 4–5 times per week; 2–3 times per week; once per week; less than above; and do not drink alcohol). The regression results with the abovementioned additional controls for (1), (2), and (3) are summarized in Table 6 Panel 2 (Model 4(a)–Model 6(b)), where the individuals who completed vocational school or higher (this designation is known as tertiary education (ISCED levels 5 to 8) by UNESCO or higher education by the World Bank) are denoted in (a), and those who completed senior high school or had a lower education level are denoted in (b) [47,48]. The corresponding regression results omitting the above controls ((1), (2), and (3)) are displayed in Panel 1 (Model 1(a)–Model 3(b)) using the same sample.

The coefficient of IHEG was −0.077 and statistically significant (Model 1(a)), whereas the coefficient was −0.060 when controlling for the additional variables (the variables were (1) satisfaction with health/medical care, (2) do not smoke dummy variable, (3) alcohol consumption dummy variable (drink alcohol every day; 4–5 times per week; 2–3 times per week; once per week; less than above; and do not drink alcohol)), and the result was statistically significant (Model 4(a)). Similar results were also found in the other models, in which the coefficients had a similar magnitude, were statistically significant, and had the same sign (Model 1(b) with Model 4(b); Model 2(a) with Model 5(a); Model 2(b) with Model 5(b); Model 3(a) with Model 6(a); and Model 3(b) with Model 6(b)). The difference between the coefficients of IHEG in Panel 1 and those in Panel 2 was small. This finding suggests that the indirect channel of the impact of parent education on adult children's health did not conflict with our main conclusions.

In addition, other factors affected health status. For example, (1) increasing levels of satisfaction with health/medical care also improved an individual's health status; (2) compared with the smoking group, the nonsmokers report better subjective health and mental health, whereas there was no great difference between these two groups regarding objective health; and (3) drinking alcohol more frequently positively affected SRH and objective health, whereas alcohol consumption negatively affected mental health.

5. Conclusions

How does an IHEG affect a married individual's well-being (e.g., SRH, mental health, and objective health) and lifestyle for sustainable development (e.g., activities to improve environmental sustainability)? This study first investigated the relationship between the gap of education levels between married couples, health status, and a sustainable lifestyle using an original international survey data collected from 32 countries on six continents. A self-rated health status index, a mental health index, and an objective health status index were utilized to assess the health statuses of individuals. Objective health was a dummy variable that was equal to 1 if an individual did not experience an illness or surgery in the past half-year. Objective unhealthy includes physical illness and mental illness. Moreover, six unique indices are used to investigate sustainable lifestyles. The instrument variable method was utilized to investigate the causal relationship between the two issues above.

The main conclusions are as follows. First, in general, compared to couples with equal education levels, couples with education gaps reported worse levels of SRH, mental health, and objective health when an individual's level of education, household income, occupation, and other factors were held constant. Second, the negative effect of IHEGs on health differed in various groups. For example, the negative effect of IHEGs on health was greater for the high-education group than for the low-education group. Moreover, for the high-education group, the negative effect of IHEGs on health was greater for women, individuals in Asian countries, and couples in middle-income countries than for their counterparts (men, individuals in Europe/North America and South America/Australia, and high-income countries). However, for the low-education group, a reduction in the education gap seemed to improve the husband's health status. In this situation, however, women's mental health deteriorated. Third, for the channels of the impact of IHEGs on health, the positive effect of income hypothesis was not supported, whereas the results supported both the negative effect of the longer working hours hypothesis and the negative effect of a couple's skill gap hypothesis. Finally, the education gap between married couples may reduce the activities of individuals in improving environmental sustainability, such as decreasing the probability of volunteering, reducing the donation of income or goods, purchasing energy-saving products, energy-saving activities, and sorting or reducing rubbish. For a highly educated individual, the education gap between married couples reduces the likelihood of charitable activities (e.g., donation). However, low-educated individuals reduce the probability of activities regarding household consumption and a sustainable lifestyle (e.g., energy-saving activities).

Table 6. Robustness check: to consider the intergenerational influence in the relation between education and health.

Panel 1	Self-Rated Health	Self-Rated Health	Mental Health	Mental Health	Objective Health	Objective Health
	Model 1(a) High Education	Model 1(b) Low Education	Model 2(a) High Education	Model 2(b) Low Education	Model 3(a) High Education	Model 3(b) Low Education
IHEG1: value	−0.077 *** (0.027)	0.040 (0.039)	−0.033 ** (0.016)	0.040 * (0.024)	0.007 (0.012)	0.045 ** (0.018)
Other controls	Yes	Yes	Yes	Yes	Yes	Yes
Panel 2	Self-Rated Health	Self-Rated Health	Mental Health	Mental Health	Objective Health	Objective Health
	Model 4(a) High Education	Model 4(b) Low Education	Model 5(a) High Education	Model 5(b) Low Education	Model 6(a) High Education	Model 6(b) Low Education
IHEG1: value	−0.060 ** (0.026)	0.029 (0.038)	−0.030 * (0.016)	0.037 * (0.022)	0.009 (0.012)	0.045 ** (0.018)
Satisfaction with health/medical care	0.134 *** (0.006)	0.139 *** (0.009)	0.081 *** (0.004)	0.081 *** (0.005)	0.016 *** (0.003)	0.011 ** (0.004)
Do not smoke	0.062 *** (0.016)	0.104 *** (0.023)	0.060 *** (0.010)	0.076 *** (0.014)	0.009 (0.008)	0.016 (0.011)
Frequency of drinking alcohol (ref. drink alcohol every day)						
4–5 times per week	−0.016 (0.038)	0.052 (0.056)	0.008 (0.023)	−0.041 (0.034)	0.010 (0.018)	−0.009 (0.026)
2–3 times per week	−0.050 (0.030)	−0.058 (0.044)	0.084 *** (0.018)	0.034 (0.026)	−0.003 (0.014)	−0.034 * (0.020)
Once per week	−0.031 (0.031)	−0.087 * (0.045)	0.095 *** (0.018)	0.081 *** (0.027)	−0.005 (0.014)	−0.065 *** (0.021)
Less than above	−0.115 *** (0.050)	−0.123 *** (0.042)	0.074 *** (0.018)	0.049 * (0.025)	−0.024 * (0.014)	−0.055 *** (0.019)
Do not drink alcohol	−0.139 *** (0.031)	−0.123 *** (0.042)	0.075 *** (0.018)	0.081 *** (0.026)	−0.030 ** (0.014)	−0.080 *** (0.020)
Other controls	Yes	Yes	Yes	Yes	Yes	Yes
Observation	12,211	6012	12,211	6012	12,211	6012
R-squared	0.162	0.159	0.101	0.086	0.039	0.010

Note: Standard errors in parentheses. *** $p < 0.01$, ** $p < 0.05$, * $p < 0.1$.

Based on the results of this study, the policy implication for improving public health (national welfare) is as follows. First, it has been shown that reducing the IHEG may also improve an individual's health status. Because a gender gap in the levels of education exists, particularly in developing countries, the gender gap in school enrollment and levels of education was high. Therefore, increasing the female school enrollment rate may contribute to reducing the IHEG, which can improve health statuses and increase the national human capital needed to increase economic growth. Second, the results showed that family-work conflicts still exist, particularly for the highly educated groups, women, individuals in Asian countries, and middle-income countries. Health improvement policies may be more important for these groups. Third, long working hours and poor support or help between couples with a higher IHEG were the main channels for the negative effects of IHEG on health. Therefore, regarding traditional gender-role consciousness, providing more support for family care by the government, implementing family-friendly systems, such as flexible work hours in the workplace, and improving communication between couples can lead to health status improvements. Finally, from the perspective of the SDGs, the United Nations recommended the sustainability development goals, which clearly linked 17 goals to sustaining human well-being. This study is related to responsible consumption and production (goal 12), good health and well-being (goal 3), quality education (goal 4), gender equality (goal 5), and reducing inequalities. The empirical study results suggest that reducing the intrahousehold education gap may positively contribute to establishing a sustainable development society by improving both individual well-being and a sustainable lifestyle.

Author Contributions: S.M. designed and conducted the website survey; X.P. and X.M. designed the research including conceptualization and methodology. X.P. analyzed the data and wrote the original paper. X.M., X.P. and C.Z. revised the paper; writing—review and editing, S.M., X.M. and C.Z.; supervision, X.M. and S.M.; funding acquisition, X.M. and S.M. All authors have read and agreed to the published version of the manuscript.

Funding: This research was supported by Specially Promoted Research through a Grant-in-Aid, grant number: 26000001 from the Japanese Ministry of Education, Culture, Sports, Science and Technology (MEXT) and Environment Research and Technology Development Fund (S-16-3, S-14-1, S-15-4) and Grant in Aid from the Ministry of Education, Culture, Sports, Science and Technology in Japan (MEXT): Grant in Aid (20H00648), Ministry of Environment, Japan (1-2001), and by a grand from Japan Society for the Promotion of Science (JSPS), grant number: 20H01512. Any opinions, findings, and conclusions expressed in this material are those of the authors and do not necessarily reflect the views of the agencies.

Conflicts of Interest: The authors declare no conflict of interest.

References

1. United Nations. Resolution A/RES/70/1. Transforming our world: The 2030 agenda for sustainable development. In Proceedings of the Seventieth General Assembly, New York, NY, USA, 15 September–2 October 2015.
2. Akenji, L.; Chen, H. *A Framework for Shaping Sustainable Lifestyles*; United Nations Environment Programme: Nairobi, Kenya, 2016.
3. Van Der Heide, I.; Wang, J.; Droomers, M.; Spreeuwenberg, P.; Rademakers, J.; Uiters, E. The relationship between health, education, and health literacy: Results from the dutch adult literacy and life skills survey. *J. Health Commun.* **2013**, *18*, 172–184. [CrossRef] [PubMed]
4. Clark, D.; Royer, H. The effect of education on adult mortality and health: Evidence from Britain. *Am. Econ. Rev.* **2013**, *103*, 2087–2120. [CrossRef] [PubMed]
5. Lee, J. Pathways from education to depression. *J. Cross-Cult. Gerontol.* **2011**, *26*, 121–135. [CrossRef] [PubMed]
6. Ding, W.; Lehrer, S.F.; Rosenquist, J.; Audrain-McGovern, J. The impact of poor health on academic performance: New evidence using genetic markers. *J. Health Econ.* **2009**, *28*, 578–597. [CrossRef] [PubMed]
7. Breslau, J.; Lane, M.; Sampson, N.; Kessler, R.C. Mental disorders and subsequent educational attainment in a US national sample. *J. Psychiatr. Res.* **2008**, *42*, 708–716. [CrossRef] [PubMed]
8. Cornaglia, F.; Crivellaro, E.; McNally, S. Mental health and education decisions. *Labour Econ.* **2015**, *33*, 1–12. [CrossRef]
9. Fletcher, J.M. Adolescent depression: Diagnosis, treatment, and educational attainment. *Health Econ.* **2008**, *17*, 1215–1235. [CrossRef]
10. Chiappori, P.-A. Collective labor supply and welfare. *J. Political Econ.* **1992**, *100*, 437–467. [CrossRef]

11. Browning, M.; Bourguignon, F.; Chiappori, P.-A.; Lechene, V. Income and outcomes: A structural model of intrahousehold allocation. *J. Political Econ.* **1994**, *102*, 1067–1096. [CrossRef]

12. Chiappori, P.-A.; Fortin, B.; Lacroix, G. Marriage market, divorce legislation, and household labor supply. *J. Political Econ.* **2002**, *110*, 37–72. [CrossRef]

13. Couprie, H. Time allocation within the family: Welfare implications of life in a couple. *Econ. J.* **2007**, *117*, 287–305. [CrossRef]

14. Lise, J.; Seitz, S. Consumption inequality and intra-household allocations. *Rev. Econ. Stud.* **2011**, *78*, 328–355. [CrossRef]

15. Cherchye, L.; De Rock, B.; Vermeulen, F. Married with children: A collective labor supply model with detailed time use and intrahousehold expenditure information. *Am. Econ. Rev.* **2012**, *102*, 3377–3405. [CrossRef] [PubMed]

16. Browning, M.; Chiappori, P.-A.; Lewbel, A. Estimating consumption economies of scale, adult equivalence scales, and household bargaining power. *Rev. Econ. Stud.* **2013**, *80*, 1267–1303. [CrossRef]

17. Tsurumi, T.; Imauji, A.; Managi, S. Relative income, community attachment and subjective well-being: Evidence from Japan. *Kyklos* **2018**, *72*, 152–182. [CrossRef]

18. Cherchye, L.; De Rock, B.; Lewbel, A.; Vermeulen, F. Sharing rule identification for general collective consumption models. *Econometrica* **2015**, *83*, 2001–2041. [CrossRef]

19. Haddad, G.K. Gender ratio, divorce rate, and intra-household collective decision process: Evidence from iranian urban households labor supply with non-participation. *Empir. Econ.* **2014**, *48*, 1365–1394. [CrossRef]

20. Groot, W.; Brink, H.M.V.D. Age and education differences in marriages and their effects on life satisfaction. *J. Happiness Stud.* **2002**, *3*, 153–165. [CrossRef]

21. Ma, X.; Piao, X. The impact of intra-household bargaining power on happiness of married women: Evidence from Japan. *J. Happiness Stud.* **2018**, *20*, 1775–1806. [CrossRef]

22. Miles, D. Gender effect on housework allocation: Evidence from Spanish two-earner couples. *J. Popul. Econ.* **2003**, *16*, 227–242. [CrossRef]

23. Bertrand, M.; Kamenica, E.; Pan, J. Gender identity and relative income within households. *Q. J. Econ.* **2015**, *130*, 571–614. [CrossRef]

24. Cronin, M.; Glenn, P. Oral communication across the curriculum in higher education: The state of the art. *Commun. Educ.* **1991**, *40*, 356–367. [CrossRef]

25. Bynner, J.; Schuller, T.; Feinstein, L. Wider benefits of education: Skills, higher education and civic engagement. *Zeitschrift für Pädagogik* **2003**, *49*, 341–361.

26. Kruss, G.; McGrath, S.; Petersen, I.-H.; Gastrow, M. Higher education and economic development: The importance of building technological capabilities. *Int. J. Educ. Dev.* **2015**, *43*, 22–31. [CrossRef]

27. Gerstein, M.; Friedman, H.H. Rethinking higher education: Focusing on skills and competencies. *Psychol. Issues Hum. Resour. Manag.* **2016**, *4*, 104–121. [CrossRef]

28. Caruso, C.C. Possible broad impacts of long work hours. *Ind. Health* **2006**, *44*, 531–536. [CrossRef]

29. Kuroda, S.; Yamamoto, I. Why do people overwork at the risk of impairing mental health? *J. Happiness Stud.* **2018**, *20*, 1519–1538. [CrossRef]

30. Virtanen, M.; Ferrie, J.E.; Singh-Manoux, A.; Shipley, M.J.; Stansfeld, S.A.; Marmot, M.; Ahola, K.; Vahtera, J.; Kivimaki, M. Long working hours and symptoms of anxiety and depression: A 5-year follow-up of the Whitehall II study. *Psychol. Med.* **2011**, *41*, 2485–2494. [CrossRef]

31. Virtanen, M.; Stansfeld, S.A.; Fuhrer, R.; Ferrie, J.E.; Kivimaki, M. Overtime work as a predictor of major depressive episode: A 5-Year follow-up of the Whitehall II study. *PLoS ONE* **2012**, *7*, e30719. [CrossRef]

32. Shove, E.A. Efficiency and consumption: Technology and practice. *Energy Environ.* **2004**, *15*, 1053–1065. [CrossRef]

33. Retamal, M. Collaborative consumption practices in Southeast Asian cities: Prospects for growth and sustainability. *J. Clean. Prod.* **2019**, *222*, 143–152. [CrossRef]

34. Zakaria, N.F.; Rahim, H.A.; Paim, L.; Zakaria, N.F. The mediating effect of sustainable consumption attitude on association between perception of sustainable lifestyle and sustainable consumption practice. *Asian Soc. Sci.* **2019**, *15*. [CrossRef]

35. Heiskanen, E.; Lovio, R.; Jalas, M. Path creation for sustainable consumption: Promoting alternative heating systems in Finland. *J. Clean. Prod.* **2011**, *19*, 1892–1900. [CrossRef]

36. Alcott, B. The sufficiency strategy: Would rich-world frugality lower environmental impact? *Ecol. Econ.* **2008**, *64*, 770–786. [CrossRef]

37. Peattie, K.; Collins, A. Guest editorial: Perspectives on sustainable consumption. *Int. J. Consum. Stud.* **2009**, *33*, 107–112. [CrossRef]

38. Hotta, Y.; Santo, A.; Tasaki, T. EPR-Based Electronic Home Appliance Recycling System under Home Appliance Recycling Act of Japan. Institute for Global Environmental Strategies, 2014. Available online: https://www.oecd.org/environment/waste/EPR_Japan_HomeAppliance.pdf (accessed on 8 April 2020).

39. Hotta, Y.; Aoki-Suzuki, C. Waste reduction and recycling initiatives in Japanese cities: Lessons from Yokohama and Kamakura. *Waste Manag. Res.* **2014**, *32*, 857–866. [CrossRef]

40. Awasthi, A.K.; Li, J. Management of electrical and electronic waste: A comparative evaluation of China and India. *Renew. Sustain. Energy Rev.* **2017**, *76*, 434–447. [CrossRef]

41. Xie, J.; Nozawa, W.; Managi, S. The role of women on boards in corporate environmental strategy and financial performance: A global outlook. *Corp. Soc. Responsib. Environ. Manag.* **2020**, 1–16. [CrossRef]

42. Tilman, D.; Clark, M. Global diets link environmental sustainability and human health. *Nature* **2014**, *515*, 518–522. [CrossRef]

43. Wooldridge, J. *Econometric Analysis of Cross Section and Panel Data*, 2nd ed.; MIT Press: Cambridge, MA, USA, 2010.

44. Bera, A.K.; Greene, W.H. Econometric analysis. *J. Am. Stat. Assoc.* **1994**, *89*, 1567. [CrossRef]

45. Stock, J.H.; Wright, J.H.; Yogo, M. A survey of weak instruments and weak identification in generalized method of moments. *J. Bus. Econ. Stat.* **2002**, *20*, 518–529. [CrossRef]

46. Chapman, A.; Fujii, H.; Managi, S. Multinational life satisfaction, perceived inequality and energy affordability. *Nat. Sustain.* **2019**, *2*, 508–514. [CrossRef]

47. ISCED. International Standard Classification of Education. Available online: http://uis.unesco.org/sites/default/files/documents/international-standard-classification-of-education-isced-2011-en.pdf (accessed on 28 May 2019).

48. Worldbank. Higher Education. Available online: https://www.worldbank.org/en/topic/tertiaryeducation (accessed on 2 June 2020).

49. Piliavin, J.A.; Siegl, E. Health benefits of volunteering in the Wisconsin longitudinal study. *J. Health Soc. Behav.* **2007**, *48*, 450–464. [CrossRef] [PubMed]

50. Luoh, M.-C.; Herzog, A.R. Individual consequences of volunteer and paid work in old age: Health and mortality. *J. Health Soc. Behav.* **2002**, *43*, 490. [CrossRef] [PubMed]

51. Detollenaere, J.; Willems, S.; Baert, S. Volunteering, income and health. *PLoS ONE* **2017**, *12*, e0173139. [CrossRef]

52. Berkman, L.F. The role of social relations in health promotion. *Psychosom. Med.* **1995**, *57*, 245–254. [CrossRef]

53. Currie, J.; Moretti, E. Mother's education and the intergenerational transmission of human capital: Evidence from college openings. *Q. J. Econ.* **2003**, *118*, 1495–1532. [CrossRef]

54. Cutler, D.M.; Lleras-Muney, A. Understanding differences in health behaviors by education. *J. Health Econ.* **2009**, *29*, 1–28. [CrossRef]

55. Dursun, B.; Cesur, R.; Mocan, N. The impact of education on health outcomes and behaviors in a middle-income, low-education country. *Econ. Hum. Biol.* **2018**, *31*, 94–114. [CrossRef]

56. Lawrence, E.M. Why do college graduates behave more healthfully than those who are less educated? *J. Health Soc. Behav.* **2017**, *58*, 291–306. [CrossRef]

57. Michael, Y.L.; Berkman, L.F.; Colditz, G.A.; Kawachi, I. Living arrangements, social integration, and change in functional health status. *Am. J. Epidemiol.* **2001**, *153*, 123–131. [CrossRef] [PubMed]

58. Malamud, O.; Wozniak, A. The impact of college on migration: Evidence from the Vietnam generation. *J. Hum. Resour.* **2012**, *47*, 913–950. [CrossRef]

59. Winkleby, M.A.; Jatulis, D.E.; Frank, E.; Fortmann, S.P. Socioeconomic status and health: How education, income, and occupation contribute to risk factors for cardiovascular disease. *Am. J. Public Health* **1992**, *82*, 816–820. [CrossRef] [PubMed]

 *sustainability*

Article

Attachment to Material Goods and Subjective Well-Being: Evidence from Life Satisfaction in Rural Areas in Vietnam

Tetsuya Tsurumi [1,*]**, Rintaro Yamaguchi** [2]**, Kazuki Kagohashi** [3] **and Shunsuke Managi** [4]

[1]   Faculty of Policy Studies, Nanzan University, Nagoya 466-8673, Japan
[2]   National Institute for Environmental Studies, Tsukuba 305-8506, Japan; yamaguchi.rintaro@nies.go.jp
[3]   Faculty of Global Liberal Studies, Nanzan University, Nagoya 466-8673, Japan; kago84@ic.nanzan-u.ac.jp
[4]   Urban Research Institute and Department of Civil Engineering, School of Engineering, Kyushu University, Fukuoka 819-0395, Japan; managi.s@gmail.com
*   Correspondence: tsurumi@nanzan-u.ac.jp

Received: 1 October 2020; Accepted: 24 November 2020; Published: 26 November 2020

Abstract: In our daily lives, some people tend to use the same material goods more extensively than other people. It would appear that people like this consume fewer material inputs, other things being equal. Our research question is whether they are also happier in terms of life satisfaction. To study this, we first hypothesized that they are happier due to the endowment effect, prosocial or pro-environmental motivations, or income and substitution effects. We show that income and substitution effects are positive for people who use products for longer. Using a reduced form model that incorporates these four effects together, and empirical data originally collected from rural areas in Vietnam, we divide consumption into material consumption and residual consumption and demonstrate that, in general, increased material consumption is not associated with increased well-being; however, for those who take better care of their possessions, this effect is reversed, and material consumption does increase well-being. Our study shows that for people who take better care of their possessions, increased consumption is linked to increased well-being. This finding has a useful policy implication for developing countries to improve their well-being by promoting economic growth alongside responsible consumption.

Keywords: mottainai; attachment; subjective well-being; life satisfaction; happiness

---

1. Introduction

There have been increasingly urgent calls to transform the current material-intensive economy to a circular economy [1,2]. There are many potential ways to achieve this transformation but reusing material goods as much as possible in the forms of eco-design, direct reuse, or recycling, is a prominent, simple way to do this, both on the demand and supply sides [3]. This study explores how attachment to goods is related to life satisfaction, an oft-cited aspect of subjective well-being (SWB). There exists vast literature on attachment theory, including psychological studies that focus on interpersonal relationships, that is, the intimate emotional bonds between people, such as those between parents and children (e.g., [4]). Research on attachment has been expanded to "places" or "neighborhoods" in sociology and human geography [5] (p. 144) and empirical studies revealed that many people (from 40% to 65%) demonstrated attachment to their neighborhoods [6] (p. 274). According to [7], place attachment can be categorized as either "social attachment", which includes institutional ties, social activity, and local intimates; or "affective attachment" in which the satisfaction is with the neighborhood itself [5] (p. 145). Following this categorization, we use the term attachment to goods in the latter sense—namely, we focus in this paper on material possession attachment [8]. In particular, we hypothesize the following four relationships between longer use of goods and SWB.

First, the tendency to use the same good for longer can increase effective income, if other things remain equal. This excess resource may be put into the further purchasing of either consumable or durable goods. In a conventional worldview, this enables us to reach higher utility, which we aim to establish in the theoretical model in the next section.

Second, in a well-known experiment, Kahneman et al. [9] compared people's willingness to accept giving up a mug that they were previously awarded. The surprising result showed that the willingness to accept was higher than the market price, implying that some subjective value was added to the owner of the mug since it was given to them. The authors' interpretation was that once you own a material good, you might feel attached to it after keeping it for a certain period of time, so you would avoid parting with it even if an equivalent or superior good could be purchased. They called this the endowment effect [10], which was demonstrated in a laboratory experiment [11–13] (p. 194). This effect, more generally called loss aversion, provides another rationale to assume that utility depends on whether consumers continue to use the same commodity. This has been partly examined in the theme of product attachment [14,15]. Recent research suggests that attachment to possessions influences disposition decisions at various stages [16,17], and previous studies have shown that strong emotional attachment to a product discourages consumers from replacing or discarding it [14,17] (p. 215).

Third, in the modern world, where there is increasing awareness of the importance of sustainability and a tendency toward a circular economy, people may be happier when they decide to keep using the same product that is still usable, even if they can afford to repurchase a new one. They find reusing to be prosocial, whether it is due to pure environmentalism or the "warm glow" effect [18]. In the warm glow theory, an agent is assumed to prefer one alternative but aspires to choose another for ethical reasons and receives psychological satisfaction (i.e., a warm glow) from acting in accordance with their aspirations [19] (p. 502). Empirical studies have revealed that the warm glow effect can be found in the context of environmental protection [20,21]. A related concept is the positive feelings associated with empathy that can reinforce prosocial behavior [22] (p. 60). From this perspective, people may regard throwing products away as immoral because it harms the environment or others. Note that this prosocial effect does not necessarily involve attachment, endowment, or loss aversion effects.

Fourth, there could be some latent factors that affect both material possession and SWB. Due to these factors, those who tend to use the same good may also tend to focus on long-term relationships, for example. An in-depth investigation of these variables is beyond the scope of the current study, but it is useful here to point to the spirit of mottainai, which is a Japanese adjective frequently used to describe the feeling when something still usable is wasted [23]. The waste from food-related activities such as grocery shopping, cooking, eating, handling surplus food, and so forth, could be reduced by following the mottainai spirit of Japanese culture [24]. Such cultural factors can affect both attachment to goods and SWB. Interestingly, this can be said about material goods, money, and even talent and opportunities.

To date, it has not been clear how the transformation into consumption based on a circular economy would affect SWB. In contrast, the vast literature has focused on the relationship between income and SWB (for the comprehensive review on this theme, see [25]). Previous studies revealed that income positively correlates to SWB within countries (e.g., [26–29]). It has been argued that income ceases to augment SWB once individual income level rises above a certain threshold [25]. This has been called the "Easterlin paradox" since it has been confirmed that the reported happiness does not rise in proportion to income increase in those countries such as the United States, Japan, and European countries [28,30–33]. Other studies include the perspective of aspiration for better lives, career achievement, and so forth, which plays an important role in explaining the relatively lower levels of SWB (e.g., [34]).

In contrast to the previous studies that question the role of income in raising SWB, there has also been growing research that provides evidence that income matters to augmenting SWB [25,35,36]. For example, using panel data to control for individual fixed effects, it has been revealed that the income of the reference group (together with their own income) significantly affects SWB [37]. Argyle [31] argues that there exists no satiation point of income, as opposed to the previous studies that explored the threshold of income, in line with the Easterlin paradox. This research shows that relative income matters and strongly affects SWB [25,38]. Recent studies revealed that relative income and consumption (such as home-ownership) significantly contribute to increasing SWB [39,40].

Donati [41] suggests that the effect of material well-being on SWB is not linear but rather diminishes with higher levels of material well-being. This is in line with the economic concept of diminishing marginal utility. Empirical studies also show diminishing marginal effects. Gokdemir [42] shows that, in Turkey, only the consumption of durable goods is correlated with life satisfaction. DeLeir and Kalil [43] shows that out of nine consumption categories, only one, leisure, is positively correlated with SWB, using U.S. data. Zhang and Xiong [44] employs 77 consumption categories and 13 SWB indicators to investigate the relationship between consumption and SWB in Japan and shows the particularly strong correlation between relational consumption and SWB. Dumludag [45] shows that the relationship between life satisfaction and each consumption category varies in accordance with the development stage. Relatedly, Pandelaere [46] reviews studies on experiential versus material consumption and suggests that even though many studies find an advantage for experiential consumption, this effect does not occur for materialists, which implies that materialists do not benefit more from material than from experiential consumption owing to unrealistic expectations. The above-mentioned literature implies that material well-being cannot be easily increased and that it depends on the contents of consumption or the development stage. However, previous studies do not fully clarify the effect of "individuals' attitudes toward material goods" on material well-being. In this study, we thus investigate the relationship between material well-being and "attachment to material goods" in the context of a rural developing country.

The rest of the paper will proceed as follows. In the next section, we theoretically explore the effects of conventional substitution and income. In Section 3, we use a theoretically reduced form that incorporates four effects and empirically test our hypothesis using survey data from rural Vietnam. Finally, we provide a discussion and conclusions in Section 4.

2. Materials and Methods

2.1. Conceptual Model

To see how longer use of goods can affect life satisfaction, an oft-cited aspect of SWB, it is useful to set up a simple model of two goods and two time periods. In doing so, we compared the consumers' utility levels in two distinct contexts. Imagine two kinds of consumer goods. Good 1 is a general, perishable good that needs to be continually repurchased, such as food. Good 2 is still a consumer good, but durable to a certain extent, so that some people purchase it every time period, while others continue to use what they bought in the previous period. Examples include clothing, personal computers, and smartphones.

In the benchmark model, the consumer purchases Goods 1 and 2 in both periods, t and t + 1. In the extended model, the consumer purchases both goods in period t, but in the subsequent period, purchases only Good 1 and continues to use Good 2 bought in the previous period. Our strategy was simply to compare the two indirect utilities achieved in both models, other things being equal.

Before diving into the details, we came up with at least four relationships between the use of goods and SWB, as described in the Introduction. Formally, life satisfaction (LS), an oft-cited aspect of SWB, may be composed of material and non-material LS (e.g., [47]):

$$LS = ML + NML.$$

ML consists of traditional economic incentives, whereas NML includes a wide variety of non-economic incentives, such as endowment and attachment effects, prosocial/pro-environmental behavior, and the spirit of *mottainai*. In the remainder of this conceptual section, we focus on the LS-equivalent of money-derived (intertemporal) utility, ML.

First, in the benchmark model, the representative consumer was assumed to maximize their (intertemporal) utility, *ML*, with regard to Goods 1 and 2 in both periods. Second, in the extended model, we wanted to see what would happen if the consumer continued to use Good 2 in the second period. As detailed in Appendix A, the comparison of the benchmark and extended model led us to our central finding:

Proposition. *In the given framework, indirect utility is strictly higher in the extended model than in the benchmark model.*

This proposition tells us that, in a basic model that strips away behavioral features, the first channel, having fewer, durable goods, is positively correlated with SWB, as mentioned in the Introduction. In particular, not having to purchase Good 2 induces both income and substitution effects. The income effect means that the money that could have been used to purchase Good 2 in the subsequent period is now freed up to enhance effective income, which increases consumption of both Goods 1 and 2. The substitution effect from/to Goods 1 and 2 depends on their relative prices as well as the discounting and interest rate.

2.2. Empirical Strategy

In the previous subsection, we demonstrated that, other conditions being equal, consumers who use the same material goods for longer may report higher intertemporal SWB. In the following, we empirically test this expectation by measuring the hypothesized effects all together, leaving separate identification of the aforementioned four effects to future research. Our empirical model is as follows:

$$LS_i = \gamma_1 + g(C_{1i}) + \sum_j \beta_j Y_{ij} + \epsilon_1 \tag{1}$$

Here, i represents the individual, and LS_i represents life satisfaction. Material consumption (C_1) is on the right-hand side of Equation (1), which is assumed to determine LS on the left-hand side. γ_1 is a constant, ϵ_1 is the uncorrelated error term, and Y_{ij} stands for other control variables with β_j as their coefficients. As control variables, we used variables that have been used in prior studies of SWB (i.e., age, gender, level of subjective health, education, marriage, having children, and number of family members). We also included residual consumption (i.e., consumption other than material consumption), $(total\_consumption_i - (C_{1i}))$ as a control variable. We intended to decompose the effect of total consumption into material consumption (C_1) and non-material consumption (residual consumption), which is in line with our theoretical model in the previous subsection.

This study followed [48]'s model, which utilizes nonparametric functions for consumption, so as to clearly and visually explore its different functional forms. As for the other explanatory variables, we used parametric functions. Thus, we applied semiparametric regression. We used generalized additive models (GAMs; [49]), in which the linear predictor depends not on a weighted sum of explanatory variables, as in linear regressions, but on unknown smooth functions, g. As such, GAMs enabled us to identify non-linear, locally diverse relationships between consumption and SWB.

2.3. Data

This study relied on a dataset originally obtained from a field survey conducted by the authors in March 2020 in two rural areas (Thieu Ngoc and Darsal) in Vietnam. The data that supported this study's findings are available from the corresponding author on reasonable request.

Due to the difficulty of obtaining a sufficient sample size using internet panels in rural areas of Vietnam, we conducted face-to-face surveys. We had the full cooperation of the local government of both Thieu Ngoc and Darsal, and the opportunity to conduct face-to-face surveys in all households in both villages. We chose these two areas because the local population is mainly engaged in primary industries, especially farming. Therefore, they are typical rural villages where people live in the traditional Vietnamese way. The main crops are rice in Thieu Ngoc, and coffee in Darsal.

Thieu Ngoc is in the Thieu Hoa district of Thanh Hóa province in the North Central Coast region, while Darsal is in the Dam Rong district of Lam Dong province in the central highlands. As shown in Table 1, per capita income in Thieu Ngoc in 2018 was US$1075, which is significantly lower than the US$2093, US$3036, and US$1615 for the whole country, urban areas, and rural areas, respectively (General statistics office of Vietnam). Thus, per capita income in Thieu Hoa is lower than in most rural areas in Vietnam. On the other hand, the per capita income in Darsal in 2018 was US$1753, which is nearly the same as all rural areas in Vietnam (US$1615).

Table 1. Official statistics of Thieu Hoa and Darsal in 2018.

	Thieu Hoa	**Darsal**
Population	6508	4485 (over 18 years old)
Per capita income (US$)	1075	1753
Area (km^2)	7.47	84.76

Note: The exchange rate was calculated at 0.000043 US$/dong, which was the average exchange rate in 2018. Data were sourced from the local governments of Thieu Hoa and Darsal.

In the face-to-face survey, we attempted to ensure the accuracy of responses through translations and multiple checks by Vietnamese native speakers, as well as by providing extensive training to our field agents, in which they received consistent instructions directly from one of the authors. We originally had participants of 1824 and 3043 from Thieu Hoa and Darsal, respectively. After eliminating the subjects who responded with "I don't know/I don't want to answer this question" to questions regarding consumption and other subjects that were deemed invalid, 1250 and 2435 eligible subjects from Thieu Hoa and Darsal remained, respectively.

To understand the general details of Thieu Hoa and Darsal residents, we conducted the following questionnaire, as shown in Table 2.

Figure 1 shows the distribution of annual household income. The average annual household income is US$6395, and the median is US$4719.

Next, Figure 2 shows material possessions per household. Most households possess two or three motorcycles, while only a few own automobiles. Concerning home appliances, more than 90% of households own televisions and rice cookers, about 74% own refrigerators, less than 50% own washing machines, and only about 20% own air conditioners. Regarding personal computers and mobile phones, about 14% of households own personal computers, while more than 95% own mobile phones.

Figure 3 shows the degree of food self-sufficiency. We found that 43% of respondents answered 3 (quite frequently) and about 25% answered 4 (very frequently). This implies that many people in the sample do not spend all of their money on food, so they are able to buy other things or put money into savings.

Figure 4 shows the frequency of bartering with other neighborhood residents. It shows that more than 34% of respondents frequently barter with other neighborhood residents. As shown in Figure 5, typical bartering goods are food (34%), furniture (16%), and clothes (12%), implying that people in the areas barter a variety of material goods.

Table 3 shows the survey questionnaire used for the main analysis. Concerning consumption, we asked questions pertaining to total monthly household consumption and material consumption. Material consumption represents the household consumption of "goods" (electrical appliances, furniture, clothes, shoes, publications, and other sundries, excluding expenditure related to housing, cars, and motorbikes). Monthly household consumption expenditure includes not only expenditures on "goods", which we have defined above, but also expenditures on housing, cars, motorbikes, medical expenses, insurance, and education, among others. As shown in Section 2.2, in our empirical model, we divided total monthly household consumption into material consumption, defined above, and residual consumption, and focused on the relationship between material consumption and SWB, while controlling residual consumption in the empirical model.

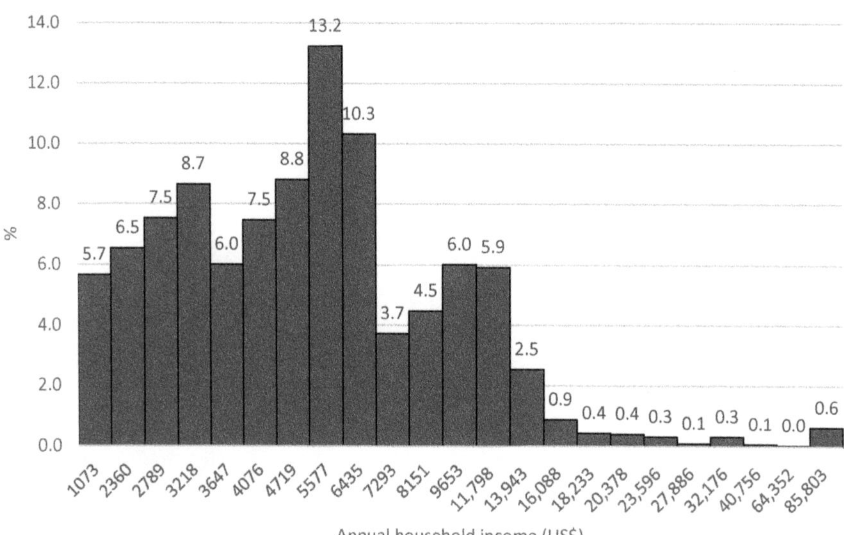

Figure 1. Annual household income in Thieu Hoa and Darsal (US$; N = 3685).

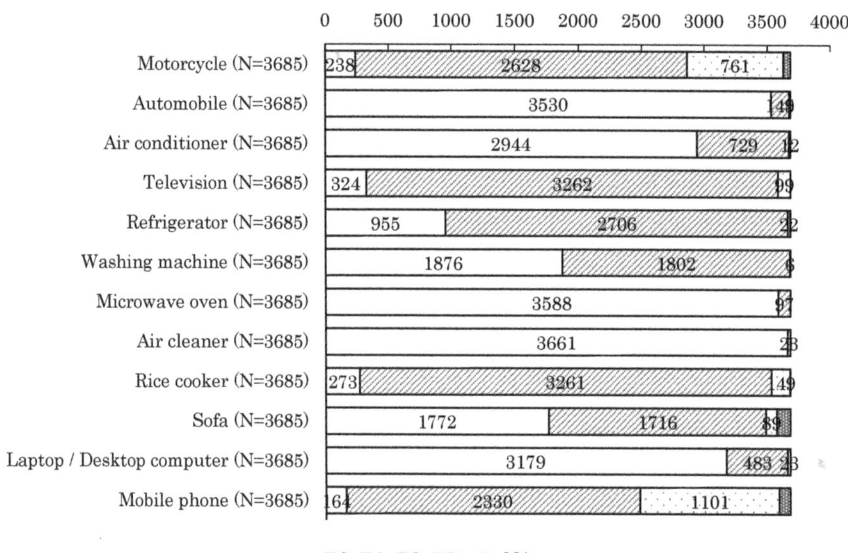

Figure 2. Material possessions per household (N = 3685).

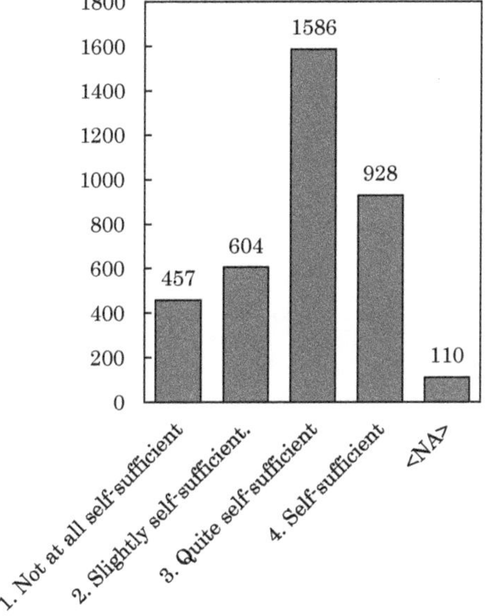

Figure 3. Food self-sufficiency (N = 3685).

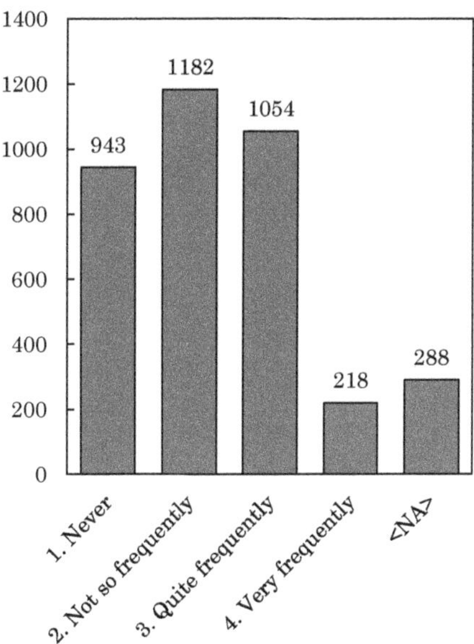

Figure 4. Frequency of bartering with neighborhood residents (N = 3685).

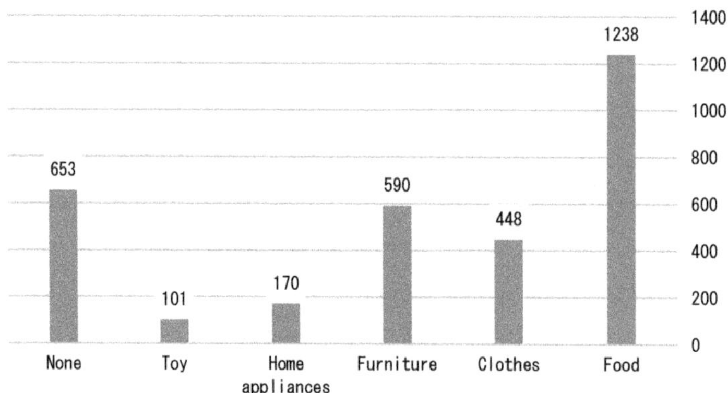

Figure 5. Type of goods bartered (N = 3685).

Table 2. Survey questionnaire to obtain basic information from the sample.

	Survey Question	**Notes**
Household Income: Unit (US$)	Please tell us your yearly household income.	
Household material possessions	How many of the following goods does your household have? * Motorcycle * Automobile * Air conditioner * Television * Refrigerator * Washing machine * Microwave oven * Air purifier * Rice cooker * Sofa * Laptop/desktop computer * Mobile phone	
Self-sufficiency	Please tell us your household's degree of food self-sufficiency.	1. Not at all self-sufficient 2. Slightly self-sufficient. 3. Quite self-sufficient 4. Self-sufficient
Frequency of barter with other neighborhood residents	How often do you barter with other neighborhood residents?	1. Never 2. Not so frequently 3. Quite frequently 4. Very frequently
Bartering goods	What kind of material goods do you barter with other neighborhood residents? (Multiple choice)	1. Food 2. Clothes 3. Furniture 4. Home appliances 5. Toys 6. None

Table 3. Survey questionnaire for the main analysis.

	Survey Question	Notes
Life satisfaction	Overall, how satisfied are you with your life?	"5: completely satisfied"; "1: not at all satisfied"; five-point scale (standardized 0 to 1).
Material consumption (US$)	What is the average monthly amount spent in your household to purchase "goods" (electrical appliances, furniture, clothes, shoes, publications, and other sundries, excluding expenditure related to housing, cars, and motorbikes)?	
Monthly household consumption expenditure: unit (US$)	Overall, approximately how much does your household spend monthly on consumption? Please see the consumption categories under "Reference" and include all these expenses before answering this question. * Reference. Consumption categories in the "Public Opinion Survey on the Life of the People" conducted by the Cabinet Office in Japan: Apparel (clothes and shoes) Food expenses (foodstuff and dining costs) Housing expenses (loans, rent, land fees, equipment repairs/maintenance expenses, construction, other housing-related services, and utility fees) Durable consumer goods (electrical appliances, furniture, bedding, automobiles/motorcycles/bicycles, etc.) Miscellaneous expenses (sundries, consumables, and hairdressing) Medical expenses Transportation expenses Social expenses Insurance Communications expenses (postage, mobile phone, and Internet) Educational expenses Entertainment expenses	
Age	Please tell us your age.	Unit: age.
Gender	Please tell us your gender.	(1: man; 0: woman)
Marriage	Are you married?	(1: married; 0: other)
Children	Do you have children?	(1: yes; 0: no)
Number of family members	Including yourself, how many people live in your household?	
Education	Please tell us your highest academic qualification (if a student, please tell us which school you graduated from most recently).	(college graduate or higher: 1; other: 0).
Attachment	Please select all items that are applicable. * I want to utilize "goods" and look after them for as long as possible.	(1: applicable; 0: not applicable)

Tables 4–6 show descriptive statistics for the survey questionnaire used for the main analysis. The average monthly household material consumptions are US$179, US$188, and US$175; these distributions are shown in Figures 6–8, respectively. We found similar descriptive statistics for variables used for analysis and the distribution of monthly household material consumption among the overall sample and Subsamples.

Table 4. Descriptive statistics for the overall sample.

	Obs.	Mean	S.D.	Min.	Max.
Darsal dummy	3685	0.661	0.474	0	1
Life satisfaction (standardized: 0 to 1)	3685	0.751	0.181	0	1
Monthly household material consumption (US$)	3685	179	124	32	965
Monthly household total consumption (US$)	3685	255	426	54	1931
Age	3685	39.9	12.8	18	101
Gender	3685	0.499	0.500	0	1
Marriage	3685	0.871	0.335	0	1
Children	3685	0.906	0.292	0	1
Number of family members	3685	4.01	1.28	1	10
Education	3685	0.044	0.206	0	1
Attachment dummy	3685	0.242	0.428	0	1

Table 5. Descriptive statistics for overall sample for Subsample 1 (Attachment dummy = 1, N = 892).

	Obs.	Mean	S.D.	Min.	Max.
Darsal dummy	892	0.646	0.479	0	1
Life satisfaction (standardized: 0 to 1)	892	0.747	0.244	0	1
Monthly household material consumption (US$)	892	188	122	32	965
Monthly household total consumption (US$)	892	259	435	54	1931
Age	892	39.4	13.6	18	100
Gender	892	0.524	0.500	0	1
Marriage	892	0.859	0.348	0	1
Children	892	0.898	0.303	0	1
Number of family members	892	4.17	1.45	1	10
Education	892	0.052	0.221	0	1

Table 6. Descriptive statistics for overall sample for Subsample 2 (Attachment dummy = 0, N = 2689).

	Obs.	Mean	S.D.	Min.	Max.
Darsal dummy	2689	0.667	0.471	0	1
Life satisfaction (standardized: 0 to 1)	2689	0.752	0.220	0	1
Monthly household material consumption (US$)	2689	175	124	32	965
Monthly household total consumption (US$)	2689	254	443	54	1931
Age	2689	40.2	12.5	18	100
Gender	2689	0.490	0.500	0	1
Marriage	2689	0.332	0.874	0	1
Children	2689	0.907	0.290	0	1
Number of family members	2689	3.96	1.22	1	10
Education	2689	0.043	0.202	0	1

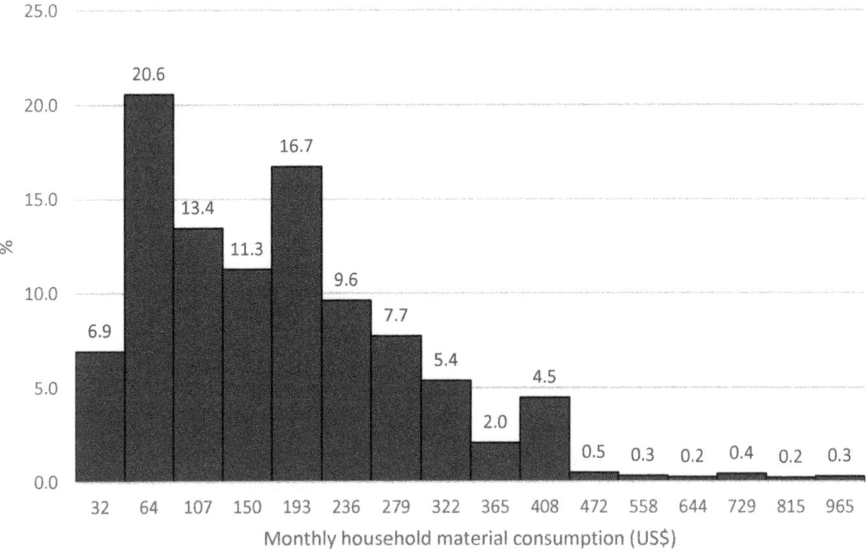

Figure 6. Monthly household material consumption for overall sample (US$; N = 3685).

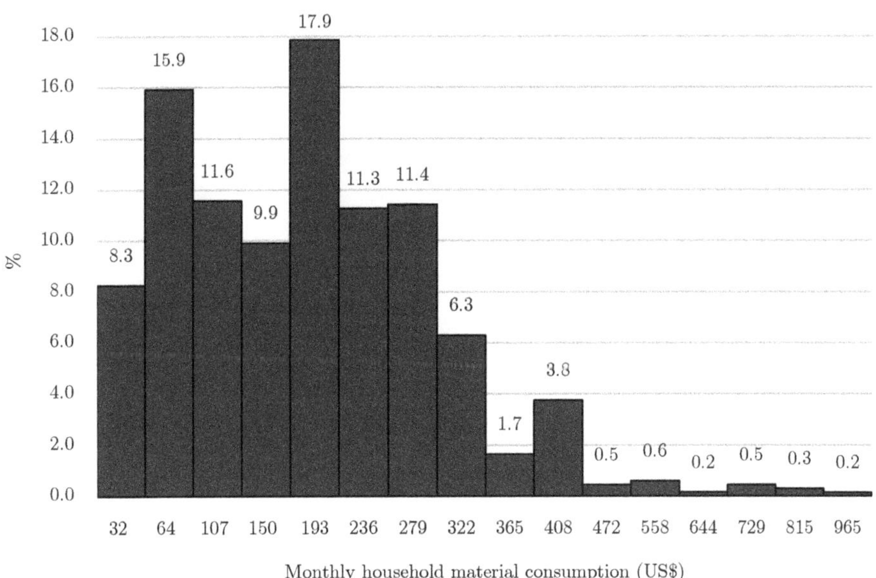

Figure 7. Monthly household material consumption for Subsample 1 (US$, Attachment dummy = 1, N = 892).

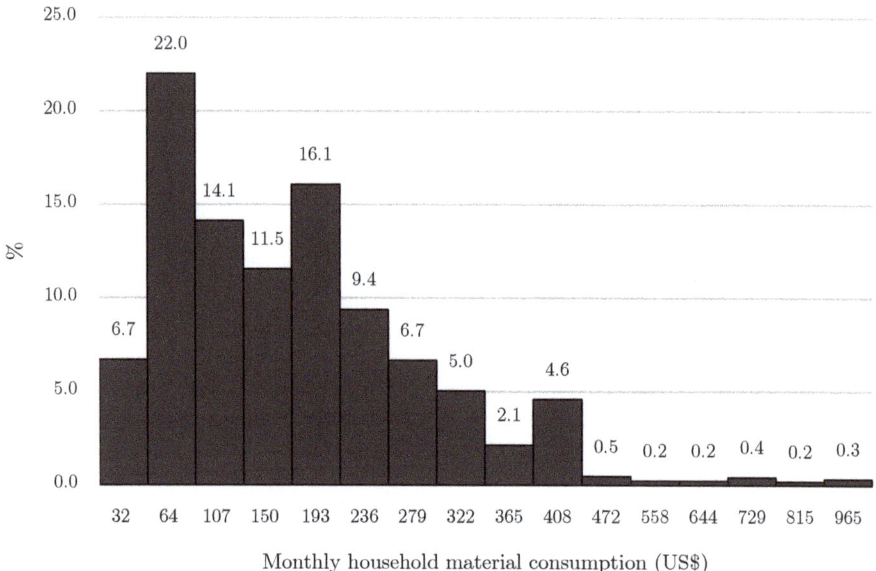

Figure 8. Monthly household material consumption for Subsample 2 (US$, Attachment dummy = 0, N = 2689).

3. Results

Figure 9 shows the estimation results of Equation (1), which includes all samples, Subsample 1 (Attachment dummy = 1), and Subsample 2 (Attachment dummy = 0). In these figures, monthly material consumption expenditure (unit: US$) is shown on the horizontal axis, while life satisfaction, which was standardized from 0 to 1, is shown on the vertical axis. The solid line is the estimated SWB function curve, while the dotted lines represent 95% confidence intervals. The "0" on the vertical axis denotes the samples' average life satisfaction. As shown in Table 7, all estimated nonparametric functions are statistically significant. We found a monotonous decreasing trend for the pooled overall sample in the upper panel of Figure 9. For a robustness check, we applied additional estimation using Subsamples A (men) and B (women). The estimation results for Subsamples were almost the same as our main result using the overall sample in Figure 9. For Subsample 1, whose attachment dummy equals 1, although the confidence interval is relatively wide due to the relatively small sample, meaning the estimation result must be interpreted with caution, we found an increasing trend from US$138 to US$468. For Subsample 1, the upper and lower confidential intervals were both more than 0 from US$138 to US$468 with regard to the vertical axis, which corresponds to the situation where the predicted contribution to life satisfaction is larger than its sample mean. The predicted contribution to life satisfaction of Subsample 1 was 0.101 for the samples' average monthly household material consumption (US$179). Although the confidence interval is too wide to interpret, we found a decreasing trend over US$468. For Subsample 2, whose attachment dummy equals 0, we found a decreasing trend.

Table 8 reports the parameter estimates for control variables. We found the expected signs for most variables, with the exception of age, gender, and marriage. Regarding age, we found an inverted-U relationship between age and life satisfaction for Subsample 2, which is not in line with the literature. The exceptions may indicate unique characteristics for rural developing countries. The negative coefficients of marriage and the number of family members can be interpreted as an increase in economic burdens in budget constraints. Furthermore, the absolute value of the negative coefficient of the number of family members for Subsample 1 is larger than that for Subsample 2. This might imply that those who have an attachment to material goods tend to have a stronger scarcity consciousness of income. There is a possibility that those who have an attachment to material goods tend also to have an attachment to family or importance of family and, thus, have higher standards of expenses for each family member, which increases the sense of income scarcity. Another possibility is that Subsample 1 is attached to material goods, so much so that the expenditures on a marginal increase of family members would be more painful than Subsample 2. The positive coefficients of residual consumption imply that non-material consumption is positively correlated with life satisfaction, which is in line with [50].

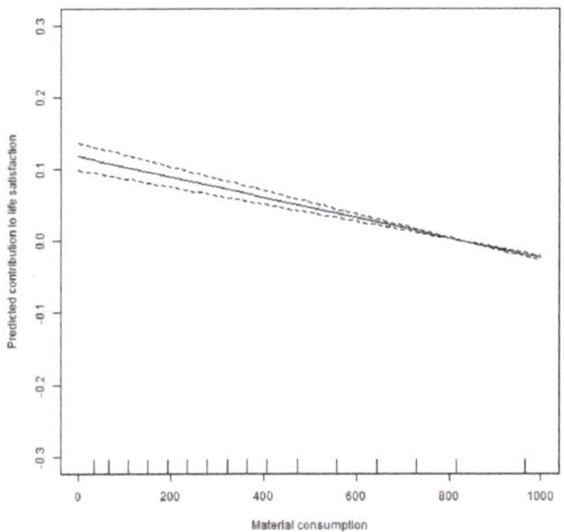

Overall sample (N = 3685).

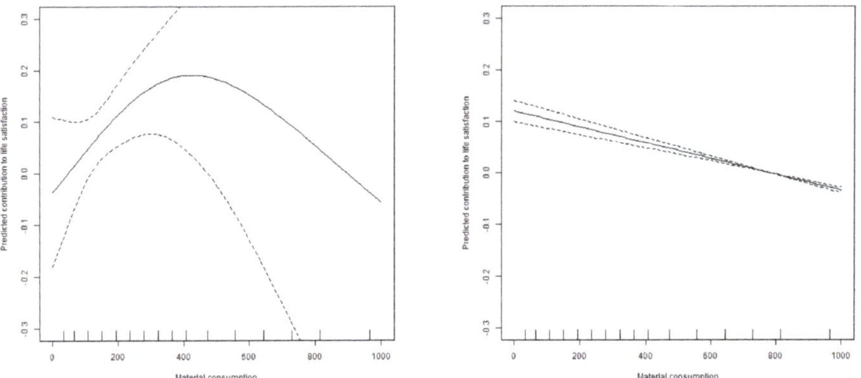

Subsample 1 (Attachment dummy = 1, N = 892); Subsample 2 (Attachment dummy = 0, N = 2689)

Figure 9. Material consumption (US$) and life satisfaction in Vietnam.

Table 7. Model fit statistics.

	Approximate Significance of Smooth Term (F Value)
Overall sample	137.8 ***
Subsample 1 (Attachment dummy = 1)	4.426 *
Subsample 2 (Attachment dummy = 0)	141.9 ***

Note: *** and * denote statistical significance at the 1% and 10% levels, respectively.

Table 8. Parametric estimation results for control variables.

Variable	Overall Sample	Subsample 1	Subsample 2
Age	0.030	−0.022 *	0.011
	(0.12)	(0.010)	(0.0073)
Age squared	−0.10	0.00022 *	−0.00016 *
	(0.15)	(0.00011)	(0.000079)
Gender (Male dummy)	0.010	−0.042	0.019 *
	(0.0069)	(0.056)	(0.0086)
Marriage	−0.042 *	−0.072	−0.088
	(0.026)	(0.10)	(0.059)
Children	−0.014	−0.016	−0.0086
	(0.016)	(0.13)	(0.076)
Number of family members	−0.11 ***	−0.084 ***	−0.029 *
	(0.025)	(0.020)	(0.013)
Education	0.018	0.23 *	0.025
	(0.017)	(0.12)	(0.080)
Residual consumption	0.023 **	0.019 *	0.021 **
	(0.011)	(0.009)	(0.010)
Darsal dummy	−0.12 ***	−0.99 ***	−0.34 ***
	(0.0079)	(0.016)	(0.037)
Constant term	0.89 ***	1.04 ***	4.33 ***
	(0.020)	(0.038)	(0.25)
Adjusted R squared	0.15	0.31	0.12
No. of observation	3685	892	2689

Note: Standardized coefficients are shown. Standard errors are in parenthesis. ***, **, and * denote statistical significance at the 1%, 5%, and 10% levels, respectively.

4. Discussion

If developing countries succeeded in emulating the prevailing consumption patterns of developed countries, the entire world would be on an equal footing in terms of consumer behavior. This possible conversion brings us on the right track to alleviating poverty in low-income countries, but without decoupling material consumption and income, it may also run the risk of exacerbating fears of increasing environmental burdens through over-consumption. In order to generate the sustainable consumption advocated for by the ongoing discourse on ecological footprints, planetary boundaries, and sustainable development goals (SDGs), current consumption styles in developing countries need to be reexamined so that we can grasp how consumption and subjective well-being interact.

In this context, Tsurumi et al. [48,50] showed that SWB saturation through material consumption can be observed not only in developed countries (like Japan) but also in developing countries (like Vietnam), in both urban and rural areas. Furthermore, ref [50] identified that a negative correlation between material consumption and SWB can be observed in rural areas in Vietnam. These findings imply that regardless of the stage of economic development, SWB saturation through material consumption can occur. In light of sustainable consumption, the situation where SWB saturation through consumption exists may correspond to be a "rampant or vain consumption" in terms of SWB. To accomplish sustainable consumption, we need to avoid meaningless consumption, which does not contribute to increasing SWB. Furthermore, if we can identify ways to increase our marginal utility from consumption, we can sustain or even increase our SWB by consuming less, thereby lowering our environmental burden.

In this study, we investigated whether attachment to material goods can interact with SWB in terms of life satisfaction. To that aim, we have constructed a two-good, two-period model and considered whether the level of utility increases if the potentially durable good is carried over to the second period. The result demonstrates that the consumer's utility does spike, due to the combined effect of substitution (between general consumption and the durable good) and income. The latter income effect arises because income that could have been used for the purchase of the durable good is now freed up. We have also discussed other oft-cited important channels: endowment effect, pro-social behavior, and latent factors such as the mottainai spirit.

Our theoretical model also relies on a number of assumptions, which can be relaxed in follow-up studies. The utility function is confined to the additive specification. Although our basic insights do not alter, this can be formally extended to the constant-elasticity-of-substitution and Cobb–Douglas types, which are more common in the literature. Sensitivity analysis of important parameters, such as the discount rate, the relative price of the durable good, and wages, may be relevant in considering what shapes less material-intensive economies. Moreover, an interesting setting would be where a consumer may choose to keep using the durable good or not, depending on its relative price, discounting, and other factors.

Consistent with this theoretical prediction, our empirical estimation results show that, while the marginal contribution of material consumption to life satisfaction is declining on average in rural Vietnam, it is increasing for people who utilize material goods possessed for a long time with good care. This implies that they obtain higher life satisfaction by less consumption of material goods than the average sample. There is a possibility that the quality of goods affects the relationship between material consumption and life satisfaction. However, in our survey areas of Thieu Hoa and Darsal, people have few options to select goods. In these villages, there are only several small shops people can use. In addition, people in the area cannot use Internet shopping services like Amazon because there is no delivery service. Furthermore, it takes a long time to reach urban areas where there are large shops and people usually do not have cars and therefore cannot bring home the goods purchased, and they have no delivery services to their village. Here, we note that, even if material consumption is low, total consumption and residual consumption (such as relational consumption shown in [50]) can be high, since our model divides total consumption into material consumption and residual consumption. Although our sample is limited to rural areas in Vietnam, and what we have demonstrated is not causality, this may point to a promising channel to sustain life satisfaction even if—or perhaps because—less material is consumed by households in developing countries. Considering that we found similar descriptive statistics for the control variables used in our analysis among the subsamples and that we did not obtain statistically significant coefficients for most of the control variables, as shown in Table 8, the observed differences between the subsamples in Figure 9 are thought not to be due to differences in control variables and that attachment to material goods is the most likely the cause of the results shown.

Consumerist motivations and relative consumption are also expected to be prevalent in developing countries. The literature recognizes that the consumption of some goods is conspicuous in rural households in developing countries (e.g., [51,52]). It is well known that the rural poor spend much on festivals and ceremonies in India [53] and in African countries, which [51] suggests they may be a substitute for the consumption of material goods, such as radio and television. Moreover, in a survey of poor rural households in India, ref [54] reports that individuals who spend more on conspicuous consumption have lower levels of SWB, while their income relative to others does not affect SWB. Thus, while the evidence is scarce and mixed, our results shown in Subsample 2 in Figure 7 do not seem to support the hypothesis that individuals with consumerist motivations and relative consumption have higher SWB. This implies that people who consume material goods less tend to be those who have stronger social capital, and they tend to barter material goods such as food. The strong social capital can significantly improve people's life satisfaction (e.g., [55]). Our result is in line with [50] and reveals the positive correlation between relational consumption based on strong social capital and life satisfaction in rural Vietnam.

We have theoretically and empirically demonstrated that using the same good longer correlates with sustained or higher SWB in terms of life satisfaction, other conditions being equal. To bridge our results with environmental sustainability, we also need to clarify how goods attachment, SWB, and environmental burden interrelate with each other. Attachment to goods does not necessarily reduce environmental burden if attachment leads to far more possession than necessary. This relationship may also be dynamically complicated by the recent emergence of sharing economies. Studying these relationships may produce an important implication for directing ourselves toward circular economies.

Other channels than those we have discussed thus far, connecting SWB and the continued use of material goods, may exist, even if they have not yet been discussed in the literature. For example, people who own less may experience higher SWB, as they may have less clutter and are able to focus on the "here and now." This may also be related to the ethics of minimalism. Minimalism stresses the importance of the non-material aspects of life and is sometimes characterized by anti-consumerism (i.e., "less is more") [56] (p. 67). A minimalist lifestyle may be deemed as environmentally friendly; however, the overall environmental impacts are still ambiguous, as those lifestyles may either trigger more throwing away or reduced purchasing at the outset. In fact, having fewer goods is becoming more common, enhanced by recent popular movements focusing on organization methods that involve disposing of things when they no longer bring joy [57]. Some studies actually point to preserving utility by retaining the memory of certain goods, rather than their physical possession [58], or by "social recycling" instead of throwing away [59].

To take another recent example, in an increasingly popular sharing economy, purchasing a durable good may be replaced by subscribing to a sharing program or purchasing a service. One direction of the mode of consumption in the sharing economy is collaborative consumption [60–64], which contrasts with individualistic consumerism and may contribute to sustainable consumption within the planetary boundaries [65]. Mobility services as a substitute for car ownership are a case in point. These recent and important discussions may be expanded upon in future research.

A limitation of the current study is that we did not consider the effect of the burden of long working hours on life satisfaction. Higher material consumption can be related to having a stressful or demanding job. Therefore, having long working hours is a potential explanation for the negative correlation between material consumption and life satisfaction observed in Figure 7. However, considering that we found a positive correlation between material consumption and life satisfaction for Subsample 1 in Figure 7, the potential negative effects of long working hours may be surpassed by the positive effects of the attachment to material goods on life satisfaction.

Another limitation of the current study is that we only show a correlation between material consumption and life satisfaction, not causality. However, looking after material goods may give people life satisfaction, or more satisfied people may tend to look after material goods better. This can be explored in future research.

Our study shows that for people who take better care of material goods, increased consumption is linked to increased life satisfaction in the study sample. This finding has a useful policy implication for developing countries to improve their well-being.

Author Contributions: Conceptualization, T.T., R.Y. and S.M.; methodology, T.T. and R.Y.; software, T.T.; validation, T.T.; formal analysis, T.T.; investigation, T.T. resources, T.T.; data curation, T.T.; writing—original draft preparation, T.T. and R.Y.; writing—review and editing, T.T., R.Y. and K.K.; supervision, T.T.; project administration, T.T.; funding acquisition, T.T. All authors have read and agreed to the published version of the manuscript.

Funding: This research was supported by the Environment Research and Technology Development Fund (S-16) of the Ministry of the Environment, Japan.

Acknowledgments: Authors thank Michikazu Kojima for his insightful comments.

Conflicts of Interest: The authors declare no conflict of interest. The funder had no role in the design of the study; in the collection, analyses, or interpretation of data; in the writing of the manuscript, or in the decision to publish the results.

Appendix A

Here we present a formal model that led us to the Proposition in Section 2.1, that utility is higher when a durable good is possessed longer, all else being equal. In the basic model, let C_{1t}, C_{2t}, C_{1t+1}, and C_{2t+1} denote Goods 1 and 2 bought at t and $t + 1$, respectively. As described in the main text, to fix ideas, imagine that Good 1 is a mix of general consumer material good, whereas Good 2 is a relatively durable, but still material consumer good. The prices of both goods are constant and certain, expressed by p_1 and p_2, respectively. Because this is an intertemporal problem, we wrote $\delta > 0$ for the utility discount rate and $r > 0$ as the interest rate. For analytical ease, the (instantaneous) utility function is further specified as additive: $U(C_{1t}, C_{2t}) = C_{1t}^2 + C_{2t}^2$. Other typical specifications such as Cobb–Douglas or the constant elasticity of substitution (CES) utility functions do not alter our basic insights. Finally, the consumer earns income

w only in t, part of which is saved for the period $t + 1$. Formally, our problem is

$$\max_{C_{1t},\, C_{2t}, C_{1t+1},\, C_{2t+1}} U(C_{1t}, C_{2t}) + \frac{1}{1+\delta} U(C_{1t+1}, C_{2t+1})$$

subject to

$$p_1 C_{1t} + p_2 C_{2t} + \frac{1}{1+r}(p_1 C_{1t+1} + p_2 C_{2t+1}) = w$$

A regular optimization exercise enabled us to solve for the consumption of two goods in both periods and to write indirect utility as a function of wage, prices, discount rate, and interest rate.

In the extended model, where Good 2 continues to be possessed, our problem changes slightly to the following:

$$\max_{C_{1t},\, C_{2t}, C_{1t+1}} U(C_{1t}, C_{2t}) + \frac{1}{1+\delta} U(C_{1t+1}, C_{2t+1})$$

subject to

$$p_1 C_{1t} + p_2 C_{2t} + \frac{1}{1+r}\, p_1 C_{1t+1} = w$$

and

$$C_{2t} = C_{2t+1}.$$

Observe that the first constraint lacks Good 2 in the second period, as it does not have to be purchased. In addition, the second constraint states that the quantity of Good 2 consumed in the next period remains the same. It is commonplace to assume that investment goods are subject to depreciation; this assumption can be applied to our example. On the other hand, we have already seen that some behavioral literature suggests a positive endowment effect can also be attained from durable goods, in which case the value of the good being studied actually appreciates for that person. In any case, we bypassed the endowment effect here, as it is contained in M in the current formulation. Thus, temporal changes in the value of Good 2 to the consumer may be either positive or negative. We relegated more general cases to our future research, and simply assumed the second constraint in the current study.

References

1. Geissdoerfer, M.; Savaget, P.; Bocken, N.M.P.; Hultink, E.J. Circular Economy Sustainability Paradigm. *J. Clean. Prod.* **2019**, *143*, 757–768. [CrossRef]
2. Korhonen, J.; Honkasalo, A.; Seppälä, J. Circular Economy: The Concept and its Limitations. *Ecol. Econ.* **2018**, *143*, 37–46. [CrossRef]
3. Kirchherr, J.; Reike, D.; Hekkert, M. Conceptualizing the circular economy: An analysis of 114 definitions. *Resour. Conserv. Recycl.* **2017**, *127*, 221–232. [CrossRef]
4. Bowlby, J. *A Secure Base: Clinical Applications of Attachment Theory*; Routledge: Abingdon, UK, 1988.

5.   Giuliani, M.V. Theory of Attachment and Place Attachment. In *Psychological Theories for Environmental Issues*; Bonnes, M., Lee, T., Bonaiuto, M., Eds.; Ashgate: Farnham, UK, 2003; pp. 137–170.

6.   Hidalgo, M.C.; Hernández, B. Place attachment: Conceptual and empirical questions. *J. Environ. Psychol.* **2001**, *21*, 273–281. [CrossRef]

7.   Gerson, K.; Stueve, A.; Fischer, C.S. Attachment to Place. In *Networks and Places: Social Relations in the Urban Setting*; Fischer, C.S., Jackson, R.M., Stueve, C.A., Gerson, K., Jones, L.M., Baldassare, M., Eds.; The Free Press: Wong Chuk Hang, Hong Kong, 1977; pp. 139–161.

8.   Kleine, S.S.; Baker, S.M. An integrative review of material possession attachment. *Acad. Mark. Sci. Rev.* **2004**, *1*, 1–39.

9.   Kahneman, D.; Knetsch, J.L.; Thaler, R.H. Experimental tests of the endowment effect and the Coase theorem. *J. Political Econ.* **1990**, *98*, 1325–1348. [CrossRef]

10.   Thaler, R. Toward a positive theory of consumer choice. *J. Econ. Behav. Organ.* **1980**, *1*, 39–60. [CrossRef]

11.   Kahneman, D.; Knetsch, J.L.; Thaler, R.H. The Endowment Effect, Loss Aversion, and Status Quo Bias. *J. Econ. Perspect.* **1991**, *5*, 193–206. [CrossRef]

12.   Knetsch, J.L. The Endowment Effect and Evidence of Nonreversible Indifference Curves. *Am. Econ. Rev.* **1989**, *79*, 1277–1284.

13.   Knetsch, J.L.; Sinden, J.A. The Persistence of Evaluation Disparities. *Q. J. Econ.* **1987**, *102*, 691–696. [CrossRef]

14.   Mugge, R.; Schoormans, J.P.L.; Schifferstein, H.N.J. Product Attachment: Design Strategies to Stimulate the Emotional Bonding to Products. In *Product Experience*; Schifferstein, H.N.J., Hekke, P., Eds.; Elsevier: Amsterdam, The Netherlands, 2008; pp. 425–440.

15.   Schifferstein, H.N.J.; Zwartkruis-Pelgrim, E.P.H. Consumer-Product Attachment: Measurement and Design Implications. *Int. J. Des.* **2008**, *2*, 1–13.

16.   Dommer, S.L.; Winterich, K.P. Disposing of the self: The role of attachment in the disposition process. *Curr. Opin. Psychol.* **2020**, *39*, 43–47. [CrossRef] [PubMed]

17.   Ting, H.; Thaichon, P.; Chuah, F.; Tan, S.R. Consumer Behavior and Disposition Decisions: The Why and How of Smartphone Disposition. *J. Retail. Consum. Serv.* **2019**, *51*, 212–220. [CrossRef]

18.   Andreoni, J. Impure Altruism and Donations to Public Goods: A Theory of Warm-Glow Giving. *Econ. J.* **1990**, *100*, 464–477. [CrossRef]

19.   Cherepanov, V.; Feddersen, T.; Sandroni, A. Revealed Preferences and Aspirations in Warm Glow Theory. *Econ. Theory* **2013**, *54*, 501–535. [CrossRef]

20.   Menges, R.; Schroeder, C.; Traub, S. Altruism, Warm Glow and the Willingness-to-Donate for Green Electricity: An Artefactual Field Experiment. *Environ. Resour. Econ.* **2005**, *31*, 431–458. [CrossRef]

21.   Taufik, D.; Bolderdijk, J.W.; Steg, L. Acting Green Elicits a Literal Warm Glow. *Nat. Clim. Chang.* **2015**, *5*, 37–40. [CrossRef]

22.   Morelli, S.A.; Lieberman, M.D.; Zaki, J. The Emerging Study of Positive Empathy. *Soc. Personal. Psychol. Compass* **2015**, *9*, 57–68. [CrossRef]

23.   Maathai, W. *Unbowed: A memoir*; Anchor: New York, NY, USA, 2007.

24.   Sirola, N.; Sutinen, U.M.; Närvänen, E.; Mesiranta, N.; Mattila, M. Mottainai!-A practice theoretical analysis of Japanese consumers' food waste reduction. *Sustainability* **2019**, *11*, 6645. [CrossRef]

25.   Clark, A.E.; Frijters, P.; Shields, M.A. Relative income, happiness, and utility: An explanation for the Easterlin paradox and other puzzles. *J. Econ. Lit.* **2008**, *46*, 95–144. [CrossRef]

26.   Diener, E. Subjective Well-Being. *Psychol. Bull.* **1984**, *95*, 542–575. [CrossRef] [PubMed]

27.   Diener, E.; Sandvik, E.; Seidlitz, L.; Diener, M. The relationship between income and subjective well-being: Relative or absolute? *Soc. Indic. Res.* **1993**, *28*, 195–223. [CrossRef]

28.   Easterlin, R.A. Does Economic Growth Improve the Human Lot? Some Empirical Evidence. In *Nations and Households in Economic Growth: Essays in Honor of Moses Abramovitz*; David, P.A., Reder, M.W., Eds.; Academic Press: New York, NY, USA, 1974; pp. 89–215.

29.   Diener, E.; Diener, M.; Diener, C. Factors Predicting the Subjective Well-Being of Nations. *J. Personal. Soc. Psychol.* **1995**, *69*, 851–864. [CrossRef]

30.   Myers, D.G.; Diener, E. The Pursuit of Happiness. *Sci. Am.* **1996**, *274*, 54–56. [CrossRef] [PubMed]

31.   Argyle, M. Causes and Correlates of Happiness. In *Well-Being: The Foundations of Hedonic Psychology*; Kahneman, D., Diener, E., Schwarz, N., Eds.; Russell Sage Foundation: New York, NY, USA, 1999; pp. 353–373.

32. Easterlin, R.A. Will raising the incomes of all increase the happiness of all? *J. Econ. Vehavior Organ.* **1995**, *27*, 35–47. [CrossRef]

33. Easterlin, R.A. Feeding the illusion of growth and happiness: A reply to Hagerty and Veenhoven. *Soc. Indic. Res.* **2005**, *74*, 429–443. [CrossRef]

34. Cheng, Z.; Wang, H.; Smyth, R. Happiness and job satisfaction in urban China: A comparative study of two generations of migrants and urban locals. *J. Econ. Vehavior Organ.* **2014**, *51*, 2160–2184. [CrossRef]

35. Stevenson, B.; Wolfers, J. Subjective well-being and income: Is there any evidence of satiation? *Am. Econ. Rev.* **2013**, *103*, 598–604. [CrossRef]

36. Stevenson, B.; Wolfers, J. Economic Growth and Subjective Well-Being: Reassessing the Easterlin Paradox. *NBER Work. Paper Ser.* **2008**, *14282*, 1–28.

37. Ferrer-i-Carbonell, A. Income and well-being: An empirical analysis of the comparison income effect. *J. Public Econ.* **2005**, *89*, 997–1019. [CrossRef]

38. Dolan, P.; Peasgood, T.; White, M. Do we really know what makes us happy? A review of the economic literature on the factors associated with subjective well-being. *J. Econ. Psychol.* **2008**, *29*, 94–122. [CrossRef]

39. Rickardsson, J.; Mellander, C. Absolute vs Relative Income and Life Satisfaction. *CESIS Electron. Work. Pap. Ser.* **2017**, *451*, 1–29.

40. Foye, C.; Clapham, D.; Gabrieli, T. Home-ownership as a social norm and positional good: Subjective wellbeing evidence from panel data. *Urban Stud.* **2018**, *55*, 1290–1312. [CrossRef]

41. Donati, P. *Relational Sociology: A New Paradigm for the Social Sciences*; Routledge: London, UK, 2010.

42. Gokdemir, O. Consumption, savings and life satisfaction: The Turkish case. *Int. Rev. Econ.* **2015**, *62*, 183–196. [CrossRef]

43. DeLeire, T.; Kalil, A. Does consumption buy happiness? Evidence from the United States. *Int. Rev. Econ.* **2010**, *57*, 163–176. [CrossRef]

44. Zhang, J.; Xiong, Y. Effects of multifaceted consumption on happiness in life: A case study in Japan based on an integrated approach. *Int. Rev. Econ.* **2015**, *62*, 143–162. [CrossRef]

45. Dumludag, D. Consumption and life satisfaction at different levels of economic development. *Int. Rev. Econ.* **2015**, *62*, 163–182. [CrossRef]

46. Pandelaere, M. Materialism and well-being: The role of consumption. *Curr. Opin. Psychol.* **2016**, *10*, 33–38. [CrossRef]

47. Levitt, S.D.; List, J.A. What do laboratory experiments measuring social preferences reveal about the real world? *J. Econ. Perspect.* **2007**, *21*, 153–174. [CrossRef]

48. Tsurumi, T.; Yamaguchi, R.; Kagohashi, K.; Managi, S. Are cognitive, affective, and eudaimonic dimensions of subjective well-being differently related to consumption? Evidence from Japan. *J. Happiness Stud.* **2020**. [CrossRef]

49. Hastie, T.J.; Tibshirani, R.J. *Generalized Additive Models*; CRC Press: Boca Raton, FL, USA, 1990; Volume 43.

50. Tsurumi, T.; Yamaguchi, R.; Kagohashi, K.; Managi, S. The role of social capital in improving subjective well-being through consumption: Evidence from rural and urban Vietnam. Under review.

51. Banerjee, A.V.; Duflo, E. The economic lives of the poor. *J. Econ. Perspect.* **2007**, *21*, 141–168. [CrossRef]

52. Roychowdhury, P. Visible inequality, status competition, and conspicuous consumption: Evidence from rural India. *Oxf. Econ. Pap.* **2016**, *69*, 36–54. [CrossRef]

53. Bloch, F.; Rao, V.; Desai, S. Wedding celebrations as conspicuous consumption signaling social status in rural India. *J. Hum. Resour.* **2004**, *39*, 675–695. [CrossRef]

54. Linssen, R.; Van Kempen, L.; Kraaykamp, G. Subjective well-being in rural India: The curse of conspicuous consumption. *Soc. Indic. Res.* **2011**, *101*, 57–72. [CrossRef]

55. Bjørnskov, C. The happy few: Cross-country evidence on social capital and life satisfaction. *Kyklos* **2003**, *56*, 3–16.

56. Dopierała, R. Minimalism—A New Mode of Consumption? *Przegląd Socjol.* **2017**, *66*, 67–83.

57. Kondō, M. *The Life-Changing Magic of Tidying up: The Japanese Art of Decluttering and Organizing*; Ten Speed Press: Berkeley, CA, USA, 2014.

58. Winterich, K.P.; Reczek, R.W.; Irwin, J.R. Keeping the memory but not the possession: Memory preservation mitigates identity loss from product disposition. *J. Mark.* **2017**, *81*, 104–120. [CrossRef]

59. Donnelly, G.E.; Lamberton, C.; Reczek, R.W.; Norton, M.I. Social recycling transforms unwanted goods into happiness. *J. Assoc. Consum. Res.* **2017**, *2*, 48–63. [CrossRef]

60. Albinsson, P.A.; Perera, B.Y. Alternative Marketplaces in the 21st Century: Building Community through Sharing Events. *J. Consum. Behav.* **2012**, *11*, 303–315. [CrossRef]
61. Belk, R. Sharing. *J. Consum. Res.* **2010**, *36*, 715–734. [CrossRef]
62. Botsman, R.; Rogers, R. Beyond Zipcar: Collaborative Consumption. *Harv. Bus. Rev.* **2010**, *88*, 30.
63. Felson, M.; Spaeth, J.L. Community Structure and Collaborative Consumption: A Routine Activity Approach. *Am. Behav. Sci.* **1978**, *21*, 614–624. [CrossRef]
64. Möhlmann, M. Collaborative consumption: Determinants of satisfaction and the likelihood of using a sharing economy option again. *J. Consum. Behav.* **2015**, *14*, 193–207. [CrossRef]
65. Steffen, W.; Richardson, K.; Rockström, J.; Cornell, S.E.; Fetzer, I.; Bennett, E.M.; Biggs, R.; Carpenter, S.R.; De Vries, W.; De Wit, C.A.; et al. Planetary boundaries: Guiding human development on a changing planet. *Science* **2015**, *347*. [CrossRef] [PubMed]

Publisher's Note: MDPI stays neutral with regard to jurisdictional claims in published maps and institutional affiliations.

 *sustainability*

Article

To See a World in a Grain of Sand—The Transformative Potential of Small Community Actions

Atsushi Watabe [1,*] and Simon Gilby [2]

[1] Institute for Global Environmental Strategies, 2108-11 Kamiyamaguchi, Hayama, Kanagawa 240-0115, Japan
[2] Independent Researcher, London SE1, UK; simongilby@gmail.com
* Correspondence: watabe@iges.or.jp

Received: 6 August 2020; Accepted: 4 September 2020; Published: 9 September 2020

Abstract: The recognition of the urgent need for more sustainable lifestyles dates from the late 20th century, originating in concerns about resource depletion and climate change. Research and policy measures have evolved since then, paying increasing attention to systemic change over individual behaviour. However, as individual behavioural change is constrained by the systems within which choices are made, more study is needed to understand better how systemic changes occur. Drawing on the experiences of the Sustainable Lifestyles and Education Programme of the UN-led One-Planet Network in collaborating with small collective actions for sustainable lifestyles, the paper analyses the needs and approaches for sustainable lifestyles and opportunities for the local actors to grow their capacities in developing ways of living sustainably. These experiences show that the pursuit of sustainable lifestyles is not a one-shot change in behaviour. It is a continuous process where actors identify and tackle locally specific opportunities for responsible and sustainable ways of living, and through a process of mutual learning and experimentation gradually shape shared visions of sustainable living. Systemic changes for sustainable living are ultimately neither about simply improving people's awareness or attitudes or replacing some components of the external systems. They are the creation of capacities and aspirations of people actively and continuously engaging to shape alternative systems of living.

Keywords: sustainable lifestyles; collective actions; One-Planet Network

---

1. Introduction

Today, our global footprint is about one and a half times the Earth's total capacity to provide renewable and non-renewable resources to humanity. If nothing changes, in 35 years, with an increasing population that could reach 9.6 billion by 2050, we will need almost three planets to sustain our ways of living. Rethinking the ways we produce, consume and exchange has become crucial to move towards a society where we can all live well within the boundaries of our planet [1].

For as long as the critical role of our lifestyles in ensuring the sustainability of the planet has been widely recognised, policymakers, practitioners and researchers have argued for the need for more sustainable lifestyles with such a statement [2–5]. Since the Sustainable Lifestyles and Education Programme of the One-Planet Network was launched in 2014, the programme's partners including the authors of this paper have also worked in-line with this thinking.

Although we believe that this thinking is still valid, by conceptualising sustainable lifestyles in such a way, we run the risk of narrowing down the scope to "environmentally friendly" individual behaviour [6,7] and ignoring the broad range of elements that constitute (un)sustainability of living [8–10]. In fact, during the past two decades, research focus has shifted from individual behavioural change to the changes in "systems" that shape the enabling and constraining contexts of lifestyles [11–14], or the entangled elements that shape the possibilities and limitations of our ways of living [15,16].

However, it is not easy to induce the changes to the systems that eventually enable the uptake of sustainable ways of living [11]. Indeed, we are not very clear about what we mean by "systemic changes", or the roles that actors in the society—either as individuals, members of organisations, or organisations—can play. This paper aims to contribute to the discussion of sustainable lifestyles through gaining a deeper understanding of the changes of "systems" that eventually propel the shift in lifestyles and draws on the lessons learned from the projects supported under the framework of the Sustainable Lifestyles and Education Programme (SLE Programme) of the UN 10-Year Framework of Programmes on Sustainable Consumption and Production Patterns (One-Planet Network). The next section looks into the shifting research focus on lifestyles in association with the common trap of individual focus and increasing attention to collective actions, followed by the introduction of the SLE Programme. The third section looks at the materials and methods used by the projects. The fourth section reflects on the critical lessons of ground-level collective efforts supported under the SLE Programme, paying attention to the challenges these initiatives addressed, approaches adopted to address these challenges, and reflections on lessons learned. Based on these reflections, the final section discusses the meaning of "systemic change" at the day-to-day level.

2. Sustainable Lifestyles and Increasing Attention to Collective Actions

2.1. Lifestyles—Allure into the Diagnostic Point of View

The sharp increase in the use of natural resources in association with an ever-increasing demand for goods and services has created significant pressure on the natural environment and the planet as a whole. The issue has become more and more pressing throughout the late twentieth century and early twenty-first century, due to the rapid economic development and continued population growth in emerging economies, resulting in the urbanisation and emergence of the new middle class. Against this backdrop, dominant arguments on sustainable lifestyles have focused on the negative impacts of our lifestyles. Mass-consumption lifestyles have been fingered as the cause of the unsustainable and severe damage to the environment by bringing about ever-increasing use of natural resources, greenhouse gas emissions, and waste and pollution [17]. Our society needs to achieve decoupling of resource use and economic growth or natural resource use and human wellbeing in this context [18,19]. Having taken this consideration as a starting point, the literature on sustainable lifestyles or living have generally referred "to using as few resources as possible, reducing carbon footprints, and reducing environmental damage" [10]. It has often been taken as a term interchangeable with sustainable consumption [20] though some have called attention to the fact that lifestyle is far broader than consumption [21]. However, such focus on the "negative impact of consumer behaviour on the sustainability of the environment" has made us incapable of capturing how lifestyles, entangled with broader elements of the society, would continue or change. Here, it is worth looking back on how the concept of lifestyles has been used in much wider contexts than environmental sustainability.

The concept of lifestyles has been developed gradually within social science since the nineteenth-century, with significant contributions being made, inter alia, by Thorstein Veblen (on conspicuous consumption of goods and leisure activities), Georg Simmel (on the dynamics of styles through trickle-down emulation in the modern urban life), and Max Weber (on the stratification of relationships based on styles rather than position in the labour market). Sociologists have used the concept to capture the dynamics of classification in modern societies through differentiation and emulation [22]. Such earlier concepts have been elaborated by Bourdieu clarifying the differences of social, economic and cultural capitals among people in different positions in the social hierarchy [23]. In the late twentieth century, against the backdrop of the flattering of the social forms and the instability of the modern forms of social membership [24], sociologists deepened the analysis of lifestyles as the process of reflexive self-identification, arguing that people express their personal views of their (ideal) position in society through their consumer behaviour, products and services they acquire and the ways they organise their time through using them [25]. However, empirical research also supports the remaining validity of class [26–29].

Many fields of science have adapted lifestyles as a key concept. For instance, health care science has identified many lifestyle diseases such as cardiovascular diseases (CVDs), stroke, diabetes and certain forms of cancer. Their risks are heavily associated with the individual patterns of actions, such as diet, smoking and physical exercise [30,31]. Marketing research has traced the changes and stratifications of consumer lifestyles by analysing the changing trends of expenditures [32–34]. It is important to note that many of these different streams of lifestyle studies share an angle: a lifestyle, understood as the pattern of individual (or group) actions, needs to be studied to address the background/direct cause of specific status of the present society. Such a diagnostic angle is valid as long as it tries to pin down the causes of modern challenges. However, it inevitably narrows down the objectives and outputs of the studies toward producing practical measures for intervening in the single behaviours performed by individuals. Spaargaren and Oosterveer warned that "lifestyle politics could mistakenly be interpreted as dealing with 'individual' human beings and their private, individual affairs primarily and exclusively" [12]. Such an angle is narrow compared to the arguments that lifestyles are not single behaviours based on the free choice of individuals, but collective constructs and social constructions that groups of people (ethnic groups, subculture groups, etc.) develop and change dynamically over time [22]. Evans and Jackson also pointed out that lifestyles as self-identification are not personal processes but "tensions between individual and collective identities" [21].

The mainstream sustainable lifestyles discourse seems to have been taken over by this narrow scope. For several reasons, this has caused difficulty in understanding how changes in lifestyles occur, and effectively inducing desirable changes. Firstly, they often pay attention to a limited range of "environmentally sustainable" actions (e.g., buycotting, boycotting ethical consumption, voluntary simplicity) in specific domains (e.g., water, food, energy, consumer goods) [9]. Because of this narrow focus, they tend to avert their eyes away from the uncomfortable fact that a significant part of our daily practices has become more and more resource-intensive [6,7]. Secondly, framing lifestyles solely as individual actions and behaviour misleads one into the well-known "mystery" of the attitude–behaviour gap [35–38]. This concept is described as when people, despite having abundant knowledge of the negative consequences of their behaviours as well as clear intentions to reduce their impacts, do not adopt such "sustainable" practices in reality [35,39,40].

2.2. Lifestyles—As a Dynamic Web of Practices

However, the gap is not a mystery. The difficulty in interpreting the gap comes from the assumption that an individual is considered to have a portfolio of values, attitudes, norms, interests and desires, and selects from them to decide on the course of action [41]. However, research has shown that such a linear assumption of attitude, behaviour and choice (ABC) is overly simplistic [42]. Practice theory offers a deeper understanding of the complex nature of inertia or changes of behaviours [43–48]. The concept of practices, originating from the earlier sociologists such as Bourdieu and Giddens, have evolved since the 1990s with particular attention to sustainable consumption. Practice comprises competence, meaning, materials, and the interplays among people, instead of the behaviours driven by the attitudes of individuals. According to practice theory, a new practice is formulated and spread in association with the changes in the systems of provision [11–14,21], or is shaped through the transfer of the competence, meaning, or materials from existing practices to another [49,50]. For instance, the spread of daily bathing was the result of the changes in the knowledge and belief on health, sanitation and appropriateness, as well as the development of the water system. Mobility patterns in urban areas not only change due to the implementation of a congestion tax, but also by the changing contexts of practices such as eating, learning and working [49].

In this perspective, people conduct actions in a web of many elements that force them to take multiple roles (e.g., consumers, family members, employees). Their actions are not conducted in a single domain but are closely linked to the needs and conditions of multiple domains (e.g., moving for shopping and working) [9,14,41]. Finally, in such a manner, behaviours and lifestyles are the constructs of collective/social interactions [22,41]. On this account, "the mystery" of the gaps between attitude and behaviours "arises from the failure of the portfolio model as a paradigm for human activity" [41]. What we observe as "changes in lifestyle" is often a superficial layer of the complex changes of the systems comprising a variety of elements that create and deliver specific values of goods or services benefiting our living, as well as the associated appropriation or abandonment of competence, meanings, and materials, shaped by the unique conditions on which individuals and groups of people are situated. For instance, it is often argued that people tend to change their behaviours at turning points in their lives. However, this conclusion has been challenged by other research as overlooking the fact that changes at turning "points" may take place over years, multiple changes in behaviours take place and influence each other, and the changes are often for the sake of meeting the needs of their families, instead of personal needs with further investigation needed to more fully understand how people transition through their lives [51]. Lifestyles often change without intention or clear design but through dynamic processes of bonding different elements into bricolage [52]. In sum, our lifestyles are organised as a part of the webs of diverse needs and competences in various domains, complex network of people, and contexts of technology and institutional settings. Researchers have proposed concepts to capture such complexity, e.g., "web of constraints" to look into the construct and constraint of practices through the changes in technology, social relations [15], systems of provision [14,20], and the prism of sustainable consumption covering the practice and the conditions and contexts of the economy and environment [53].

2.3. Collective Actions on Lifestyles

However, such focus on the systems or webs poses additional questions. How do systems change? How can we initiate changes in systems? In paying attention to systems, we are tempted to expect that powerful actors such as governments may replace a few of their vital parts to quickly change the entire systems and induce desirable changes of behavioural patterns. Such a view is not entirely untrue if we look back on some past cases, such as how the development of the subway network changed the mobility of the citizens. However, such an engineering point view keeps us in the trap of the narrower focus on the systemic causes of "environment-(un)friendly" behaviours of individuals. While lifestyles are shaped in the web of elements, we would need to pay attention to the changes of "relations between elements" rather than jumping on to the replacement of components through technical fixes. Moreover, given the radical alteration to our current systems required to enable sustainable ways of living, top-down approaches are insufficient in terms of the reductions required and ineffective when attempting to force people to adopt new ways of living without their full participation and consent. Such an approach would deprive people of opportunities for critically understanding the conditions and consequences of their ways of living, and, thus, may lead to backlash and failure.

Here, collaborative actions, including grassroots or micro-scale initiatives deserve attention [54–57]. Such initiatives are often carried out in protective spaces [58] where actors are relatively free from the ordinary constraints such as political/legal requirements, dominant actors of the market, and social norms that force them to follow business as usual. Through engaging in and driving the alternative actions and the enabling conditions thereof, people grow and exercise their capacities to create alternative contexts of living that enable them to live more sustainably [59–63]. Additionally, micro-scale collective actions have some features that may help us explore the key questions of this paper—how actors engage in systemic changes for sustainable lifestyles.

1. Collaborative actions take place in specific local settings where actual challenges of (un)sustainable living are dynamically shaped. All "unsustainability" issues take concrete forms in the particular contexts of certain spaces—rural, suburban or urban areas. For instance, in his argument on the role of the cities in shaping grassroots niches, Marc Wolfram [64] pointed out "the implications of the cities for the way in which citizens and local civil society actors get involved in the spatially

embedded reproduction of socio-technical regimes and/or creation of sustainability innovations", referring to Bulkeley et al. and Baker et al. [65,66]. While unsustainability issues emerge at certain spaces, people also address them at specific sites, sometimes resulting in new forms of cooperative relations. For this reason, examination of collective initiatives opens up opportunities to learn how unsustainability challenges arise in the real contexts of specific spaces [67] as well as how individual/organisational capacities to address them are formulated and often unevenly distributed [54].

2. Facing the dynamics of (un)sustainability, collective actions may not be able to induce ground-level changes if they simply transfer existing knowledge or skills on "sustainable behaviours" to the "beneficiaries" who are assumed to lack them. Instead, knowledge and skills are co-created through the collaboration and mutual learning of the participants with different backgrounds [68], including the reshaping of the relations among them [54,62]. Interestingly, collective actions do not only help to build alternative relationships between human actors. The creation of new knowledge and skills often means that new relations are established between human and nonhumans constituting the web of practices through various forms of actions [69], for potentially changing their status in the global socioeconomic systems. Thus, we could learn from collective actions how such knowledge and relations are built in-between different elements and support the actors altering the current constraints of living.

3. Collaborative actions frequently do not go as planned in the real-world context. They face a variety of unforeseen challenges making actors review what steps they should carry out to achieve their goals. Sometimes they are urged to consider their objectives—what kind of lifestyles or societal contexts that they wish to make. In doing so, they attach meanings to the conditions they face—why the current conditions are unacceptable and what kind of alternative they want to create [59]. In other words, they do not only engage in the actions toward pre-set goals but participate in the transitions "in-the-making" where the issues, models of participation, and attending public are dynamically formed [70]. For such a reason, by looking at or participating in collective actions, we can learn how various actors learn from reality and dynamically shape their transitions.

In short, the (un)sustainability of lifestyles emerges and dynamically evolves in the locally specific contexts of living, knowledge of (un)sustainability and capacities to address it are formed in the dynamic relations among people and surrounding conditions and actors learn from reality and dynamically shape the transitions. We can tentatively posit that changes in the technical or institutional elements of the systems and changes in the intentions and competences of the people are adjacent and entangled with each other, and, thus, are not replaceable separately, and that systemic changes may not occur (only) through the replacement of elements to drive the current unsustainable living to the pre-defined sustainable living, but through the process in which actors engage in the reshaping of the webs. Further investigation into these points would, therefore, guide us in gaining a deeper understanding of what the changes of "systems of provisions" are or the entanglements shaping the contexts of living, and how people and organisations, as well as material elements, interact in guiding such changes. On such a consideration, the latter half of the paper takes a closer look at the three critical points for gaining insights on how actors initiate and engage in the systemic changes.

1. How did actors identify the locally specific challenges of (un)sustainable living?
2. How did actors create and share the knowledge and skills to address them?
3. How and what did actors learn to dynamically and continuously shape their transitions?

To this end, we draw on some of the ground-level collective actions that we have collaborated with or learned from under the SLE Programme.

3. Materials and Methods

To gain the insights into the three critical points of the changes in systems or webs of elements shaping our lifestyles, the paper draws mainly on some of the projects that the SLE Programme has selected through open calls for project proposals and supported from 2016 to 2019. There was a total of four calls for project proposals from developing countries and countries with emerging economies that address unique challenges for (un)sustainable lifestyles and contribute to reducing greenhouse gas (GHG) emissions through their activities. The SLE Programme selected 24 projects in total. The SLE Programme has given financial assistance ranging from USD 50,000 to 400,000 depending on the proposals, utilising funds from the Government of Japan, as well as providing technical support covering procedural or contract issues, suggestions regarding activities and needed skills/capacity development or resource persons. Programme coordinators visited project sites as necessary for consultation and provided support as needed, for instance, negotiating with the local government or assisting with remedial action. A significant part of the information referred to in this paper was collected through such collaboration between the programme coordination desk and the project implementers.

In addition, the paper also refers to some of the ground-level initiatives which the SLE Programme studied through a scan of practices and policies for sustainable lifestyles during 2016–2017. More specifically, these cases were examined under the project entitled Envisioning Future Low-Carbon Lifestyles and Transitioning Instruments (2017–2019) that scanned transformational policies and ground-level initiatives globally through desk-based research and a call for case studies. The SLE Programme mostly learned from these ground-level initiatives through online interviews and also collaborated on a case study report published in 2019 [71]. Table 1 gives a summary of the open call for projects and scan of policies and instruments. See Appendix A for the list of projects and cases.

Table 1. Sustainable Lifestyles and Education (SLE) Programme's Open Call for Projects and Scan of Policies and Instruments.

	Open Call for Projects	Scan of Policies and Instruments
Selection	Four open calls for projects (2016–2019 24 projects selected out of more than 600 proposals	A call for case studies (2017) 30 cases of policies and civil society initiatives identified out of 120 submissions and desk-based research
Main criteria	Activities in developing countries and countries with emerging economies Completing in 12 to 24 months Addressing local challenges for sustainable lifestyles Contribution to greenhouse gas (GHG) emissions reduction	Transformational policies and instruments supporting pathways toward low-carbon and sustainable lifestyles Covering both developing and developed countries
SLE Programme's Collaboration	Financial support (USD 50,000 to 400,000) Online support (e.g., contract issues, specific skills and knowledge, elaboration of action plans, GHG monitoring) On-site support (e.g., attending training workshops, negotiation with local stakeholders)	Awarding of a few best cases Online interviews to capture the detail of the contexts, activities and outcomes Publication of a case study report with the implementers
Information source	Online and on-site consultation Project stories	Online interviews with project implementers Case study reports Desk-based research including peer-reviewed and grey literature

The authors of the paper were engaged in the process of selecting and supporting the projects. After the selection of the projects, they attempted to accompany the process of these projects in planning, implementing and monitoring, rather than conduct the summative evaluation with pre-designed criteria. This enabled the authors to elaborate the perspectives on the three critical points, i.e., (a) challenges of (un)sustainable living, (b) approaches taken to address the challenges and (c) learning obtained, through the collaboration and communication with the project implementers. The authors admit that a variety of conditions, such as the authors' positions as the programme coordinators, the capacities of the project implementing teams and the external conditions that affected the project implementation, made considerable influences of the following analysis. However, precisely because of such positions, the authors could come closer to the ground-level innovations in the making.

4. Results: Ground-Level Collective Actions for Sustainable Lifestyles

In what follows, we will look back on the challenges of sustainable living addressed, the approaches taken, and the insights obtained through collaboration with partners, based on the experience of our cooperation and communication with the projects.

4.1. Challenges of (Un)Sustainable Living

Twenty-four projects supported under the SLE Programme and 30 cases of initiatives we learned about through the call for submissions and desk-based research aimed at enabling sustainable lifestyles in one or more "domains" of living, such as energy consumption, water use, wasting and recycling, food production and consumption, housing (including heating and cooling houses), purchasing of consumer goods and livelihoods. However, we should pay attention to the fact that our behaviours and the surrounding contexts enabling and constraining them to comprise a mixture of conditions which cannot be separated into those "domains". On account of the crosscutting nature of lifestyles and the diversity of the purposes of the collective actions, we would need to pay more attention to the "why" of these actions. In other words, we would need to understand the challenges of the current patterns of living or the contexts associated.

All projects addressed the challenges of living or lifestyles which cause negative impacts on environmental sustainability. Among them, 17 of 24 projects dealt with the consequences caused by increased consumption and production in the context of rapid economic growth and urbanisation. This was not surprising considering the criteria of the calls for project proposals and case studies. The SLE Programme's calls for projects asked for submissions of proposals contributing to low-carbon lifestyles based on the local needs and opportunities. The call for case studies also requested the applicants address unsustainability issues in association with overconsumption. Projects thus aimed to reduce environmental impact through actions such as waste reduction and recycling, energy-saving and water-saving, more sustainable production of food or textiles and effective land management.

However, such negative impacts on the environment are not the only challenges for sustainable lifestyles. Even societies isolated from growth often suffer from negative environmental effects emerging in geographical, political or economic conditions. Furthermore, societies with economic growth, as well as those with a stagnant economy, may also suffer from a diversity of threats to stable livelihoods and consumption patterns. Many projects tried to create contexts of living whereby people could live more stable, secure lives, while at the same time mitigating negative environmental impacts. Detailed examples follow in the boxes below, starting with Box 1, which further illustrate the arguments made in this article.

Box 1. Case 1: Empowerment of Armenian Rural Community with Solar Power.

A project based in Armenian rural communities dealt with saving electricity and gas use in families and public facilities. The project team worked with community leaders and local governments in introducing some simple tools for utilizing abundant solar power to support local lives. However, the primary issue was not increased environmental impacts, but the energy mix of the local society, which was dependent on imported gas. While household income is limited due to economic stagnation, a hike in gas prices has put financial pressure on households and various local organisations, including schools and kindergartens which were forced to shut down in the winter. Additionally, because rural highland areas get dark early in the evening, people were reluctant to go out and join social activities. This indicates that increasing energy costs posed a threat to socioeconomic opportunities.

The project, thus, planned to introduce solar cookers and dry-fruit makers to families to reduce energy cost, solar water-heaters at schools and kindergarten to support them to operate in the winter and solar streetlights to enable people to participate in social activities safely. The project, however, needed to make a couple of modifications to their activities of installing solar tools. They found that the solar cookers for households were not practical in the winter due to low temperatures. They gave up the plan and instead established a large solar dry-fruit maker at the women's centre. This allowed the community people to organise training sessions for villagers as well as people from the neighbouring areas. They also received requests from the parents of the kindergarten kids where they installed a solar water heater. The instalment of the water heater was originally intended to reduce operational costs so that the kindergarten can open longer in the winter. However, the parents asked if the project can set up a warm swimming pool using the heated water. In such a way, the cancellation of the solar cooker, and the changes in the purpose of using warm water, enabled the project team and the participants to explore a broader range of opportunities for utilizing the solar power in improving their living conditions than initially planned.

Additionally, the project's proposal to set up the solar-photovoltaic-powered streetlights was not favourably accepted by the local government in the beginning. In the male-dominated rural societies in Armenia, the village government did not trust the project team comprised only of women. However, the repeated visits enabled them to gradually gain a shared understanding of the challenges of the community—due to the high altitude of the area, even the central sections of the village become dark at 6 pm, making villagers feel unsafe participating in social activities. The local government suggested the streetlights be installed in the street in front of the village's community sports centre so that more villagers can come out. Moreover, the government promised to secure the budget to expand the lighting after the project period.

In short, challenges of unsustainable lifestyles comprise two issues, namely, (a) an increase in the negative impacts of our behaviours on the environment, economy and society, and (b) the destabilisation or vulnerability of our lifestyles due to changes in environmental, economic and social conditions. Efforts towards supporting the shift to more sustainable lifestyles encompass the pursuit of situations where people can adopt responsible living that minimise the negative outgoing impacts and a reliable living which provides people with the capacity to prepare for, withstand and recover from external shocks and stresses. Importantly, these two elements are interconnected with each other in several ways. First, specific economic conditions, namely the dependency on a high-cost energy source, endangered both household economies and health, as well as limiting their participation in essential services, thus, leading to the high impacts, are shown in a few cases such as Armenia. Second, instability of livelihoods makes it more difficult for the local people to choose alternative options for livelihoods or participate in collective actions to conserve and improve their natural environment. Third, increasing overconsumption or the sharp rise in demand for goods or services among specific groups in society often caused insecurity or instability for other vulnerable groups. For instance, the rapid growth of the tourism sector in Da Nang, Vietnam caused a sharp rise in water demand in coastal resort development zones (Box 2). This resulted in an increasingly unstable water supply for the residents living in inland areas. Both those who have lived in the city for many years and the newcomers who tend to live in the hilly zones are equally affected.

Box 2. Case 2: Water-Smart Lifestyles in a Growing City of Vietnam.

A project in Vietnam worked on the increasing demand for water in the city of Da Nang. The expanding tourism sector in Da Nang has driven economic growth, as well as a construction boom in the seafront area, with other areas of the city also seeing a population influx. This has pushed water demand close to the limit of the current water supply system. As a result, a significant number of hotels have opened in coastal areas and require an enormous amount of water to provide services to their guests. This leap in water demand in the tourism sector destabilised the water supply to citizens, including those living in high-altitude areas. Furthermore, the increasing tendency of severe flooding and storms pushed the city to develop a more stable water system with both supply-side and demand-side measures.

The project organised a kick-off workshop in October 2017, inviting various participants concerned in the issue of water security. Many participants pointed out that it would be most useful to provide active learning programmes for younger children to support the families in changing water-using behaviours since families are positive in taking up new knowledge their children obtained. The project, thus, built a partnership with schools and kindergartens to prepare and conduct relevant programmes. Later on, the learning programmes evolved into a city-wide campaign of water-saving.

Note, however, that their understanding of the current "challenges" of unsustainable living is subject to change over time. As we will see in the later sections, they often face unforeseen conditions during the implementation phase. At such occasions, they reconsider their actions and roles for changing the contexts with a clearer understanding of the background causes of these challenges.

4.2. Approaches Taken—To Unlock the "Web"

Projects took actions combining some of the activities as follows to address these two-fold challenges of unsustainable living and the conditions.

4.2.1. Visualisation of the Impacts of the Current Patterns of Living and Benefits of Alternatives

Firstly, almost all the projects worked to take those living conditions or associated impacts which were not visible to the stakeholders, and make them visible. Once people have a clear vision of what they are doing currently, how much they pay and what impacts these behaviours will cause, they may have the chance to reconsider or even redirect some of their behaviours. Thus, visualisation works as the first step to create an alternative connection between people, resources or practices that may stabilise their living conditions with fewer negative impacts. One of the projects in Thailand, for instance, developed a database of the energy use in urban households and provided families with a home energy audit clarifying the status of energy use in comparison with their neighbours, with clear suggestions on no- or low-cost measures for energy saving. Another project in India launched "Food-Info-Marts" where urban food consumers purchase organic produce grown in the surrounding farm areas and also obtain knowledge about healthy diets. A third project in Ballina, Ireland (Box 3) measured household ecological footprints and worked with them to develop means of lowering their impacts, through storytelling.

Box 3. Case 3: Community Footprinting in Ballina, Ireland.

Working with small community groups in a small town in Ireland, this project managed to achieve reductions of 28% in recipients' ecological footprints. This was achieved through accurate footprinting so that recipients understood their environmental impact and storytelling, which included both short case studies developed by the participants themselves, and as well as slogans created through local competitions in schools to reinterpret technical messages. Through providing the necessary tools and placing the recipients at the centre of the project through the co-creation of footprint reduction activities and asking them to co-create messages to encourage others to join, the project was able to achieve high levels of engagement and maintain interest over the multi-year implementation time span. The project was rooted in what was tangible and interesting for the participants by focusing on transport, energy, water and waste. This participant-centred approach demonstrates the impact that can be achieved through co-development and small collective actions.

4.2.2. Introduction of Tools/Facilities That Enables People to Connect to "Elements" Differently

Visualisation of the current patterns could be more supportive for people in creating alternative practices when specific suggestions or guidance follows, or when people can take part in collective actions for experimenting with alternative behaviours. Some projects introduced tools, equipment, or facilities that enable changes to the current patterns of behaviour that were visualised. Introduction of small tools and more extensive facilities were carried out on many projects. For instance, the project in Vietnam introduced information and guidance on water-saving measures to selected households in the city. Following that, the project also installed several simple tools supporting water-saving actions, such as water-saving taps in the pilot households. The tools made it easier for people to take concrete steps in changing their water-using behaviours. The project in Armenia is another case, where it introduced cheap tools to turn sunshine into energy, and provided other benefits for local living. Another project in Chile established two facilities utilising geothermal energy, namely, a firewood drier and a greenhouse producing leafy vegetables. These facilities brought together local farmers in the experimental production of firewood and vegetables.

Such actions have several effects. Physical tools work as the interface between individuals or families and those resources that can meet their needs. The tools provided in the aforementioned cases in Armenia and Vietnam enabled people to adopt cleaner ways of using water, food or energy, or even turn something they have wasted into a useful product. Information tools, on the other hand, connect people with different knowledge and resources through the facilitated exchange of information on, for instance, surplus food, available water or recyclable paper, and enable them to avoid waste, reduce costs and gain additional income.

4.2.3. Setting up of Spaces for Collaboration and Co-Creation of Knowledge

Furthermore, tools and facilities can be even more effective when they are managed in cooperation with local people and organisations. Tools or facilities can work for local people as a centre for knowledge creation through which they can share about their daily experiences, challenges or concerns, and find excellent opportunities. This then contributes to the development of capacity and aspiration of the local people in collectively creating alternative ways of living.

Projects set up a variety of spaces for collaboration and co-creation of knowledge. Some of them formed groups of participants who seek productive usages of physical tools or facilities. In contrast, others took advantage of the existing organisations of sites such as schools, kindergartens, offices or housewife groups, such as the Green Office project in Viet Nam (Box 4). They often provided training sessions to these groups or encouraged them to take collective action to make tangible changes in their particular contexts, such as the workplace. Some projects introduced physical (e.g., experimental fields) or virtual spaces (e.g., mobile apps) where people can meet and exchange their different needs, offers, skills and knowledge.

Box 4. Case 4: Upscale and Mainstream Green Office (GO) Lifestyles in Vietnam.

With Vietnam undergoing rapid urbanisation and economic growth, the country is seeing the emergence of a white-collar middle class. Faced with a deteriorating environment, many workers and companies are becoming increasingly interested in sustainability. In response to this, the Green Office project was established by AITVN under the One-Planet Network in order to develop green offices in key consumption areas such as energy, water, waste, paper and office equipment leading to reductions of 20% in CO2 emissions. To support this, a Green Office (GO) Lifestyle toolkit and GO Standards for a public audience were developed. The project was run in a total of ten offices across three cities—Hanoi, Ho Chi Minh City and Da Nang. Following implementation, the average emissions per capita was reduced by 8.9%, with the highest reduction being 22.5%. Over a thousand people were introduced to the Green Office approach, and the development of the toolkit and standards has ensured the sustainability of the project in the longer term.

Combining these activities is useful in producing alternative sets of competence, materials and meanings. Competence here is not limited to one's ability to do something one is taught in the training workshop. It also includes the ability to create alternative connections between oneself with materials—such as water, sunlight, waste and food to bring out their values to sustain the living or improve the living conditions. It also covers the competence to communicate, negotiate, and collaborate with other people—family members, neighbours, colleagues, public or private organisations—in creating the local socioeconomic contexts where they can continuously engage in the creation of new values. Thus, these approaches were not just the replacement of the parts of the systems. The above methods of visualisation, introduction of tools and setting up of spaces for co-creation fostered the creation of alternative connections among the elements of the systems such as people, material and skills and created new competences and meanings.

4.3. Learning by Doing—How Actors Grew through Moving forward the Innovations in the Making

Even if a project addresses a definite challenge and identifies the practical approaches to address it, it is still not possible to create a perfect implementation plan which anticipates everything. Thus, it is crucial to understand at which opportunities the actors are urged to reconsider their actions, and how they are informed to review their activities or purposes.

4.3.1. Learning Opportunities

Projects were often urged to reconsider the contexts and change some of the planned actions.

- Difficulty in building relationships with partners and participants: Projects were able to smoothly build cooperative relationships with partners when they were already well aware of the necessity to change current living patterns. However, if this was not the case, projects found it hard to find participants and identify and build partnerships with the key organisations to collaborate with and gaining active participation of the local people.
- Compatibility of the knowledge, skills and tools with the local cultures or environments: Following the identification of key partners, the projects took further steps to organise participants and initiate training courses or introduce tools. At this stage, projects often had difficulty in delivering the knowledge and skills of the participants or supporting them in using the tools introduced. Sometimes projects discovered that the tools or knowledge they tried to apply are not useful in the local context.
- Unforeseen changes in the external conditions: For projects required to generate outcomes in only a few years, sudden changes in external conditions can be a severe challenge jeopardising plans. The activities in the project in Chile were delayed due to the changes in the local government during the early stages of the project. The project in Zimbabwe with farmers' organisations had to cancel some of the activities due to a drought, currency crisis and cholera outbreak.

4.3.2. Reflections on Lessons Learned

These challenges enabled them to review and modify their actions, partnerships and even the goals. They adjusted, for instance, the contents of the training programmes, types of tools and timings and locations of activities to fit the local contexts or unforeseen situation. Many of them carried out modifications more than once.

In some cases, the project team reconsiders the activities with a recognition that they need to take some additional measures to achieve their intended outcomes. The additional measures would be, for instance, the introduction of different tools or skills; the organisation of participants for growing their skills and motivation or the introduction of policies, infrastructure, or education programmes that would enable conditions for behaviour changes.

Such additional measures often required the project team to collaborate differently with partners. The case in Armenia shows how project partners, such as the local government and parents, shifted their ways to engage with the project from the potential beneficiaries into the collaborators. The water-saving project in Da Nang, Vietnam added new activities of the active learning programmes in collaboration with additional partners, namely schools and kindergartens. Likewise, a few projects engaged with policymakers to deliver policy recommendations out of their actions in search of more substantial support to ensure continued benefits after the completion of the project period. Moreover, participants of a few projects redesigned their group structures during the implementation period to make the best use of the enthusiasm or advanced skills of some of the key persons of the local society in growing the capacities and aspirations of the local participants such as farmers' clubs in Zimbabwe (Box 5). Enthusiastic or skilled "participants" become "trainers", "masters" or "leaders" in the later stage of the project implementation and play crucial roles in creating and spreading the skills they developed, marketing their products for securing their livelihoods and connecting with additional partners. Through these "additional" actions, participants developed their capacities and assumed more essential roles in causing the changes.

Box 5. Case 5: Farmers' Clubs in Zimbabwe.

<table><tr><td>

The project in Zimbabwe formed farmers' groups and established experimental fields where the groups provided knowledge and skills for conservation agriculture, horticulture and small livestock production. From the previous implementation in other areas, the project team identified these means as measures for helping farmers mitigate their impacts on the natural environment and, at the same time, become resilient to external shocks such as flood and droughts, or an economic crisis. In addition to training on production, the project also covered nutrition, sanitation and health and marketing to strengthen farming households and communities in a holistic manner.

The project started amid the currency fluctuation that limited farmers' capacity to purchase necessary farm inputs. Participants saw that they could stabilise their household economy with small additional incomes from these activities. However, the project team and participants gave up one of their planned activities of establishing communal saving as it was challenging to develop and manage it stably due to continuous currency risk. Another unforeseen crisis hit the project when a cholera outbreak occurred in the region. During this crisis, participating farmers moved quickly to visit neighbouring areas to disseminate what they learned about health and sanitation as well as the importance of having stable livelihoods. Through these unforeseen events, the farmers grew their capacity to play a more significant role in the local society in creating resilient living conditions.

</td></tr></table>

Unforeseen situations often hinder the smooth implementation of the project activities as planned. However, they also give them opportunities for the project teams and partners in reconsidering their roles in creating the alternative contexts of living and grow their competencies to more proactively move forward. Furthermore, with more proactive roles and competencies, project teams and partners gain deeper considerations of the purposes—what they are collaborating on, creating alternative contexts of living. Thus, ground-level transitions in the making does not mean that they continuously modify their activity plans toward the pre-determined goals. The purposes of the transitions or the visions of the desired future lifestyles or their contexts continue to evolve.

5. Discussion and Conclusions

Thus far, we have analysed some of the common points learned from the small- and micro-scale collaborative actions which the Sustainable Lifestyles and Education Programme of the One-Planet Network has collaborated on from 2016 to 2019. We have learned that the efforts to achieve sustainable lifestyles are, in a nutshell, collective actions of creating the contexts where local people can achieve more responsible and reliable living. Local initiatives have addressed a wide variety of locally specific challenges, namely, the increasing negative impacts associated with growing consumption, and unstable livelihoods and consumption. These two challenges are deeply entangled, as was shown from the above-cited cases 1 and 2. Thus, most of them needed to aim for responsible and reliable living at the same time. To this end, they visualised the current status of living and the associated impacts as illustrated in cases 2 and 3, introduced physical tools or facilities to better utilise the locally available resources in meeting the day-to-day needs (cases 1, 2, 4) and set up spaces for collaboration where local actors learn from each other (cases 1–5) to create alternative connections between human actors and material objects, or individuals and organisations. However, we have also seen that none of these collective actions went as planned despite detailed preparation. They faced difficulties in establishing partnerships (cases 1, 4), applying knowledge and skills (cases 1, 3) and unforeseen events caused by political and economic conditions or natural disasters (cases 1, 5). Facing such challenges, local actors modified their plans. In doing so, they gained a deeper consideration of the drivers of their unsustainable conditions, what kind of alternatives they envision and what roles they can play in their efforts to achieve them. In other words, local actors co-create competencies and the meaning of (un)sustainable lifestyles in collaborative spaces.

The above lessons indicate that we would need to reconsider the meanings and conditions of systemic changes that enable sustainable lifestyles. Sustainable lifestyles at the local level will require changes to the systems enabling and constraining the local living contexts. However, such systemic changes are not realised only by replacing a few of the parts or elements to "make it work with less waste" or to "teach people to make a wise choice." We have seen that, for instance, visualisation of the negative impacts of living and introduction of the tools or facilities to enable people to reduce energy, water or waste may work when local actors play active roles in such visualisation or introduction/operation of the tools. The systems [11,12,44] or entanglements [15,16] where our lifestyles are situated is not about engineering elements of materials and tools that would potentially contribute to meeting our needs with "reduced negative impacts". They comprise materials and tools, knowledge and skills to utilise them and meanings or aspirations of people in creating positive relations with knowledge, skills, tools and materials. Likewise, systemic changes are not about replacing some of its parts but are the process where actors explore the potentials of alternative ways of living, and the roles they can play in pursuit of such alternatives. The transition to sustainable lifestyles is not a direct and one-time shift from current unsustainable patterns to pre-defined sustainable patterns. A change in lifestyles needs systemic changes which do not take place somewhere distant from individuals or groups of people but arise as the growth of skills, tools, intentions, competences and aspirations through the collective efforts in response to locally specific challenges. It is more of a collaborative and continuous exploration [54–56,59,70] into the locally specific meaning of (un)sustainable living conditions and creation of competences that connect people and people or people and knowledge, skills, tools and materials in different ways.

With the above points in mind, we would also need to reconsider the ways in which governments, business and civil society can support or promote the shift in lifestyles. A broad range of policy measures and business models are already tested and are contributing to the uptake of more sustainable lifestyles. The business sector is, for example, providing information on the negative impacts associated with the production and use of specific goods and services so that consumers can make wiser choices or introduce products or services that enable consumers to meet their needs with fewer negative impacts. Governments are supporting such measures through, for instance, offering economic incentives, setting up standards of goods and services, criteria for information provision or providing certificates [71]. Such policies could potentially be more effective when they can encourage proactive interpretation and adaptation by local actors toward their collective efforts in addressing particular issues of the local living conditions. By way of an illustration, the business sector can set up a "laboratory" type of collaborative space together with local businesses or citizens' groups for developing and delivering innovative goods or services that reduce negative impacts and at the same time generating income opportunities for the local society [72,73]. By doing so, they do not only contribute to responsible and reliable living but also enlarge the prospects for further innovations driven by the local actors.

Moreover, we should also pay attention to the dynamic natures of innovations of lifestyles and living contexts in the making, instead of trying to make a one-time shift to the pre-determined goal. Monitoring and evaluation of the progress of activities "as planned" is necessary but not sufficient for supporting dynamic collaboration. Monitoring activities can be counter-productive when they are done to make sure the 100% achievements of the planned objectives, since they may kill the chances for the participants of learning from the reality to develop their competencies and to generate their own unique meaning of sustainable living. To support the exploration of the local actors, we—as governments, civil society actors or researchers—would need to accompany them in responding to unforeseen challenges and eventually creating their competencies for and meanings of sustainable living. Recently, Developmental Evaluation has gained traction with a similar line of thinking [74–76]. We would further need to elaborate such innovative ways of supporting the ground-level collaborative actions, taking account of the dynamics of the conditions of activities, roles of the stakeholders and the goals of the initiatives.

This paper aims to propose a deeper understanding of how people engage in the "systemic changes" of lifestyles, building on the research on lifestyles that focus on how lifestyles are entangled in the systems of provision comprising various elements. The work draws on the results of the work undertaken as part of the SLE Programme. The analysis is, therefore, post-hoc and was developed from discussions regarding the results and the commonalities that were found between projects that were very culturally and geographically diverse. This poses two specific limitations: (a) the analytical approach used by this paper was not integrated into the projects and the project implementers did not consider this specific approach when implementing their projects, and (b) the analysis here does not pay sufficient attention to the pathways of the projects and participants after their initial period. In order to deepen understanding of the transition to sustainable lifestyles and the ways in which stakeholders interact with each other and alter practices, it would be necessary to improve our ways of working with ground-level innovations, integrating this approach into project planning and collaborating for longer years. As the future development of the Global South will be critical to whether humanity is able to avoid the worst effects of climate change, a greater understanding of the means by which we can learn to live well within planetary boundaries will be needed.

More than 30 years have passed since international society recognised the necessity to shape more sustainable lifestyles. A multitude of trials for shifting lifestyles exist globally and aim for alternative ways of achieving wellbeing with fewer resources and energy, giving more and more insight into the need to and possibilities of shaping alternative ways of living. However, we need to bear in mind that the potentials of such trials are not limited to their achievements of pre-determined goals, such as reduced material or energy use, through the one-shot actions of tweaking one or two elements of lifestyles. Instead, they need to be fostered as inseparable steps of endless co-creation of new meanings of alternative ways of living and competence to realise them. We, thus, need to continue our exploration to better learning from and collaborating with these ground-level innovations.

Author Contributions: The concept and structure of this article was developed jointly by the authors. A.W. is the lead author, and developed the initial draft, to which S.G. added case studies. Final development of the paper was then conducted jointly by the authors. All authors have read and agreed to the published version of the manuscript.

Funding: The development of the paper is supported by the Policy Design and Evaluation to Ensure Sustainable Consumption and Production Patterns in Asian Region (PECoP-Asia) project funded by the Environmental Research and Technology Development Fund, Japan.

Acknowledgments: The authors are thankful for the implementers of the initiatives under the Sustainable Lifestyles and Education Programme of the One-Planet Network (10-Year Framework of Programmes on Sustainable Consumption and Production Patterns).

Conflicts of Interest: The authors have been involved in the processes of evaluating and supporting the 24 projects under the Sustainable Lifestyles and Education Programme and the selection of case studies cited in the paper.

Appendix A

Table A1. Projects selected under the Open Call for Projects and Policies and Instruments Studied.

Open Call for Projects	Scan of Policies and Instruments
Africa	
Promoting Environmental Practices through Music, Cameroon	Car Free Day Marrakech, Morocco
Sustainable Lifestyles among Rural Families in Zimbabwe: Small-scale Conservation Farming to Change Lifestyles in Africa and Beyond, Zimbabwe	Green Belt Movement, Kenya
Food Waste in South Africa: Capacity Building through Research, and Trial of a Cellular Phone Application, to Reduce On-farm Food Waste and Increase Food Redistribution (Food for Us Project), South Africa	iShack, South Africa
Showing the Sustainable Lifestyle Behaviour and Technologies for Efficient Households, Zambia	Support for Women in Agriculture and Environment (SWAGEN), Uganda
Polycentric Infrastructure and Community Development Paradigm for Sustainable Urban Transitions (PICD-SUT), Malawi	
Establishing a South Africa Plastics Pact (small-scale)	
ASIA & THE PACIFIC	
SCRIPT (Sustainable Consumption and Recycling Interventions for Paper and Textiles) for Reducing Urban Climate Footprints, India and Bangladesh	Energy Efficiency in the Indian Building Sector
Upscaling Sound Food Waste Management Practices through Youth and Community Education in Schools and Institute of Higher Learning, Malaysia	Fifth Environmental Basic Plan, Japan
Upscale and Mainstream Green Office Lifestyles in Vietnam	Gross National Happiness, Bhutan
A New Approach of Reducing Greenhouse Gas (GHG) Emissions through Changing Lifestyles toward Water and Electricity-saving in Urban Households in Da Nang, Vietnam	Indonesian City Walkability (Bandung and Bogor)
Asia Pacific Low-Carbon Lifestyles Challenge	MUNI Meetups, Philippines
Promoting Household Energy Conservation through Feedback Services and Home Energy Audit on Residential Sustainable Lifestyle Programs, Thailand	National Work-Life Balance Policy, South Korea
Sustainable Urban Food Production and Connected Ecological Rural Farming for Reducing Climate and Environmental Impacts of Food Demand, India	Oki Town, Japan
Sustainable livelihoods within sustainable landscapes in Papua New Guinea	Reverse Migration, India
Active City-Community Engagement to Leverage Emissions Reduction through Activities that Transform Energy-use (ACCELERATE), Philippines	San Carlos City, Philippines
	Shu Shi (WildAid), China
	Sino-Singapore Tianjin Eco-City, China/Singapore
	Using Local Resources in Building Construction, India
	Zero Waste Activities, Dumaguete, Philippines

Table A1. *Cont.*

Open Call for Projects	Scan of Policies and Instruments
North and Latin America and The Caribbean	
Education for Sustainability and Consumption, Brazil	Costa Rica's Biodiversity Law
Direct Use of Geothermal Energy for the Promotion of Sustainable Production Model in Rural Areas in Chile: Implementation of Pilot Projects in Firewood Drying and Greenhouse for Agricultural Farming, Chile	Edukatu, Brazil
How Emerging Urban Youth can be an Engine for More Low-carbon, Sustainable Lifestyles: Beginning in Bogota, Colombia	Montréal's Multi-Model Transportation Mix, Canada
Better by Design—Replicating Promising Practices, Tools and Methodologies to Support and Enable Companies in Latin America to Improve the Sustainability of their Food and Beverage Products, Peru, Nicaragua and Honduras	Patagonia, United States
Sustainable Lifestyles in the Workplace, Morocco & Colombia	Rizoma Field School, Uruguay
Solar energy for improved rural livelihoods in Peru	Sidewalk Toronto, Canada
The recovery of traditional rice and wheat cultivation for food sovereignty in integrated agroecological production systems, Colombia	
Europe	
Solar Energy for Low-Carbon Sustainable Lifestyles in Solak, Aygavan and Malishka Rural Communities of Armenia	Ballina Eco District, Ireland
Encouraging young specialists to power the agri-food value chains and building sustainable business models, Armenia and Chile	BioSzentandás, Hungary Incredible Edible Todmorden, UK
	London Bans Junk Food Ads on the Transport for London Network, UK National Consumption Strategy, Sweden National Loneliness Policy, UK Nudge in a Green Direction, Belgium Ruby Cup, UK Sieben Linden Ecovillage, Germany

References

1. Institute for Global Environmental Strategies. *The 10YFP Programme on Sustainable Lifestyles and Education*; Institute for Global Environmental Strategies: Hayama, Japan, 2017; pp. 1–5.
2. Del Pino, S.P.; Metzger, E.; Drew, D.; Moss, K. *The Elephant in the Boardroom: Why Unchecked Consumption Is Not an Option in Tomorrow's Markets*; World Resources Institute: Washington, DC, USA, 2017; pp. 1–36.
3. Scott, K. A literature review on sustainable lifestyles and recommendations for further research. *Stock. Environ. Inst. Proj. Rep.* **2009**, *42*. [CrossRef]
4. Lorek, S.; Spangenberg, J.H. Sustainable consumption within a sustainable economy-Beyond green growth and green economies. *J. Clean. Prod.* **2014**, *63*, 33–44. [CrossRef]
5. Meroni, A.; Corubolo, M.; Piredda, F.; Jegou, F.; Schäfer, B.; Salvador, G. Exploring Pathways towards Sustainable Lifestyles. In Proceedings of the Global Research Forum on Sustainable Consumption and Production Workshop, Rio de Janeiro, Brasil, 13–15 June 2012; pp. 1–15.
6. Moloney, S.; Strengers, Y. "Going Green"? The limitations of behaviour change programs as a policy response to escalating resource consumption. *Environ. Policy Gov.* **2014**, *24*, 94–107. [CrossRef]
7. Gatersleben, B.; White, E.; Abrahamse, W.; Jackson, T.; Uzzell, D. Values and sustainable lifestyles. *Arch. Sci. Rev.* **2010**, *53*, 37–50. [CrossRef]
8. Barr, S.; Shaw, G.; Coles, T. Sustainable lifestyles: Sites, practices, and policy. *Environ. Plan. A* **2011**, *43*, 3011–3029. [CrossRef]

9. Rakic, M.; Rakic, B. Sustainable lifestyle marketing of individuals: The base of sustainability. *Amfiteatru Econ.* **2015**, *17*, 891–908.

10. Cohen, S. Understanding the Sustainable Lifestyle. Available online: https://www.europeanfinancialreview. com/understanding-the-sustainable-lifestyle/ (accessed on 8 September 2020).

11. Evans, D.; Abrahamse, W. Beyond rhetoric: The possibilities of and for "sustainable lifestyles". *Environ. Polit.* **2009**, *18*, 486–502. [CrossRef]

12. Spaargaren, G.; Oosterveer, P. Life(style) Politics for Sustainable Consumption Analysing the Role of Citizen-Consumers in Global Environmental Change. In Proceedings of the European-American workshop on "Climate Change Mitigation: Considering Lifestyle Options in Europe and the US", Berkeley, CA, USA, 1 May 2009; pp. 1–20.

13. Spaargaren, G. Theories of practices: Agency, technology, and culture. Exploring the relevance of practice theories for the governance of sustainable consumption practices in the new world-order. *Glob. Environ. Chang.* **2011**, *21*, 813–822. [CrossRef]

14. Welch, D.; Warde, A. Theories of practice and sustainable consumption. *Handb. Res. Sustain. Consum.* **2014**, 84–100. [CrossRef]

15. Dijk, M.; Backhaus, J.; Wieser, H.; Kemp, R. Policies tackling the "web of constraints" on resource efficient practices: The case of mobility. *Sustain. Sci. Pract. Policy* **2019**, *15*, 62–81. [CrossRef]

16. Backhaus, J.; Wieser, H.; Kemp, R. Disentangling practices, carriers, and production–consumption. In *Putting Sustainability into Practice: Applications and Advances in Research on Sustainable Consumption*; Kennedy, E.H., Cohen, M.J., Krogman, N.T., Eds.; Edward Elgar Publishing: Cheltenham, UK, 2015; pp. 109–133. ISBN 9781784710590.

17. United Nations Environment Programme. *Visions for Change Recommendations for Effective Policies on Sustainable Lifestyles*; United Nations Environment Programme: Paris, France, 2011.

18. United Nations Environment Programme. *Decoupling Natural Resource Use and Environmental Impacts from Economic Growth Using Less Resources and Reducing the Environmental Impact*; United Nations Environment Programme: Paris, France, 2011; ISBN 9789280731675.

19. Jackson, T. The challenge of sustainable lifestyles. In *2008 State of the World*; Worldwatch: Washington, DC, USA, 2008; pp. 45–60.

20. Spaargaren, G.; Vliet, B. Van Lifestyles, consumption and the environment: The ecological modernization of domestic consumption. *Environ. Polit.* **2000**, *9*, 50–76. [CrossRef]

21. Evans, D.; Jackson, T. *Towards a Sociology of Sustainable Lifestyles*; Resolve Working Paper; University of Surrey: Guildford, UK, 2007; pp. 1–24.

22. Holt, D.B. Poststructuralist lifestyle analysis: Conceptualizing the social patterning of consumption in postmodernity. *J. Consum. Res.* **1997**, *23*, 326. [CrossRef]

23. Bourdieu, P. *Distinction-A Social Critique of the Judgement of Taste*; Harvard University Press: Cambridge, MA, USA, 1984; ISBN 0674212770.

24. Slater, D. *Consumer Culture and Modernity*; Polity Press: Cambridge, UK, 1997.

25. Giddens, A. *Modernity and Self-Identity Self and Society in the Late Modern Age*; Stanford University Press: Stanford, CA, USA, 1991.

26. Roberts, K. *Leisure in Contemporary Society*; CABI Publishing: Wallingford, UK, 1999.

27. Veal, A.J. Leisure, Culture and LifestyleLoisir, culture et mode de vieDiversión, cultura y estilo de vida. *Loisir Soc.* **2012**, *24*, 359. [CrossRef]

28. Petev, I.D. The association of social class and lifestyles: Persistence in American sociability, 1974 to 2010. *Am. Sociol. Rev.* **2013**, *78*, 633–661. [CrossRef]

29. Tomlinson, M. Lifestyle and social class. *Eur. Sociol. Rev.* **2003**, *19*, 97–111. [CrossRef]

30. Christiansen, C.H.; Matuska, K.M. Lifestyle balance: A review of concepts and research. *J. Occup. Sci.* **2006**, *13*, 49–61. [CrossRef]

31. Tabish, S.A. Lifestyle diseases: Consequences, characteristics, causes and control. *Cardiol. Curr. Res.* **2017**, *9*, 00326. [CrossRef]

32. Lawson, R.; Todd, S. Sumer lifestyles: A social stratification perspective. *Mark. Theory* **2002**, *2*, 295–307. [CrossRef]

33. Keister, L.A.; Benton, R.; Moody, J. Lifestyles through expenditures: A case-based approach to saving. *Sociol. Sci.* **2016**, *3*, 650–684. [CrossRef]

34. Sathish, S.; Rajamohan, A. Consumer behaviour and lifestyle marketing. *Int. J. Mark. Financ. Serv. Manag. Res.* **2012**, *1*, 152–166.

35. Terlau, W.; Hirsch, D. SUSTAINABLE consumption and the attitude-behaviour-gap phenomenon-causes and measurements towards a sustainable development. *J. Food Syst. Dyn.* **2015**, *6*, 159–174.

36. Vermeir, I.; Verbeke, W. Sustainable food consumption: Exploring the consumer "attitude–behavioral intention" gap. *J. Agric. Environ. Ethics* **2006**, *19*, 169–194. [CrossRef]

37. Sniehotta, F.F. Bridging the intention–behaviour gap: Planning, self-efficacy, and action control in the adoption and maintenance of physical exercise. *Psychol. Health* **2007**, *20*, 143–160. [CrossRef]

38. Carrington, M.J.; Neville, B.A.; Whitwell, G.J. Lost in translation: Exploring the ethical consumer intention–behavior gap. *J. Bus. Res.* **2014**, *67*, 2759–2767. [CrossRef]

39. Boulstridge, E.; Carrigan, M. Do consumers really care about corporate responsibility? Highlighting the attitude-behaviour gap. *J. Commun. Manag.* **2000**, *4*, 355–368. [CrossRef]

40. Shaw, D.; McMaster, R.; Newholm, T. Care and commitment in ethical consumption: An exploration of the 'attitude–behaviour gap'. *J. Bus. Ethics* **2016**, *136*, 251–265. [CrossRef]

41. Welch, D. Behaviour change and theories of practice: Contributions, limitations and developments. *Soc. Bus.* **2017**, *7*, 241–261. [CrossRef]

42. Shove, E. Beyond ABC: Climate change policy and theories of social change. *Environ. Plan. A* **2010**, *42*, 1273–1285. [CrossRef]

43. Schatzki, T.R.; Knorr-Cetina, K.; von Savigny, E. *The Practice Turn in Contemporary Theory*; Routledge: London, UK, 2001; ISBN 0415228131.

44. Reckwitz, A. Toward a theory of social practices: A development in culturalist theorizing. *Eur. J. Soc. Theory* **2002**, *5*, 243–263. [CrossRef]

45. Warde, A. Consumption and theories of practice. *J. Consum. Cult.* **2005**, *5*, 131–153. [CrossRef]

46. Ropke, I. Theories of practice-New inspiration for ecological economic studies on consumption. *Ecol. Econ.* **2009**, *68*, 2490–2497. [CrossRef]

47. Shove, E. *Stuff, image and skill: Towards an integrative theory of practice. Designing and Consuming Objects, Practices and Processes Workshop*; Durham University: Durham, UK, 11 January 2005.

48. Sahakian, M.; Wilhite, H. Making practice theory practicable: Towards more sustainable forms of consumption. *J. Consum. Cult.* **2014**, *14*, 25–44. [CrossRef]

49. Shove, E.; Walker, G. Governing transitions in the sustainability of everyday life. *Res. Policy* **2010**, *39*, 471–476. [CrossRef]

50. Hargreaves, T.; Longhurst, N.; Seyfang, G. *Understanding Sustainability Innovations: Points of Intersection between the Multi-Level Perspective and Social Practice Theory*; 3S Working Paper 2012-03; University of East Anglia: Norwich, UK, 2012; pp. 1–25.

51. Burningham, K.; Venn, S. Are lifecourse transitions opportunities for moving to more sustainable consumption? *J. Consum. Cult.* **2017**, *20*, 102–121. [CrossRef]

52. Lorenzen, J.A. *Going Green: The Process of Lifestyle Change*; Lorenzen, A.L., Ed.; Wiley Stable: Hoboken, NJ, USA, 2012; Volume 27, pp. 94–116. Available online: http://www.jstor.org/stable/41330915 (accessed on 8 September 2020).

53. Spangenberg, J.H.; Lorek, S. Sufficiency and consumer behaviour: From theory to policy. *Energy Policy* **2019**, *129*, 1070–1079. [CrossRef]

54. Anantharman, M. Critical sustainable consumption: A research agenda. *J. Environ. Stud. Sci.* **2018**, *8*, 553–561. [CrossRef]

55. O'Neill, K.J.; Clear, A.K.; Friday, A.; Hazas, M. "Fractures" in food practices: Exploring transitions towards sustainable food. *Agric. Hum. Values* **2019**, *36*, 225–239. [CrossRef]

56. Seyfang, G.; Haxeltine, A.; Hargreaves, T.; Longhurst, N. *Energy and Communities in Transition-towards a New Research Agenda on Agency and Civil Society in Sustainability Transitions*; Working Paper; Center for Social Economic Research on the Global Environment, University of East Anglia: Norwich, UK, 2010; pp. 1–21.

57. Seyfang, G.; Longhurst, N. Desperately seeking niches: Grassroots innovations and niche development in the community currency field. *Glob. Environ. Chang.* **2013**, *23*, 881–891. [CrossRef]

58. Smith, A.; Raven, R. What is protective space? Reconsidering niches in transitions to sustainability. *Res. Policy* **2012**, *41*, 1025–1036. [CrossRef]

59. Hagbert, P.; Bradley, K. Transitions on the home front: A story of sustainable living beyond eco-efficiency. *Energy Res. Soc. Sci.* **2017**, *31*, 240–248. [CrossRef]

60. Marchand, A.; Walker, S. Insights for sustainable lifestyles and communities: Ecocitizens as key informants. Proceedings of International Conference on Adequate & Affordable Housing for All, Toronto, ON, Canada, June 24–27 2004; pp. 1–7.

61. Miller, E.; Bentley, K. Leading a sustainable lifestyle in a 'non-sustainable world': Reflections from Australian eco-village and suburban residents. *J. Educ. Sustain. Dev.* **2012**, *6*, 139–149. [CrossRef]

62. Anantharaman, M. When Do Consumers Become Citizens? In Proceedings of the Global Research Forum on Sustainable Consumption and Production Working paper, Rio Janiero, Brazil, 13–15 June 2012; pp. 1–18.

63. Kennedy, E.H. Rethinking ecological citizenship: The role of neighbourhood networks in cultural change. *Environ. Polit.* **2011**, *20*, 843–860. [CrossRef]

64. Wolfram, M. Cities shaping grassroots niches for sustainability transitions: Conceptual reflections and an exploratory case study. *J. Clean. Prod.* **2016**. [CrossRef]

65. Bulkeley, H.; Castan Broto, V.; Maassen, A. Low-carbon transitions and the reconfiguration of urban infrastructure. *Urban Stud.* **2014**, *51*, 1471–1486. [CrossRef]

66. Baker, S.; Mehmood, A. Social innovation and the governance of sustainable places. *Local Environ.* **2015**, *20*, 321–334. [CrossRef]

67. Rogers, Z.; Bragg, E. The power of connection: Sustainable lifestyles and sense of place. *Ecopsychology* **2012**, *4*, 307–318. [CrossRef]

68. Winkler, B.; Maier, A.; Lewandowski, I. Urban gardening in Germany: Cultivating a sustainable lifestyle for the societal transition to a bioeconomy. *Sustainability* **2019**, *11*, 801. [CrossRef]

69. Wieser, H.; Backhaus, J.; Kemp, R. Exploring Consistencies and Spillovers across Food Shopping Practices. In Proceedings of the 5th International Sustainability Transitions Conference, Utrecht, The Netherlands, 26–29 August 2014; pp. 1–26.

70. Chilvers, J.; Longhurst, N. Participation in transition(s): Reconceiving public engagements in energy transitions as co-produced, emergent and diverse. *J. Environ. Policy Plan.* **2016**, *18*, 585–607. [CrossRef]

71. Gilby, S.; Mao, C.; Koide, R.; Watabe, A.; Hengesbaugh, M.; Appleby, D.; Nugroho, S.B.; Kamei, M.; Liu, C.; Chepelianskaia, O.; et al. *Sustainable Lifestyles Policy and Practice: Challenges and Way Forward*; Institute for Global Environmental Strategies: Hayama, Japan, 2019.

72. Tellioğlu, H.; Wagner, M.; Habiger, M.; Mikusch, G. Living Labs Reconsidered for Community Building and Maintenance. In Proceedings of the 9th International Conference on Communities & Technologies-Transforming Communities, Vienna, Austria, 3–7 June 2019; pp. 154–159.

73. Almirall, E. Living Labs: Arbiters of mid-and ground-level innovation. *Technol. Anal. Strateg. Manag.* **2011**, *1*, 87–102. [CrossRef]

74. Gamble, J.A.A. *A Developmental Evaluation Primer*; The J.W. McConnell Family Foundation: Montreal, QC, Canada, 2008.

75. Patton, M.Q. *Developmental Evaluation. Applying Complexity Concepts to Enhance Innovation and Use*; The Guilford Press: New York, NY, USA, 2010.

76. Dozois, E.; Langlois, M.; Blanchet-Cohen, N. *A Practitioner's Guide to Developmental Evaluation*; The J.W. McConnell Family Foundation and the International Institute for Child Rights and Development: Montreal, QC, Canada, 2010.

sustainability

MDPI

Article

Implementation of Accelerated Policy-Driven Sustainability Transitions: Case of Bharat Stage 4 to 6 Leapfrogs in India

Aditi Khodke [1,2,*], Atsushi Watabe [1] and Nigel Mehdi [2]

[1] Institute for Global Environmental Strategies, Kanagawa 240-0115, Japan; watabe@iges.or.jp
[2] Kellogg College, University of Oxford, Oxford OX2 6PN, UK; nigel.mehdi@conted.ox.ac.uk
* Correspondence: khodke@iges.or.jp

Abstract: In the face of pressing environmental challenges, governments must pledge to achieve sustainability transitions within an accelerated timeline, faster than leaving these transitions to the market mechanisms alone. This had led to an emergent approach within the sustainability transition research (STR): Accelerated policy-driven sustainability transitions (APDST). Literature on APDST asserts its significance in addressing pressing environmental and development challenges as regime actors like policymakers enact change. It also assumes support from other incumbent regime actors like the industries and businesses. In this study, we identify the reasons for which incumbent industry and business actors might support APDST and whether their support can suffice for implementation. We examine the actor strategies by drawing empirical data from the Indian national government policy of mandatory leapfrog in internal combustion engine (ICE) vehicle emission control norms, known as Bharat Stage 4 to 6. This leapfrogging policy was introduced to speed up the reduction of air pollutants produced by the transport sector. A mixed-methods approach, combining multimodal discourse analysis and netnographic research, was deployed for data collection and analysis. The findings show that unlike the status quo assumption in STR, many incumbent industry and business actors aligned with the direction of the enacted policy due to the political landscape and expected gains. However, the degree of support varied throughout the transition timeline and was influenced by challenges during the transitioning process and the response of the government actors. The case suggests we pay more attention to the actors' changing capacities and needs and consider internal and external influences in adapting the transition timelines. This study contributes to the ongoing discussion on the implementation of APDST, by examining the dynamism of actor strategies, and provides an overview of sustainability transitions in emerging economies.

Keywords: accelerated policy-driven sustainability transitions; Asian sustainability transitions; cleaner vehicle technology; urban air pollution

Citation: Khodke, A.; Watabe, A.; Mehdi, N. Implementation of Accelerated Policy-Driven Sustainability Transitions: Case of Bharat Stage 4 to 6 Leapfrogs in India. *Sustainability* **2021**, *13*, 4339. https://doi.org/10.3390/su13084339

Academic Editor: Jose Navarro Pedreño

Received: 4 March 2021
Accepted: 8 April 2021
Published: 13 April 2021

Publisher's Note: MDPI stays neutral with regard to jurisdictional claims in published maps and institutional affiliations.

1. Introduction

In the face of pressing environmental challenges, transitioning away from unsustainable consumption and production patterns as fast as possible is a necessity. Governments play a stewarding role in addressing these challenges, and they are increasingly pledging to achieve sustainability transitions within a pre-determined timeline. For example, the People's Republic of China has indicated to be carbon neutral by 2060 [1], and Japan and the European Union by 2050 [2]. Various cities across the C40 network have pledged to meet the World Health Organization (WHO) air quality guidelines by 2030 [3]. Notably, the timelines envisaged by governments for achieving such transitions tend to be faster than can be achieved through market mechanisms alone.

The academic community working in the field of sustainability transition research (STR) refers to government's stewarding role in accelerating sustainability transitions as an emergent approach [4,5]. Despite the limited empirical evidence on the success of accelerated policy-driven sustainability transitions (APDST), many prominent scholars in the field of STR assert the significance of such transitions in addressing pressing environmental and development challenges [6,7].

STR loosely defines an actor as an entity, be it an individual, institution, organisation, or a collective, related to transition [8]. Different actors are often grouped based on their analytical hierarchy or their timeline of being operational in socio-technical systems such as incumbents linked with regimes versus emergent actors connected with niches [9]. Multiple studies on socio-technical transition, driven by market mechanism, have underscored the resistance of incumbent regimes towards change due to their lock-in and path dependency [10]. For APDST, regime actors like policymakers and the governments enact the change through policy mechanisms. However, enacting accelerated policies alone is insufficient for transitioning, as the implementation of transition entails support from different actors [11].

The dynamic interaction of policymakers and incumbent regimes shapes and formulates the courses of transitions. So far, there is limited understanding of the response of other incumbent actors, particularly from the businesses and industries towards APDST. The studies that assert the significance of APDST largely take the support from other incumbent regime actors for granted, as a matter of compliance [12] or guided selection [7]. The urgent need to understand actor strategies is evidenced in the STR literature [5].

This research revisits the reasons for which the incumbent actors other than policymakers support APDST, if at all, and examine if support is sufficient for implementing the APDST. Comprehending the strategies of these actors enables the examination of the implementation of transition as well as to critically assess the assertion of the significance of APDST.

In this study, we examined the implementation of APDST by gathering empirical data from the Indian national-government-led mandatory leapfrog in the internal combustion engine (ICE) vehicle emission control norms, Bharat Stage 4 (BS4) to Bharat Stage 6 (BS6). We assessed the strategies of incumbent automotive industry actors and policymakers in response to these APDST by looking into the historical and present contexts and their discursive activities on social media; and identified reasons for incumbent industry actor's support, including their volatility of support; and their struggles in transitioning. We find that actor strategies eventually determine the directions and limitations of the accelerated transition.

The next section provides an overview of the transition scholarship and locates the novelty of APDST in the STR literature. Section 3 further elaborates on the Indian case and Section 4 explains the methodology for data collection and analysis. Section 5 sets out the findings, and Section 6 discusses our findings and compares them with the assumptions of STR. Section 7 concludes the article and suggests direction for further research.

2. Literature Review

2.1. Key Concepts in Transition Studies

The transition studies propose frameworks to harness sustainable development through technologies, practices, and governance [10]. Since the transition studies gained traction in the 1990s, they have taken holistic approaches towards comprehending change, accumulating insights about changes across the socio-technical systems, and providing insights to bring about transformative systemic shift [13].

The underpinning and seminal works of Kemp [14] and Rip and Kemp [15] argued that because technologies are embedded in societal systems, any technology change is socio-technical in nature. Changes in societal systems accompany a change in technological systems. To delineate the mechanisms of changes in socio-technical transitions, Rip and Kemp [15] defined an analytical hierarchy in socio-technical systems, namely niche, regime, and landscape, which can be interpreted as micro, meso, and macro levels. Geels [16] elaborated on this hierarchy to propose the multi-level perspectives (MLP) framework.

During the past three decades, transition theories have matured through gaining additional insights from multiple theoretical approaches: Industrial and evolutionary economics, science and technology studies, political science, and cultural studies [17], and proposed pertinent frameworks with differing foci and objectives, namely the MLP [16,18], Strategic Niche Management (SNM) [19], Technological Innovation Systems (TIS) [20,21], and transition management [22]. Moreover, researchers are proactive in advancing existing frameworks [23] and proposing new frameworks as necessary [7]. However, despite such developments, the conceptual underpinnings of transitions being socio-technical, and their analytical hierarchies, remain prevalent in transition studies.

Niches are considered incubators of new technologies and innovations. In contrast, regimes are the dominant intertwining of culture, institutional structures, actors, networks and practices that resist change [11], and landscape is where technologies become a norm and widespread [16]. This assumption has allowed researchers to consider the different positions and powers of actors in a specific system. The micro, meso, and macro levels offer different forms of stabilities where actors, networks, and their alignment determines the change in the socio-technical systems leading to the adoption of a certain technology or spread of an everyday practice [24,25].

2.2. Change in Application of Transition Studies: Development of STR

From predominately focusing on the uptake of individual technologies [18], transition studies have since focused on the uptake of sustainability-oriented innovation, technologies, and governance [26]. In particular, post-2000s, policymakers have actively applied the insights of transition studies to guide socio-technical transitions. An example of its application in addressing sustainability challenges is the Dutch government's national environmental policy plan (NM4) launched in 2001. This plan extensively used transition studies to set long-term orientation and short-term policies for addressing sustainability challenges like climate change, biodiversity loss, and exploitation of resources [27]. Academic works involving sustainability-oriented transitions are referred to as sustainability transition research (STR) [28].

2.3. Assumptions on the Pace of Transition

Though interdisciplinary crossovers can be seen [29] as the scholarship matured, transition research emerged in Northern Europe, and to date, remain dominated by Western scholars, particularly from the Netherlands. Most of them cover Northern European case studies, while the transitions outside Europe remain relatively unexplored [30].

Pioneers of transition scholarship concur that socio-technical transitions are multidecadal, long-term processes of change [10,14,16]. There are a few reasons for such unequivocal affirmation of temporality. Firstly, past transitions, for technical advancements in technology [18,31], particularly for improvements in design and performance and therefore in user-friendliness [14], mainly in the 19th and 20th centuries, required multiple decades. Secondly, as technology takes hold in a market, its price reduces, making it more affordable and further contributing to its uptake [14].

Such assumptions derived from technological transitions in developed economies do not capture globalisation's influence. Assessing transition timelines in the context of globalisation is particularly significant. Importantly, as many Asian countries have radically transitioned from import substitution development models in post-colonial times to export-led economies in the 1980s and 1990s [32], the interaction with global markets has accelerated the pace of technology development [33]. Factors like technology transfer, international knowledge and learning networks, and reduction in technology costs through outsourcing in emerging economies all accelerate the pace of technology transition; thus, the transition timeline should not be taken for granted [6].

Moreover, the Euro-centric focus has arguably diverted researchers' attention to the potentially different socio-economic contexts of transitions. For example, studies have found that the transition processes in Asian contexts deviate from transition studies [33] due to the governance structures that are yet to be fully democratic [34], often having ineffective regulatory policies and inadequate support from the private sector and civil society actors [35]. The governments often lead transitions in close collaboration with a handful of actors from the private sector [32], while many actors across the unorganised sectors are often not represented in transition processes [34,35]. The historical contexts in these countries also influence actors' configuration and roles from local, national, and international societies, as is shown in the energy transitions in Thailand and the Philippines [36].

2.4. Limited Attention and Assumptions on the Role of Actors

STR loosely defines an actor as any entity, be it an individual, institution, organisation, or a collective, related to transition [8]. Different actors are often grouped based on their analytical hierarchy or their timeline of being operational in socio-technical systems such as incumbents linked with regimes versus emergent actors connected with niches [9]. However, the definitions of niche, regime, and landscape have often been ambiguous, lacking established indicators to describe those [37]. This lack of conceptual clarity makes it harder to delineate the hierarchical boundaries of one level from another and the interlinkages among them while a particular technology becomes widely accepted [38].

With such a loose grouping of the actors in three levels, transition theorists argue that technological transitions emerge on a small scale and gain legitimacy through market mechanisms [13]. These assumptions are drawn from observations of the uptake of new technologies. STR rely on the overarching assumptions that incumbent actors are unlikely to change [5,10] and that emergent actors are necessary to enact change [28]. Incumbent actors are also referred to as 'dominant actors' [39] or 'existing actors' [22], and are considered to be locked-in unsustainable socio-technical regimes. Due to their vested interests, they resist change [10]. This resistance to change is referred to as the 'inertia'. These actors are assumed to change only through external pressure but not through their willingness to change [12]. On the other hand, emergent actors are referred to as 'new actors', 'outside actors', and 'niche actors' [40]. The 'frontrunners' among these emergent actors are referred to as 'change agent' and 'champion' [41]. As these words suggest, they are usually understood as the actors initiating experimentation and radical innovations in the protective spaces [42].

The categorisation of actors into broad groups like incumbent and emergent provides an overview of transitions. As a result, STR has provided limited attention to individual actor strategies [43]. Despite the significance of actors [8], the role of individual actors in transition processes has received relatively little attention in the existing literature [9,41].

However, due to socio-technical systems' highly entrenched nature, particularly in urban settings, where applying analytical hierarchies poses challenges [44]. Studies during the last decade revealed that not only niche actors but also policy and incumbent business at the regime level engage in the discursive practices to negotiate the creation of the socio-technical visions associated with the transitions [45] in specific sectors like transportation management [46] or renewable energy [47]. Actors involved in a particular transition process do not just lead, follow, or resist it. A more careful observation of actors' strategies will help us capture how they participate in the sense-making of a transition to take the best advantage while minimising the negative impacts in transitioning.

2.5. Accelerated Policy-Driven Sustainability Transitions (APDST)

The differing strategies of actors are worth further attention, particularly in the context of the faster transitions initiated or strongly supported by the governments. Sustainable technological transitions are often aligned with the mandate of governments to address urgent challenges to sustainable development. Pressure from multilateral organisations and international coalitions further create an urgency to address development challenges [4]. Therefore, government actors are equally interested in mainstreaming sustainable technologies. This push from government actors through directed policies could accelerate the pace of transitions compared with market-driven transitions [4]. The guided selection of sustainable technologies could overcome the time-intensive process before becoming mainstream [20]. Although there is limited empirical evidence that policy-driven transitions accelerate the pace of transition, Kern and Rogge's (2016) compelling argument [4] is shared by other scholars. Hekkert et al. [7] propose a framework for mission-oriented transitions, or the government-led accelerated policy-driven sustainability transition (APDST), where regimes guide the socio-economic systems toward a desired future within a specific time, like a mission. Whereas the empirical evidence of their success and implementation are both recognised areas for further research [4,48], APDST cases would enable researchers and policymakers to broaden their scopes beyond the market-led (or niche-driven) transitions of a sector that typically take decades. Moreover, research on implementation strengthens the understanding of the necessary implementation support and can contribute to better policy design. The assessment of actor strategies among policymakers and recipients [48] serves this purpose.

2.6. Need for Further Research on the Power Relations in APDST

The cases where governments took significant roles in transitions suggest that we pay more nuanced attention to the dynamics of positions of and relationships among the actors at all levels or, in other words, the dimension of politics [49]. Bulkeley et al. [50] and Wolfram and Frantzeskaki [51] argue that because political landscapes influence processes of socio-technical change, transition studies needs to be coupled with the lens of political ecology. Their remark is in line with broader criticisms of transition studies in overlooking the significance of political processes [52]. It contributes to Geels et al.'s suggestion [29] of the need for further research on political aspects of transitions. Even transition management research that emphasises interactions among actors through debates across strategic activities and negotiations at the tactical activities [22] is not free from this criticism [53], as it often assumes that all actors are inherently equal. Interests of different actors align with one another [27].

A more careful examination of power relations enables us to consider the dynamics of shaping the directions, paces, and even meanings of transitions, combined with the previous point about the different actors' strategies and the accelerated transitions led by governments.

Accelerated transitions [7] look as if governments make actors go straight for the pre-determined goals in the fixed timeline on the surface. However, past transition cases indicate that actors with differing needs and visions negotiate and gradually shape goals, timelines, and even meanings of the transitions over time [47]. Such dynamics should also apply to accelerated transitions. Still, some actors may be kept out of the collective sense-making of the forced goals in a short time and thus abide by an uncomfortable share of benefits and costs. Therefore, particularly in the study of accelerated transitions, we need to pay more attention to the cases where actors having different powers and strategies interact with others so that the ongoing transitions are most beneficial to them. For example, political actors may be interested in maintaining power relations in favour of specific incumbent actors while guiding the transitions [49]. Accelerated transitions can potentially be "successful" in achieving tangible "changes" in technologies or practices at the cost of placing burdens on or even excluding some of the actors having weak powers to influence the market or policies [54].

The remainder of the paper describes how APDST are enacted, applied, and responded to through analysing the discursive practices of the actors in the case of leapfrog in vehicle emission standards in India.

3. Case Study

The booming cities in India are facing many pressing environmental challenges. One is the alarming levels of air pollution [55]. The severity of this problem can be assessed through the World Health Organization's listing, which identified 14 Indian cities among the world's 20 most-polluted cities considering ambient air quality [56]. Within Indian cities, vehicles are one of the main reasons for the poor ambient air quality [55,57].

In India, the number of privately owned vehicles is one of the root cause of air pollution [58]. Each year, over 2 million cars and two-wheeled vehicles are sold within the country [59]. The majority of these vehicles are internal combustion engine (ICE)-driven. Inefficient ICE powertrains emit high levels of pollutants like hydrocarbons, carbon monoxide, nitrogen oxides, sulphur oxides, and particulate matter [60]. Poor air quality is the cause of 4.2 million premature deaths worldwide [61]. In India, the public health vulnerabilities due to air pollution from vehicles are listed in Table 1.

Table 1. Health impact of pollutants from vehicles.

Vehicular Air Pollutants [60]	Health Impacts from Air Pollutants [62]	Ambient Air Quality Index at Anand Vihar, New Delhi on 15 January 2020 [63]
Hydrocarbons	Irritation in eye, nose, and throat	Not monitored
Carbon monoxide	Damage to central nervous system	Satisfactory
Nitrogen oxides	Damage to lungs	Satisfactory
Sulphur oxides	Respiratory diseases	Good
Particulate Matter (PM)	Asthma, bronchitis, increased risk of preterm birth and morbidity rate	PM 2.5: Poor PM 10: Moderate

Considering the severity of the health crisis [62], addressing air pollution from vehicles is a priority for urban sustainability in India [55]. The Fossil Fuel Free Streets Declaration, signed by 34 cities across the Global South and North [64], is a testimony that the challenges of urban transport, including vehicular pollution, are not limited to Indian cities. Many cities across the world face similar challenges to different degrees.

Cities respond to the challenge of air pollution from vehicles by banning older ICE vehicles from the city centre, or even prohibiting their use altogether [65]. Other solutions include production side measures, like improving vehicle technology through the interventions of the national government [66].

The Indian National Ministry of Road Transport and Highways; Ministry of Heavy Industries and Public Enterprises; Ministry of Environment, Forest and Climate Change; and the Ministry of Petroleum and Natural Gas, in 2016, choose to improve the vehicle technology by introducing a draft policy for mandatory leapfrog from Bharat Stage 4 (BS4) vehicle emission control standard to Bharat Stage 6 (BS6) vehicle emission control standard [67]. They decided to skip the Euro5-equivalent BS5 emission control norms and proposed the introduction of BS6 emission control norms by 2020 to curb pollution from vehicles [57]. Applicable to both petrol and diesel vehicles, the BS6 norms were expected to reduce emissions of nitrogen, sulphur oxides, and particulate matter from new vehicles.

The BS4 to BS6 leapfrog included three sub-transitions: First, a restriction on the sales of new BS4-compliant vehicles after 1 April 2020; second, the manufacturing and sales of BS6-compliant vehicles and auto-components by 1 April 2020; and third, the availability of BS6-compliant fuel in parallel with the vehicle launch.

The pace of BS6 transition timeline was four years ahead of the former political regime's, the United Progressive Alliance (UPA), planned timeline in 2024, and one year before the incumbent political regime's, the National Development Alliance (NDA), initially planned timeline of 2021 [67,68]. The BS4 norm is equivalent to the European emission control standard Euro4, and BS6 complies with Euro6B and part of Euro6D [69]. The Euro4-to-Euro6B transition occurred over a span of nine years, whereas the BS4 to BS6 leapfrog was introduced in 2016, finalised in 2018, and was expected to be completed in 2020.

The next section discusses the research methods used to identify the actors and the assessment of the implementation through actor strategies.

4. Research Method

This research examined the implementation of APDST by assessing the actor strategies in response to the enacted the BS4 to BS6 vehicle emission control transition in India. Before examining the actor strategies towards the concerned transition, examining the political landscape for the development of emission control norms was perceived essential, in line with Section 2.6. We first started with the historical analysis of vehicle emission control norms in India. Then, following STR's prevalent research method of aggregating multimodal data to reconstruct transition trajectories, we examined policy documents on the BS4 to BS6 transition, and related news articles. This method, combined with the historic analysis of vehicle emission control standard, enabled us to identify automotive regime actors that can be representative of the incumbent automotive industry supply chain. This method posed limitations in collecting procedural data, which is crucial for assessing actor strategies. Hence, we sought the additional research method of social media data collection and analysis.

To summarise the research method involved four steps: Step 1: Assessing the history of vehicle emission control norms in India; Step 2: Aggregating multi-modal data on the BS4 to BS6 transition; Step 3: Social media analysis for data collection on actor strategies; and Step 4: Social media data analysis.

4.1. Step 1: Assessing the History of Vehicle Emission Control Norms in India

Political ecology is a significant yet understudied aspect of STR. We examined the development of vehicle emission control standard against the political landscape from its inception in India in the early 1990s. This inquiry led to further analysis of the development of automobile manufacturing industry in India in post-colonial times, which ultimately resulted in assessing the timeline from 1947 to 2018. We referred to research articles, reports, and policy documents. This analysis provided an overview of the critical junctures in India's incumbent automobile manufacturing regime, political landscape, its influence on the development of vehicle emission control standard, and automotive industry actors that can be representative of the incumbent regime. The result of this step is detailed in Section 5.1.

4.2. Step 2: Aggregating Multi-Modal Data on the BS4 to BS6 Transition

In STR, a commonly used research method involves multimodal data analysis to reconstruct the trajectories of transitions and to establish a causal relationship within the sequence of key events and transition processes [17,46]. Transition studies rely on the literature of research articles, policy documents, news, business reports, and books [16,43].

The BS4 to BS6 leapfrog was introduced in a draft policy document published by the Ministry of Road Transport and Highways in February 2016. Automobile manufacturers association contested the draft policy to negotiate the transition timeline by approaching the Supreme Court of India. The Supreme Court of India passed the final verdict in October 2018 to finalise the transition timeline, mandating to be completed before April 2020. This verdict underlined the support from two incumbent automobile manufacturers for stringent vehicle emission control norms. News articles reporting this policy change provided details on the involved political actors from the national ministries [67,68].

Based on Sections 4.1 and 4.2, 18 actors from the national ministries, individual automobile manufacturers, and auto industry associations were identified. Actors were selected considering the suggestion to identify incumbent actors from the supply chains to include both upstream and downstream actors [21] (details provided in Section 5.2).

4.3. Step 3: Social Media Analysis for Data Collection on Actor Strategies

This research was conducted between 2018 to 2020, concomitant with the implementation of the BS4 to BS6 transition. Published research articles on this topic were very limited (See [70]). Similarly, limited news outlets focusing on the automotive industry reported updates on this transition, and attention from the mainstream news media was limited.

Available data indicated the outcome of the ongoing discussion between the automotive industry regime actors and the political actors, but seldom elaborated on procedural information, which was perceived necessary for examining the process of policy implementation and changes in actor strategies.

Previous studies in STR assessed actor strategies by interviewing actors to gather information on procedural data after the transition was realised [24,71]. This research required real-time and procedural data on actor strategies, which would entail frequent and multiple interviews with the identified actors. Due to the influential positions of the identified 18 actors, frequent and multiple interviews were not feasible. Moreover, part of this research was conducted amidst the COVID-19 pandemic, which affected the availability of actors for interviews. An alternative research method of social media analysis was selected to obtain real-time procedural data.

Data collection via social media is a novel, since 2008, but increasingly popular research method in social science, business and management, environment, and multidisciplinary researches [72]. In STR, so far, only a limited number of studies have used social media for data collection, such as Henshilwood et al. [73]. Such studies have used online ethnographic research methods, also known as netnography, which allows the study of communities created via digitally mediated social interactions [74]. Use of social media for data collection enables gathering user-generated data, through multiple methods like participant observation, actively participating in online activities and interacting with users [73], or by combining qualitative and quantitative methods through text mining [75].

Here, we used text mining due to its effectiveness in extracting large volumes of data, overcoming research biases due to the researchers' self-identity, saving time required for data collection, and providing easy access to user-generated data in a pre-determined timeframe [75,76].

The selection of social media platforms depends on their user base, permission to access data from the platform, and the type of data each platform can provide [75]. To assess actor strategies, a preferred platform was one that is used to express opinions on socio-political subjects, with an assumption that it would include opinions about the BS4 to BS6 leapfrog. Facebook and Twitter are used to share opinions [75]. Between these two platforms, Twitter users are more likely to share opinions on political matters [77]. Twitter's cap on text volume, 280 characters, makes it manageable to engage with the data through text analysis.

Despite the preference for Twitter, it was crucial to verify whether the selected Indian policy and automotive industry actors actually use it. In India, the use of social media platforms by politicians to share their opinions is fairly recent, starting from 2014; this is unlike the U.S., which heavily used social media platforms like Twitter in the 2008 elections [78]. Across the NDA-led national government, Twitter is the most-used social media platform by almost all the national ministers and ministries [77]. Researchers manually verified if other identified actors use Twitter or not, and most were avid users. Only official accounts managed directly by the selected actors were considered.

A total of 25,758 tweets between October 2018 to April 2020 were collected using R programming's 'twitteR' and 'Rtweet' packages. On average, between 500 and 3200 past tweets were extracted from the selected actors. All the collected data were exported to Microsoft Excel. Though the draft BS4-to-BS6 policy was introduced in 2016, it was only finalised in October 2018. Hence, the timeline from 2018 to 2020 was selected.

4.4. Step 4: Social Media Data Analysis

Kozinets et al. [74] cautioned against text mining combined with data analysis software, which overshadows the researcher's ability to engage with the data. This research used text mining as a data extraction tool. The extracted data were carefully organised following a systematic search query [76]. The systematic search query was conducted based on the recurring terms identified related to the Bharat Stage emission control standard. The identified keywords were BS-VI, BSVI, BS6, BSIV, Bharat Stage, BS-IV, and BS4.

The critique on the authenticity of social media data [79] was addressed by the selection of Twitter, as identities of the actors are not hidden [80], secondly by collecting data from official accounts [81], thirdly by manually curating truncated data, and lastly by data triangulation against other information sources like news, speeches of the identified actors at automobile industry events, blogs, and industry reports, where available [82].

Twitter has tweet and retweet functions. A tweet is the user-generated data, whereas a retweet further distributes already-tweeted data [78]. The differences between the two are ambiguous, as users may include additional information when retweeting. Both tweets and retweets reflect a user's position when broadcasting textual discourse [83]. Broadcasting a text discourse can enable engagement with a wider audience, deliberate mobilising information, and can create a chain reaction [78]. We analysed both tweet and retweet functions.

The tweets were analysed using Gee's toolkit to identify relevant text-analysis strategies [84]: Particular attention was paid to language in use, which helped in identifying the targeted audience of the tweet; persuasive discourses [85]; and the lexical styles that reflect the power dynamics and the relative position of the actor in a wider societal sphere.

In carrying out this research, the ethical implications were carefully considered. Only publicly available published data were collected and analysed. Twitter data were collected with approval, using a standard application-programming interface (API).

5. Results and Findings

This section provides further information on how different incumbent actors from the automotive industry supply chain and the government responded to the mandatory and accelerated policy-driven transition. Here, we share findings from the historical analysis on vehicle emission control norms, introduce the selected actors related to the BS4 to BS6 transition and their analysed Twitter timelines, the key events in the BS4 to BS6 transition trajectory, and the actors' strategies in response to those.

5.1. Development of Vehicle Emission Control Standard in India

The pace of motorisation in India has been swift. This rapid pace, coupled with inefficient vehicle technology, has led to alarming levels of air pollution in Indian cities [36]. However, addressing vehicular emission was not only an environmental and public health challenge, but also a policy tool to advance foreign collaboration, export potential, and industrial competitiveness.

Emissions from vehicles have reportedly been a concern for Indian cities since the 1970s. The national development model of protectionist strategies and the inward-looking growth of the automotive industry through import substitution led to inferior quality vehicles that generated 2–3 times higher emissions, resulting in high levels of air pollution in cities [76]. According to the Japanese automobile manufacturer Suzuki Motors, the Indian automotive industry in the 1970s was technologically 30 years behind the world's most recent technology [86].

The 1980s remain the most significant decade in India for addressing air pollution from vehicles and the development of vehicle technology. The collaboration with Japanese automobile manufacturers in the 1980s led to improved fuel efficiency and technology among Indian automobile and auto component manufacturers [87]. Soon after this collaboration, India's Air Act was enacted in 1981, which identified emissions from vehicles as one of the causes of air pollution. In 1986, India's first comprehensive environment policy was enacted, providing for the first time limits on permissible emissions from vehicles [88]. The 1980s also remain significant for environmental activism, particularly after the 1986 Bhopal gas plant tragedy that caused over 3000 deaths, half a million injuries, and created deformities among new-born children in subsequent decades. The first public interest litigation against air pollution caused by vehicles was filed in 1985 [88].

Despite the attention paid to air pollution from vehicles and deliberate efforts to improve the technology through foreign collaborations, India's old vehicle fleet, poorly maintained vehicles, and the high number of two-stroke vehicles were identified as reasons for poor air quality in urban areas [89].

The Indian government launched their first vehicle emission control norms in 1990 [90]. This was also when the European Union adopted and mandated the Euro1 emission control norms for all member states. In 1991, India adopted a New Economic Policy, which changed the discourse of India's inward-looking import-substitution economy toward an economy based on the principles of liberalisation, globalisation, and privatisation.

In post-colonial times, the United Progressive Alliance (UPA) mostly led the Indian national government. In 1998, for the first time, India was governed by the majority party the New Development Alliance (NDA). The NDA government expanded the UPA government's liberalisation policies, particularly by taking active measures to promote the export potential of Indian industries [91], as well as encouraging foreign investment and promoting joint ventures between foreign and Indian automobile manufacturers. The number of automobile manufacturers in India grew as well. Within this context, sustaining the growth of the automobile manufacturing industry by only catering to the domestic market would have posed a challenge; hence, increasing export potential of the automobile manufacturing industry was one of the co-benefits of adopting the globally recognised Euro vehicle emission control norms in 2000. The national government mandated the nationwide adoption of Euro1, domestically known as India 2000 norms [60]. These norms were later renamed the Bharat Stages (BSs).

The BS2, BS3, and BS4 norms were adopted in a phased approach: First in metropolitan cities, and gradually across rest of the country in 2005, 2010, and 2017, respectively [60]. The B6 norms were initially planned to be adopted in 2024 but their adoption was advanced and finalised for 2020, citing the urgency to address air pollution from vehicles as an intergenerational equity issue.

5.2. Selected BS4 to BS6 Transition Actors

5.2.1. Policy Actors

The BS4 to BS6 draft policy to leapfrog the emission control standard was introduced in 2016 at a joint meeting among the national Ministry of Road Transport and Highways; Ministry of Heavy Industries and Public Enterprises; Ministry of Environment, Forest and Climate Change; and the Ministry of Petroleum and Natural Gas. The Supreme Court of India finalised and mandated this policy in 2018 and played a crucial role in addressing grievances of the automotive industry actors during the implementation phase. Hence, we selected the policy actors as the four ministries and the apex court.

It was observed that some ministers from the above-mentioned national ministries were in charge of multiple other national ministries. Hence, the ministers and ministries were treated as two separate sources of data.

5.2.2. Incumbent Automotive Regime Actors

After 1947, the growth of the domestic automobile manufacturing industry was largely aided by the Indian government's protectionist policies. They drove foreign automobile manufacturers out of the country [87] and retained a handful of automobile manufacturing companies, all led by Indian entrepreneurs who had newly diversified their businesses from steel manufacturing to automobile manufacturing [92]. These companies included both commercial and passenger vehicle manufacturers. Passenger vehicle manufacturers that were operating since 1947 and continued to be in operation in 2020 included Mahindra & Mahindra, TATA Motors, Bajaj Auto, and Hindustan Motors.

The 1950s–1970s was the formative phase of the vehicle and component manufacturing industry. The formation of industry associations like the Automotive Component Manufacturers Association (ACMA) in 1959, the Society of Indian Automobile Manufacturers (SIAM) in 1960, and the Federation of Automobile Dealers Associations (FADA) in 1964 are indicative of the consolidation of the Indian automotive industry.

Toward the end of the 1970s and early 1980s, the Indian national government, led by the United Progressive Alliance (UPA), opened India's inward-looking automotive industry to collaborate with technologically advanced Japanese automobile and component manufacturers. Indian-government-owned car manufacturing company Maruti Udyog [86] and Suzuki Motors established a joint venture, Maruti Suzuki, which still continues to be India's largest automobile manufacturer by market share.

The 1970s and 1980s saw an increased demand for two-wheeled vehicles for the growing middle class population [93]. TVS Motors launched India's first two-wheeled mopeds and, later, in collaboration with Suzuki Motors, ventured into the motorcycle manufacturing business.

The number of actors within the Indian automobile manufacturing industry has continuously grown between 1947 and 2020. Distinguishing between incumbent and emergent actors remains difficult [20]. Regardless, TATA Motors, Mahindra & Mahindra, and Maruti Suzuki remain among India's top five car manufacturers, whereas Bajaj Auto and TVS Motors are among the top two-wheeler manufacturers. In the Supreme Court's verdict, TVS Motors and Bajaj Auto were identified as the incumbent automobile manufacturers who supported stringent vehicle emission control norms. The three industry associations, ACMA, SIAM, and FADA, continue to be the source of the automobile industry's collective voice. ACMA represent 800+ auto component manufacturers, SIAM represent 40+ large automobile manufacturers operating in India, and FADA represent 15,000 dealers and 30+ regional dealer's associations [94–96]. These three industry associations, together with Government of India established the Automotive Skills Development Council (ASDC) in 2019, which is expected to play a key role in the capacity-building of the automobile industry towards new technological transitions [97].

To summarise, five prominent incumbent automobile manufacturers, and four industry associations were selected to represent the response of incumbent automobile regime actors towards the BS4 to BS6 leapfrog. These actors are referred to as incumbents based on the status quo definition of incumbency in STR literature [9].

5.3. Twitter Analysis

The Ministry of Heavy Industries and Public Enterprises appears not to have a Twitter account. Similarly, due to the Minister of Petroleum and Natural Gas's frequent Tweeting, the extracted data using Twitter's standard API could only collect Tweets between July 2020 and May 2020, as this timeline was beyond the scope of this research, so the position of the minster could not be taken into account.

For the automotive industry actors, it was observed that the automobile manufacturer Mahindra & Mahindra frequently referred to their chief's, Mr. Anand Mahindra, Twitter account, so the data collection was expanded to include the Tweets from the chief. Similarly, we considered including the tweets from the presidents of SIAM, ACMA, FADA, ASDC, and other automobile manufacturers; but the industry associations in most cases retweeted the positions of their presidents, which were collected from the association's account. None of the other automobile manufacturers referred to the company presidents or any particular person.

This resulted in data collection from 17 actors from the ministries, ministers, industry associations, automobile skill enhancement organisation, and automobile manufacturers. The list of all the selected actors and their analysed Twitter timelines is indicated in Table 2.

Table 2. Selected actors and their analysed Twitter timeline.

Type of Actor	Name of the Actor	Date of Retrieved Tweets (DD/MM/YYYY)
Policymakers	Ministry of Heavy Industries and Public Enterprises	None
	Minister of Heavy Industries and Public Enterprises and Minister of Environment, Forest and Climate Change	01/11/2019 to 01/04/2020
	Ministry of Road Transport and Highways	30/09/2018 to 02/04/2020
	Minister of Road Transport and Highways	10/03/2019 to 01/04/2020
	Ministry of Environment, Forest and Climate Change	01/10/2018 to 01/04/2020
	Ministry of Petroleum and Natural Gas	26/01/2019 to 01/04/2020
	Minister of Petroleum and Natural Gas	None
Automotive Industry Actors	SIAM	01/10/2018 to 01/04/2020
	ACMA	01/10/2018 to 01/04/2020
	FADA	01/10/2018 to 03/04/2020
	ASDC	01/10/2018 to 01/04/2020
	Maruti Suzuki India Limited	07/12/2019 to 01/04/2020
	Tata Motors Limited	09/03/2020 to 01/04/2020
	Mahindra & Mahindra Limited	03/09/2018 to 05/04/2020
	Chief of Mahindra & Mahindra Limited	22/10/2018 to 01/04/2020
	Bajaj Auto Limited	23/01/2019 to 01/04/2020
	TVS Motors	15/10/2019 to 01/04/2020

The Minister of Environment Forest and Climate Change, who also led the Ministry of Heavy Industry and Public Enterprises, did not Tweet on either BS4 or BS6 emission control norms. The Ministry of Road Transport and Highways mostly tweeted in Devnagari script, which was beyond the scope of this research. SIAM tweeted most frequently about the BS4 to BS6 transition, followed by the Ministry of Petroleum and Natural Gas, FADA, and TVS Motors. The frequency of tweets is indicated in Figure 1.

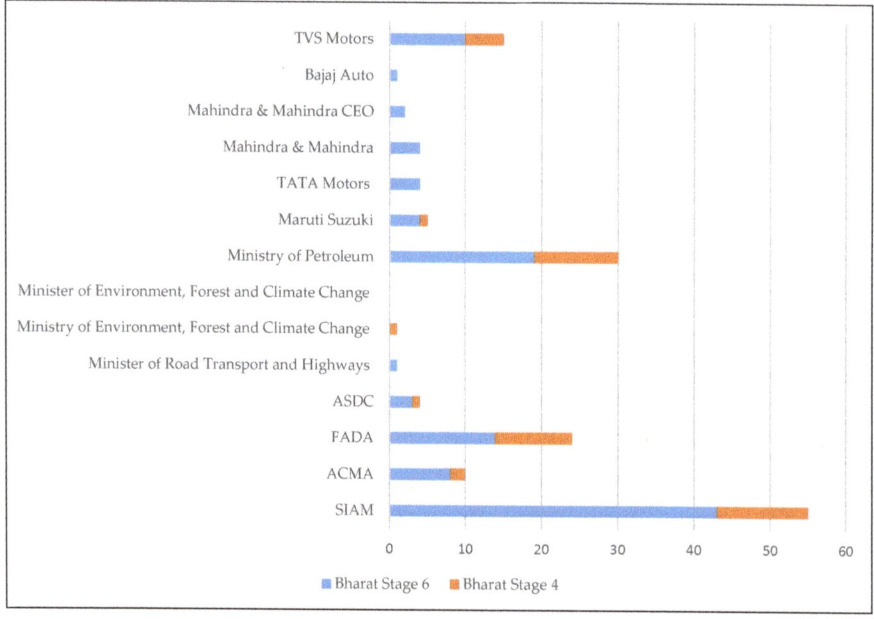

Figure 1. Frequency of tweets on BS4 and BS6.

5.4. BS4-to-BS6 Transition: Key Events

Based on aggregating the analysed Twitter data and its linked news articles, speeches of actors at automobile industry events, blogs, industry reports, and the policy documents, the following key events were identified. Some of these events were initiated by the identified actors, while others were actor's reactions to externalities. Figure 2 provides a quick overview of key events.

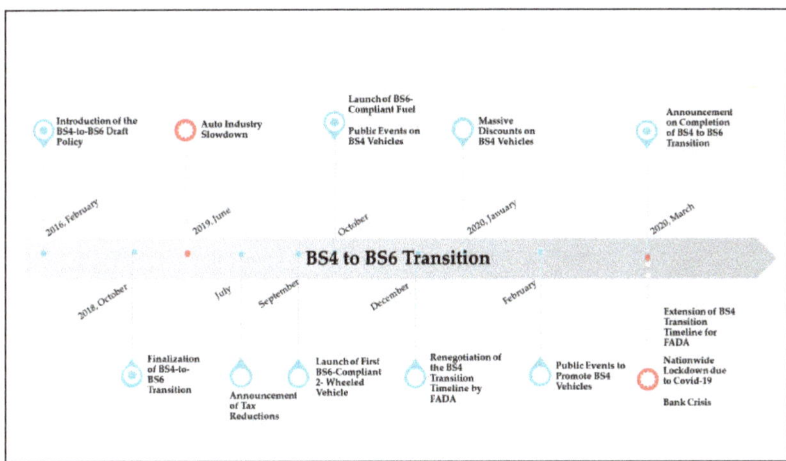

Figure 2. Overview of the BS4 to BS6 transition timeline.

5.4.1. Introduction of the BS4-to-BS6 Draft Policy (February 2016)

The draft policy for BS4-to-BS6 leapfrog, introduced in 2016 as an outcome of an inter-ministerial meeting, was contested by the Society of Indian Automobile Manufacturers (SIAM). SIAM appealed to the Supreme Court of India to negotiate the timeline on the ban on the sales of new BS4 vehicles. They underlined that after the introduction of BS6-compliant fuels, it will take their members three to six months to completely shift to BS6 vehicles and clear BS4 inventory; hence, the timeline for fuel availability and BS4 vehicle ban shall be one after the another [98]. SIAM represented its members, comprising about 40 Indian and foreign automobile manufacturers operating in India.

5.4.2. Finalisation of BS4-to-BS6 Transition (October 2018)

The Supreme Court of India rejected SIAM's appeal in October 2018 in a seminal judgement that supported the national government's decision to adopt BS6 from 2020 and prohibiting the sales of new BS4 vehicles beyond 31 March 2020. The court cited an urgency to act against air pollution and framing air pollution as an intergenerational equity issue [98].

5.4.3. Automobile Industry Slowdown (June 2019)

In 2019, the Indian automobile industry experienced one of the worst and unforeseen slowdowns. SIAM and FADA shared about the slowdown on their Twitter accounts. The plunge of 31% in the sales of passenger vehicles compared to the previous year indicates the magnitude of the crisis [99].

5.4.4. Announcement of Tax Reductions (July 2019)

The National Ministry of Finance addressed some of SIAM's and FADA's demands by reducing tax and easing the corporate social responsibility expenditure to include R&D activities [100]. After these measures, SIAM became optimistic for the growth in sales.

5.4.5. Public Events on BS4 (September 2019)

Along with the industry slowdown, the confusion among consumers about the validity of BS4-compliant vehicles throughout their registration period was a concern. This lowered the sales of BS4 vehicles. News about these events was shared on Twitter by SIAM, FADA, and the Minister of Road Transport and Highways.

5.4.6. Launch of First BS6 Compliant Vehicle (September 2019)

Honda Motors was the first to launch BS6 compliant two-wheeled vehicles, almost six months ahead of the national government's mandated timeline. The Minister of Road Transport and Highways shared about the launch on his Twitter account.

5.4.7. Launch of BS6-Compliant Fuel (October 2019)

BS6-compliant fuel was introduced in October 2019. The Ministry of Petroleum and Natural Gas shared this news on their Twitter account. This assured the automobile manufacturers and helped them plan for their BS6 inventories.

5.4.8. Renegotiation of the BS4 Transition Timeline by FADA (December 2019)

Members of FADA shared on Twitter that they had high levels of unsold BS4 vehicle inventory due to the auto industry slowdown and approached Supreme Court to renegotiate the timeline.

5.4.9. Massive Discounts on BS4 Vehicles (January 2020)

Promotion activities for BS4 vehicles surged. BS4 vehicles were promoted as technically on par with BS6 vehicles and lighter on the (customers) pockets. Individual automobile manufacturers like Maruti Suzuki, Tata Motors, TVS Motors, FADA, and ASDC shared about the discounts and promoted BS4 vehicles on their Twitter accounts.

5.4.10. Advancement of Transition Timeline (February 2020)

Some sub-national governments and financial institutes set their own timelines to stop registration and lending for the BS4 vehicles ahead of the government's timeline. This was shared by SIAM in their Twitter feed.

5.4.11. Bank Crisis (March 2020)

One of the leading private banks in India was placed under a moratorium, limiting regular banking operations. Indian consumers rely up to 74% on external finance when purchasing automobiles [101], which further affected the BS4 inventory. This was tweeted by FADA.

5.4.12. Covid-19 National Lockdown (24 March 2020)

The Indian government announced a countrywide lockdown due to the COVID-19 pandemic, which further reduced the days available to liquidate BS4 inventory. The two-wheeled dealers, which are often small and medium-sized businesses, had relatively higher BS4 inventories.

5.4.13. Extension of the Sale of BS4 Vehicles: (27 March 2020)

Due to the COVID-19-induced countrywide lockdown, the Supreme Court of India allowed the members of FADA to sell their unsold vehicle inventory affected by the pandemic [102].

5.4.14. Announcement on Completion of the BS4 to BS6 Transition (1 April 2020)

ACMA, SIAM, ASDC, and the Ministry of Petroleum and Natural Gas announced the successful completion of transition on Twitter.

5.4.15. Evaluation of Transition Timeline Extension (July 2020)

The Supreme Court Reflected upon the extension to sell BS4-compliant vehicles and concluded that some dealers misused the extension; as a result, more BS4 vehicles were sold than anticipated [102].

5.5. Actor Strategies

As explained in Section 3, the BS4-to-BS6 leapfrog involved three sub-transitions: Availability of BS6-compliant fuel and vehicles and the ban on the sales of new BS4-compliant vehicles after 31 March 2020. The actors' strategies in response to key events are elaborated in the following sections. Figure 3 provides an overview of the dynamic positioning of actors.

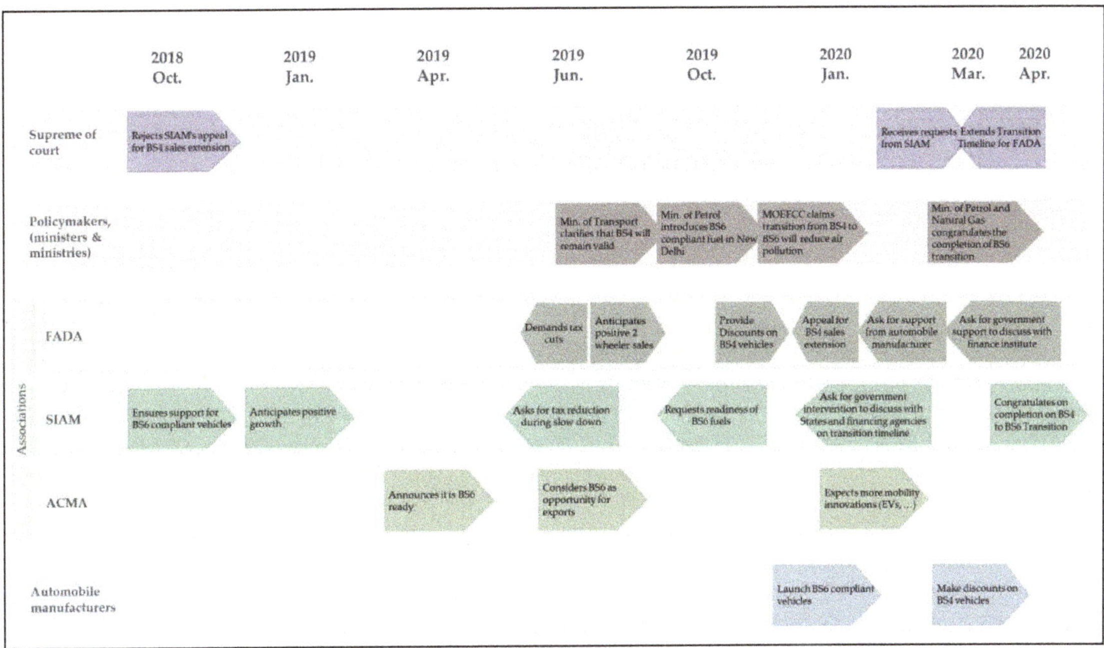

Figure 3. Dynamic positioning of actors.

5.5.1. Introduction of the BS4-to-BS6 Draft Policy (February 2016)

- SIAM, as a collective, resisted the transition timeline for the ban on the sales of BS4 vehicles.
- Many of SIAM's members were already manufacturing and exporting BS6-compliant vehicles; the economic gains of continuing with BS4 vehicles were higher [98].

5.5.2. Finalisation of BS4-to-BS6 Transition (October 2018)

- Anticipating the rejection of the appeal, SIAM stated its support for the BS4-to-BS6 transition even before the Supreme Court of India's final judgement in October 2018 on their Twitter account.
- SIAM indicated their support for the transition by assuring the launch of BS6-compliant vehicles by the national government's proposed timeline.
- ACMA, FADA, and ASDC expressed their support on Twitter for the transition during the implementation process.

5.5.3. Automobile Industry Slowdown (June 2019)

Despite the sluggish growth from the beginning of 2019 until June 2019, SIAM and FADA anticipated positive sales.

- SIAM and FADA demanded tax reductions on Twitter from July 2019 to promote vehicle sales for both BS4 and upcoming BS6 vehicles.
- On Twitter SIAM expressed conditionality on meeting the BS6 timeline due to the lack of clarity for the availability of BS6-compliant fuel from the national government.
- ACMA was the first among the automobile industry supply chain to tweet readiness for the BS4-to-BS6 transition in June 2019.

5.5.4. Announcement of Tax Reductions (July 2019)

- Members of SIAM and FADA tweeted their appreciation for the tax reductions and indicated optimism toward growth in vehicle sales because of the festive season in India.

5.5.5. Launch of First BS6 Compliant Vehicle (September 2019)

- Individual members of SIAM, like Honda Motors, introduced the first BS6-compliant two-wheeled vehicle.

5.5.6. Public Events on BS4 Vehicles (September 2019)

- Government actors like the National Minister of Road Transport and Highways clarified the validation of BS4 vehicles post March 2020.
- SIAM organised sessions to clarify the technical differences between BS4 and BS6 vehicles.

5.5.7. Launch of BS6-Compliant Fuel and Vehicles (October and November 2019)

- The National Ministry of Petroleum and Natural Gas introduced BS compliant fuel in the National Capital Region of India. They underscored the following environmental benefits of the BS4 to BS6 transition.

"Leapfrogging from BS-IV to BS-VI is a testimony of our Govt's commitment to reduce vehicular pollution and ensure a cleaner India" (Ministry of Petroleum and Natural Gas, 15 November 2019, Twitter).

- SIAM appreciated the efforts of the Ministry and remained optimistic toward the BS4-to-BS6 leapfrog.
- SIAM shared the benefits of the BS4-to-BS6 leapfrog and accentuated the importance of this transition toward reducing air pollution.
- Mahindra & Mahindra, TVS Motors, TATA Motors, Maruti Suzuki, and Bajaj Auto launched BS6-compliant vehicles ahead of the transition time.

5.5.8. Renegotiation of the BS4 Transition Timeline by FADA (December 2019)

- FADA approached the Supreme Court of India to seek an extension for the sales of BS4-compliant vehicles until the stocks of BS4 vehicles purchased before 31 March 2020 were over.

5.5.9. Massive Discounts on BS4 Vehicles (January–February 2020)

- Members of FADA and SIAM, like TVS Motors, gave massive discounts on BS4 vehicles.
- FADA organised public discussion sessions to clarify technical confusion regarding BS4 and BS6 vehicles.

5.5.10. Advancement of Transition Timeline (February 2020)

SIAM approached the Supreme Court of India to ensure all the sub-national governments and financial institutes follow the nationally determined transition timeline.

5.5.11. Bank Crisis (March 2020)

FADA sought intervention from the Minister of Road Transport and Highways and Minister of Finance to help them unload the BS4 inventories.

5.5.12. Covid-19 National Lockdown (24 March 2020)

Members of SIAM continued to supply BS4 vehicles to FADA despite their multiple requests.

- Some members of SIAM exported their unsold BS4 inventories.
- Many members of FADA could not sell the BS4 vehicles.

5.5.13. Extension of Transition Timeline (27 March 2020)

- The Supreme Court granted extension to FADA to sell the unsold BS4 stocks, manufactured before 31 March 2020.
- Many dealers registered the vehicles under their employees' names and sold them beyond the transition timeline as second-hand stock.

5.5.14. Announcement on Completion of the BS4 to BS6 Transition (1 April 2020)

- SIAM, ACMA, and the Ministry of Petroleum and Natural Gas indicated completion and success of the BS4-to-BS6 leapfrog.
- FADA requested the help of the National Minister of Road Transport.
- FADA did not succeed in selling its BS4 inventory before 31 March 2020.

6. Discussion

Our findings confirm Kern and Rogge's assumption that regime actors, like policymakers, can speed up the pace of transition by addressing sustainability challenges through directed policies [4], which, if left to market mechanisms, could take multiple decades to achieve [14]. It also supports the observation in policy studies that any enacted policies achieve some degree of success [103]. The findings also illustrate Berkhout's [104] assertion that in industrial economies where both production and consumption supply chains are linked, environmental problems are addressed through technological solutions.

STR assumes that incumbent socio-economic regimes and their related actors are path-dependent and locked-in to unsustainable practices, so they resist change [10]. A few studies, like Penna and Geels [12], Berggren et al. [105], and Shiroyama and Kajiki [106], identified that there are deviations to this assumption. Penna and Geels [12] associated the support from incumbent actors with external pressures such as those from the policymakers, which are not necessarily due to the actors' own strategy. This research contributes to the deviation. For the Indian case, the introduction of policy did play a role in aligning all the actors toward the directed sustainability transition at an accelerated pace through guiding the socio-technical system towards a pre-determined future. Yet, the incumbent actors did not just submissively follow the transition policy but eventually adjusted their positions to it after the transition was finalised and during its implementation, as was shown by some actions, such as initial contestation to the timeline, the production of BS6-compliant vehicles before the transition timeline, communication of the benefits of the transition, and the responses in the face of the lack of compatible fuel and economic crisis.

Support from incumbent regime actors is dynamic, and the degree of support can vary throughout the transition timeline. This dynamism was in response to externalities and key events in the transition trajectory. It identifies that through negotiation with the other actors and addressing some externalities, actors can reposition themselves either as those proactively moving the transitions or those opportunistically resisting.

The next sub-sections elaborate further on the reasons for support to APDST and the dynamic positioning that influenced the support.

6.1. Reasons for Support from Incumbent Actors

The deviation in the positions of incumbent actors from the mainstream assumption in the STR is thought to be due to two factors: The socio-political landscape in India, and the anticipated co-benefits of policy transitions.

In India's post-colonial development trajectories, different national governments have collaborated with indigenous entrepreneurs in establishing a domestic automobile manufacturing industry and in realising national development strategies. Despite the change in the political regime, the national government worked closely with incumbent automotive industry actors [87,92], and have mostly enacted policies to foster growth of the automobile manufacturing regime. A similar relationship between the political regime and incumbent actors was observed across other Asian countries like Korea, Thailand, and Japan [32]. This observation underlines the significance of considering the political landscape and development trajectories [52] before assuming the position of the incumbent actors.

The Bharatiya Janata Party led New Development Alliance (NDA) national government enacted the APDST by leapfrogging the BS5 vehicle emission norms. Before enactment of this policy, the NDA government's 2014 election manifesto indicated making India a 'Global Manufacturing Hub' and creating an 'innovation and technology driven society', among others. The Indian automobile manufacturing industry is one of the key sectors identified by the NDA government to realise this manifesto. Moreover, the NDA government projected the Indian automobile manufacturing industry to be world's third largest automobile manufacturing base by 2026 [107]. To realise these aspirations, adopting one of the world's most contemporary vehicle emission control norms, on par with the European manufacturers, had anticipated economic and symbolic benefits, while banning the BS4 vehicles and fuel had environmental benefits.

The support from incumbent automobile manufacturers was crucial for the success of the BS4 to BS6 transition as they are among the most powerful actors of the incumbent regime whose strategies directed the component manufacturers and dealers. As most of the analysed automobile manufacturers were active in exports, transitioning to BS6 emission norms would further strengthen their export potential. Hence, the sub-transition to manufacture BS6 vehicles was well received by the incumbent regime due to their anticipated benefits. However, the resistance came from giving upon the immediate economic gains through domestically selling BS4 vehicles, for which the pressure from the actor like the Supreme Court helped in gaining support from the incumbents.

Depending on the political landscape and the anticipated benefits of transitions, the incumbent regime's actors are likely to support directed transitions. Similarly, the actors positioned themselves in the expectation of future gains. Even though the transitions were enacted as part of a policy to address air pollution, there were multiple benefits associated with achieving this socio-technical transition. This observation is congruent with the anticipated co-benefits associated with sustainable development, particularly in addressing the problem of air pollution [108]. This also supports the idea that sustainability transitions have multiple co-benefits, which, when they are quantified and communicated effectively [109], create an impetus to address sustainability challenges [108]. In line with Berkhout [104], this confirms that incumbent actors could align with transition visions as long as the benefits of the transitions are communicated amongst all actors.

The following subsections delves further into the reasons for dynamic positions of incumbent actors and the key considerations for APSDT implementation.

6.2. Dynamic Position of Incumbent Actors

In STR, actors are often grouped based on the analytical hierarchy they belong to, or the time they have been operational in those analytical hierarchies, for example, regime vs. niche actors or incumbent vs. emergent actors. Such terms deflect our attention from dynamism in actors' positions during the implementation period. The dynamic positioning of incumbent actors in response to the key events during the transition timeline was revealed by assessing the actor strategies. Despite the support from different actors in the automobile industry supply chain and various national ministries, the implementation of the BS4-to-BS6 leapfrog encountered different challenges, and accordingly, the actor's degree of support for the transition varied. Moreover, actors like members of the FADA could not transition within the pre-determined timeline. The key reasons are discussed below.

6.2.1. Systemic Interdependencies

In addition to the changes in vehicle technology, the BS4-to-BS6 transition depended on the availability of BS6-compliant fuels. This mandate came under the Ministry of Petroleum and Natural Gas. India has both publicly owned and private oil refineries. Since the introduction of the BS4-to-BS6 draft policy, SIAM expressed concerns about the transition timeline. According to them, for the implementation of the transition, BS6-complaint fuels should be made available three to six months before the automobile manufacturers shift their production line to BS6 vehicles. However, the lack of information on the availability of BS6 fuel until October 2019 created challenges for the manufacturers to plan for BS6 vehicles, and they continued with surplus BS4 vehicles.

As opposed to the transition timeline planned by the national government, some sub-national governments and financing agencies set their own transition timelines, prohibiting the registration and lending for BS4 vehicles before 31 March 2020. To address the lack of coordination between national and sub-national governments, SIAM sought the Supreme Court's intervention.

The burden of the BS4 vehicle surplus and the changes in transition timeline was first borne by the members of SIAM and more so by the members of FADA.

6.2.2. Unforeseen Externalities

Externalities can hamper the implementation of accelerated policy transitions. During the transition timeline for BS4 to BS6, the slowdown in auto industry sales further created a surplus of BS4 vehicles in the supply chains for both the manufacturers and the dealers. SIAM's support for transition weakened, and they demanded tax reductions for vehicles. FADA supported SIAM's demand. However, the support from SIAM was regained after the national Ministry of Finance introduced tax cuts.

Members of FADA, being downstream actors of the supply chain, suffered more and could not unload the BS4 inventories; they approached the Supreme Court to seek extension for the sales of BS4 vehicles. Toward the end of the transition timeline in March 2020, the surplus was exacerbated by the enactment of a countrywide lockdown by the national government due to the COVID-19 pandemic. The Supreme Court granted a last-minute extension, three days before the end of transition timeline.

6.2.3. Non-Negotiable Timeline Revealed Power Asymmetries

The pre-determined and non-negotiable timeline underscored the power asymmetries within the automotive industry. Actors with limited capacities to negotiate suffered during the transition, as powerful actors succeeded to transition by imposing the cost of transition. In 2018, the Supreme Court of India mandated the BS4-to-BS6 transition to be completed by 1 April 2020. Though the timeline for sales and registration was common for all actors, meeting this timeline depended on mutual support among different actors along the supply chain. Upstream actors like auto components manufacturers, members of the ACMA, were well-prepared for the transition and delivered the BS6-compliant parts ahead of the transition timeline. However, despite repeated requests from the dealers to the automobile manufacturers, the dealers received stocks of BS4 vehicles until March 2020, which they were unable to sell. The two-wheel vehicle dealers, which are often small- and medium-size businesses, had limited capacity to negotiate with automobile manufacturers. They continued to receive BS4 vehicle stocks until the end of the transition timeline, making them unable to get rid of old stocks, and pleading support from policymakers.

The above examples indicate that though different incumbent regime actors supported the transition, their degree of support varied across the transition timeline. It was influenced by the challenges during the implementation process and the response of government actors in addressing those challenges. These challenges revealed the lack of coordination and power asymmetries. It underscores that enacting policy-driven transition is a multi-level and multi-actor process; sustained support for the accelerated policies requires deliberate efforts throughout the implementation process [110]. The leapfrogging from BS4 to BS6 looked successful with the support from the incumbent industry. However, it was a sleek success putting costs to actors in weaker positions and forced them to find out loopholes such as the registration of BS4 vehicles in their employees' names and selling it later as a second-hand stock.

7. Conclusions

Existing literature in STR postulates the resistance from incumbent regime actors to transition, APDST, the emergent approach in STR, assumes the support from incumbent regime actors as policymakers enact the change. This research elaborates further on what makes the incumbent regime actors, mainly from the industry and business, support APDST, and how their strategies influence each other in the face of externalities hindering the transition

This study reinstates the significance and potential of APDST in the addressing pressing environmental challenge.

We find that support from incumbent actors is not only due to the compliance towards enacted policies. The past and present relationship between industrial-political actors and anticipated co-benefits through transitioning influences actor strategies in response to APDST. Yet, sustaining the support from incumbent regime actors needs continued efforts from the government due to the challenges incurred in transitioning.

Though policy implementation challenges are often regarded as outcomes of failed policy coordination and coherence [5,110], this research finds that if systemic interdependencies, unforeseen externalities, and non-negotiable transition timelines revealing power asymmetries are not addressed sufficiently then, some actors could not fulfil the requirements within the pre-determined timeline, and can make the success of APDST questionable.

We recommend accounting for the dynamic positioning of the regime actors to deepen the understanding of the challenges of implementing APDST. This can enable a more reflexive approach towards sustainability transition that revisits the changing capacities and needs of the actors, internal and external influences, and can adapt the transition timelines accordingly. In the absence of a carefully planned APDST, the success of transition can be questionable.

Along with contributing to the recognized research gap on the implementation of APDST, this research provides empirical evidence on the dynamism of actors during transition trajectories, which hitherto remain largely unaccounted. The research method of Twitter data analysis provides an alternative to study the process of transitioning in real-time. Moreover, this study also contributes to the geographic diversity of the STR by drawing insights of APSDT in emerging economies.

The limitation of this research is twofold. Firstly, conclusions are primarily drawn based on the stated position of actors rather than their actual positions. The study partly addressed such a limitation through data triangulation, but due to the specificity of the empirical case, a limited number of data were available for triangulation. Secondly, the study only assessed the positions of well-recognised actors mentioned in the literature. In particular, the actor strategies of informal sector actors are under-represented.

As a way forward, studies in STR considering using social media analysis can benefit from combining it with interviews to verify the actor's stated position. Combining other forms of data collection to include informal sector actors [34] is highly recommended. We relied on Twitter's standard API for data collection, which capped the number of data collected. A more extensive database could contribute to strengthening these research findings.

Author Contributions: Conceptualisation, A.K.; methodology, A.K. and N.M.; data collection and analysis, A.K.; writing—original draft preparation, A.K. and A.W.; writing—review and editing, A.K., A.W., and N.M.; visualisation, A.K.; funding acquisition, A.W. All authors have read and agreed to the published version of the manuscript.

Funding: The Environment Research and Technology Development Fund (S-16-3: JPMEERF16S11630) of the Environmental Restoration and Conservation Agency of Japan, funded the development of this manuscript.

Institutional Review Board Statement: Ethical review and approval were waived for this study. The data used in this research were based on publicly available published data collected from secondary government sources, news articles and from Twitter. Twitter data was collected with approval, using a standard application programming interface (API). Apart from Twitter, no permissions were needed or sought, and there are no ethical issues raised by this work as defined by the University of Oxford's Research Ethics.

Informed Consent Statement: Not applicable.

Data Availability Statement: No new data were created or analyzed in this study. Data sharing is not applicable to this article.

Conflicts of Interest: The authors declare no conflict of interest.

References

1. Myers, S.L. China's Pledge to Be Carbon Neutral by 2060: What It Means-The New York Times. Available online: https://www.nytimes.com/2020/09/23/world/asia/china-climate-change.html (accessed on 27 March 2021).
2. Tsukimori, O. Japan Faces High Costs in Achieving Suga's 2050 Carbon Neutrality Target I The Japan Times. Available online: https://www.japantimes.co.jp/news/2020/11/09/business/japan-2050-carbon-neutrality-costs/ (accessed on 27 March 2021).
3. C40 35 Cities Unite to Clean the Air Their Citizens Breathe, Protecting the Health of Millions. Available online: https://www.c40.org/press_releases/35-cities-unite-to-clean-the-air-their-citizens-breathe-protecting-the-health-of-millions (accessed on 27 March 2021).
4. Kern, F.; Rogge, K.S. The pace of governed energy transitions: Agency, international dynamics and the global Paris agreement accelerating decarbonisation processes? *Energy Res. Soc. Sci.* **2016**, *22*, 13–17. [CrossRef]
5. Markard, J.; Geels, F.W.; Raven, R. Challenges in the acceleration of sustainability transitions. *Environ. Res. Lett.* **2020**, *15*, 081001. [CrossRef]
6. Sovacool, B.K. Energy Research & Social Science How long will it take? Conceptualizing the temporal dynamics of energy transitions. *Energy Res. Soc. Sci.* **2016**, *13*, 202–215. [CrossRef]
7. Hekkert, M.P.; Janssen, M.J.; Wesseling, J.H.; Negro, S.O. Mission-oriented innovation systems. *Environ. Innov. Soc. Transit.* **2020**, *34*, 76–79. [CrossRef]
8. Farla, J.; Markard, J.; Raven, R.; Coenen, L. Technological Forecasting & Social Change Sustainability transitions in the making: A closer look at actors, strategies and resources. *Technol. Forecast. Soc. Chang.* **2012**, *79*, 991–998. [CrossRef]
9. Fischer, L.B.; Newig, J. Importance of Actors and Agency in Sustainability Transitions: A Systematic Exploration of the Literature. *Sustainability* **2016**, *8*, 476. [CrossRef]
10. Köhler, J.; Geels, F.W.; Kern, F.; Markard, J.; Onsongo, E.; Wieczorek, A.; Alkemade, F.; Avelino, F.; Bergek, A.; Boons, F.; et al. An agenda for sustainability transitions research: State of the art and future directions. *Environ. Innov. Soc. Transit.* **2019**, *31*, 1–32. [CrossRef]
11. Roberts, C.; Geels, F.W.; Lockwood, M.; Newell, P.; Schmitz, H.; Turnheim, B.; Jordan, A. The politics of accelerating low-carbon transitions: Towards a new research agenda. *Energy Res. Soc. Sci.* **2018**, *44*, 304–311. [CrossRef]
12. Penna, C.C.R.; Geels, F.W. Multi-dimensional struggles in the greening of industry: A dialectic issue lifecycle model and case study. *Technol. Forecast. Soc. Chang.* **2012**, *79*, 999–1020. [CrossRef]
13. Geels, F.W.; Schot, J. Typology of sociotechnical transition pathways. *Res. Policy* **2007**, *36*, 399–417. [CrossRef]
14. Kemp, R. Technology and the transition to environmental sustainability. The problem of technological regime shifts. *Futures* **1994**, *26*, 1023–1046. [CrossRef]
15. Rip, A.; Kemp, R. Technological change. *Hum. Choice Clim. Chang.* **1997**, *2*, 327–400.
16. Geels, F.W. Technological transitions as evolutionary reconfiguration processes: A multi-level perspective and a case-study. *Res. Policy* **2002**, *31*, 1257–1274. [CrossRef]
17. Caniëls, M.C.J.; Romijn, H.A. Strategic niche management: Towards a policy tool for sustainable development. *Technol. Anal. Strateg. Manag.* **2008**, *20*, 245–266. [CrossRef]
18. Geels, F.W. The Dynamics of Transitions in Socio-technical Systems: A Multi-level Analysis of the Transition Pathway from Horse-drawn Carriages to Automobiles. *Technol. Anal. Strateg. Manag.* **2005**, *17*, 445–476. [CrossRef]

19. Kemp, R.; Truffer, B.; Harms, S. Strategic Niche Management for Sustainable Mobility. In *Social Costs and Sustainable Mobility. ZEW Economic Studies*; Rennings, K., Hohmeyer, O., Ottinger, R.L., Eds.; Physica: Heidelberg, Germany, 2000; Volume 7, pp. 167–187.

20. Hekkert, M.P.; Suurs, R.A.A.; Negro, S.O.; Kuhlmann, S.; Smits, R.E.H.M. Functions of innovation systems: A new approach for analysing technological change. *Technol. Forecast. Soc. Chang.* **2007**, *74*, 413–432. [CrossRef]

21. Bergek, A.; Jacobsson, S.; Carlsson, B.; Lindmark, S.; Rickne, A. Analyzing the functional dynamics of technological innovation systems: A scheme of analysis. *Res. Policy* **2008**, *37*, 407–429. [CrossRef]

22. Loorbach, D. Transition Management for Sustainable Development: A Prescriptive, Complexity-Based Governance Framework. *Governance* **2010**, *23*, 161–183. [CrossRef]

23. Elzen, B.; Van Mierlo, B.; Leeuwis, C. Anchoring of innovations: Assessing Dutch efforts to harvest energy from glasshouses. *Environ. Innov. Soc. Transit.* **2012**, *5*, 1–18. [CrossRef]

24. Wolfram, M. The Role of Cities in Sustainability Transitions: New Perspectives for Science and Policy. In *Quantitative Regional Economic and Environmental Analysis for Sustainability in Korea*; Springer: Singapore, 2016; pp. 3–22. [CrossRef]

25. Hargreaves, T.; Longhurst, N.; Seyfang, G. *Understanding Sustainability Innovations: Points of Intersection between the Multi-Level Perspective and Social Practice Theory*; Science, Society and Sustainability (3S) Research Group Working Paper; University of East Anglia: Norwich, UK, 2012.

26. Markard, J.; Raven, R.; Truffer, B. Sustainability transitions: An emerging field of research and its prospects. *Res. Policy* **2012**, *41*, 955–967. [CrossRef]

27. Kemp, R.; Rotmans, J.; Loorbach, D. Assessing the Dutch energy transition policy: How does it deal with dilemmas of managing transitions? *J. Environ. Policy Plan.* **2007**, *9*, 315–331. [CrossRef]

28. Loorbach, D.; Frantzeskaki, N.; Thissen, W. Technological Forecasting & Social Change Introduction to the special section: Infrastructures and transitions. *Technol. Forecast. Soc. Chang.* **2010**, *77*, 1195–1202. [CrossRef]

29. Geels, F.W.; Hekkert, M.P.; Jacobsson, S. The dynamics of sustainable innovation journeys. *Technol. Anal. Strateg. Manag.* **2008**, *20*, 521–536. [CrossRef]

30. Temenos, C.; Nikolaeva, A.; Schwanen, T.; Cresswell, T.; Sengers, F.; Watson, M.; Sheller, M. Ideas in motion: Theorizing mobility transitions an interdisciplinary conversation. *Transfers* **2017**, *7*, 113–129. [CrossRef]

31. Turnheim, B.; Geels, F.W. Regime destabilisation as the flipside of energy transitions: Lessons from the history of the British coal industry (1913-1997). *Energy Policy* **2012**, *50*, 35–49. [CrossRef]

32. Rock, M.; Murphy, J.T.; Rasiah, R.; Van Seters, P.; Managi, S. Technological Forecasting & Social Change A hard slog, not a leap frog: Globalization and sustainability transitions in developing Asia. *Technol. Forecast. Soc. Chang.* **2009**, *76*, 241–254. [CrossRef]

33. Berkhout, F.; Angel, D.; Wieczorek, A.J. Sustainability transitions in developing Asia: Are alternative development pathways likely? *Technol. Forecast. Social Chang.* **2009**, *76*, 215–217. [CrossRef]

34. Zusman, E.; Lee, S.Y.; Nakano, R.; Sano, D.; Lualon, U.; Nugroho, S.B. Governing Sustainability Transitions in Asia: Cases from Japan, Indonesia and Thailand. 환경사회학연구 *ECO Transl. Environ. Sociol. Res.* **2014**, *18*, 115–150.

35. Mah, D.; van der Vleuten, J.M.; Ip, J.C.M.; Hills, P. Governing the transition of socio-technical systems: A case study of the development of smart grids in Korea. *Energy Policy* **2012**, *45*, 133–141. [CrossRef]

36. Marquardt, J.; Delina, L.L. Reimagining energy futures: Contributions from community sustainable energy transitions in Thailand and the Philippines. *Energy Res. Soc. Sci.* **2019**, *49*, 91–102. [CrossRef]

37. Bilali, H. El The Multi-Level Perspective in Research on Sustainability Transitions in Agriculture and Food Systems: A Systematic Review. *Agriculture* **2019**, *9*, 74. [CrossRef]

38. Berkhout, F.; Verbong, G.; Wieczorek, A.J.; Raven, R.; Lebel, L. Sustainability experiments in Asia: Innovations shaping alternative development pathways? *Environ. Sci. Policy* **2010**, *13*, 261–271. [CrossRef]

39. Loorbach, D.; Oxenaar, S. Counting on Nature Counting on Nature. Available online: https://drift.eur.nl/wp-content/uploads/2018/02/Counting-on-Nature.-Transitions-to-a-natural-capital-positive-economy.pdf (accessed on 10 April 2020).

40. Geels, F.W. Ontologies, socio-technical transitions (to sustainability), and the multi-level perspective. *Res. Policy* **2010**, *39*, 495–510. [CrossRef]

41. Wittmayer, J.M.; Avelino, F.; Van Steenbergen, F.; Loorbach, D.; van Steenbergen, F.; Loorbach, D. Environmental Innovation and Societal Transitions Actor roles in transition: Insights from sociological perspectives. *Environ. Innov. Soc. Transit.* **2017**, *24*, 45–56. [CrossRef]

42. Smith, A.; Raven, R. What is protective space? Reconsidering niches in transitions to sustainability. *Res. Policy* **2012**, *41*, 1025–1036. [CrossRef]

43. Geels, F.W. Low-carbon transition via system reconfiguration? A socio-technical whole system analysis of passenger mobility in Great Britain (1990–2016). *Energy Res. Soc. Sci.* **2018**, *46*, 86–102. [CrossRef]

44. Nææss, P.; Vogel, N. Sustainable urban development and the multi-Level transition perspective. *Environ. Innov. Soc. Transit.* **2012**, *4*, 36–50. [CrossRef]

45. Späth, P.; Rohracher, H. 'Energy regions': The transformative power of regional discourses on socio-technical futures. *Res. Policy* **2010**, *39*, 449–458. [CrossRef]

46. Pel, B. Trojan horses in transitions: A dialectical perspective on innovation 'capture'. *J. Environ. Policy Plan.* **2016**, *18*, 673–691. [CrossRef]

47. Jørgensen, U. Mapping and navigating transitions — The multi-level perspective compared with arenas of development. *Res. Policy* **2012**, *41*, 996–1010. [CrossRef]

48. Kern, F.; Rogge, K.S.; Howlett, M. Policy mixes for sustainability transitions: New approaches and insights through bridging innovation and policy studies. *Res. Policy* **2019**, *48*, 103832. [CrossRef]

49. Meadowcroft, J. Engaging with the politics of sustainability transitions. *Environ. Innov. Soc. Transit.* **2011**, *1*, 70–75. [CrossRef]

50. Bulkeley, H.; Castan Broto, V.; Maassen, A. Low-carbon Transitions and the Reconfiguration of Urban Infrastructure. *Urban Stud.* **2014**, *51*, 1471–1486. [CrossRef]

51. Wolfram, M.; Frantzeskaki, N. Cities and Systemic Change for Sustainability: Prevailing Epistemologies and an Emerging Research Agenda. *Sustainability* **2016**, *8*, 144. [CrossRef]

52. Lawhon, M.; Murphy, J.T. Socio-technical regimes and sustainability transitions: Insights from political ecology. *Prog. Hum. Geogr.* **2012**, *36*, 354–378. [CrossRef]

53. Voß, J.; Bornemann, B. The Politics of Reflexive Governance: Challenges for Designing Adaptive Management and Transition Management. *Ecol. Soc.* **2011**, *16*, 9–32. [CrossRef]

54. Chilvers, J.; Longhurst, N. Participation in transition(s): Reconceiving public engagements in energy transitions as co-produced, emergent and diverse. *J. Environ. Policy Plan.* **2016**, *18*, 585–607. [CrossRef]

55. Shrivastava, M.; Ghosh, A.; Bhattacharyya, R.; Singh, S.D. Urban Pollution in India. In *Urban Pollution*; Charlesworth, S.M., Booth, C.A., Eds.; John Wiley & Sons, Ltd.: Chichester, UK, 2018; pp. 341–356.

56. Mahato, S.; Ghosh, K.G. Short-term exposure to ambient air quality of the most polluted Indian cities due to lockdown amid SARS-CoV-2. *Environ. Res.* **2020**, *188*, 109835. [CrossRef] [PubMed]

57. Ganguly, T.; Selvaraj, K.L.; Guttikunda, S.K. National Clean Air Programme (NCAP) for Indian cities: Review and outlook of clean air action plans. *Atmos. Environ. X* **2020**, *8*, 100096. [CrossRef]

58. Shekhar, S. Vehicular Pollution in India: Laws and Reactions. *SSRN Electron. J.* **2015**. [CrossRef]

59. SIAM Automobile Domestic Sales Trends. Available online: https://www.siam.in/statistics.aspx?mpgid=8&pgidtrail=14 (accessed on 27 March 2021).

60. Govindaraj, E.; Karthikeyan, D.; Muthu, K. History of Emission standards in India-A Critical review. *Int. J. Res. Anal. Rev.* **2019**, *6*, 28–35.

61. WHO Ambient Air Pollution-a Major Threat to Health and Climate. Available online: https://www.who.int/airpollution/ambient/en/#:~:text=Ambientairpollution-amajorthreattohealthandclimate&text=Ambientairpollutionaccountsfor,qualitylevelsexceedWHOlimits (accessed on 27 March 2021).

62. Barman, S.C.; Singh, R.; Negi, M.P.S.; Bhargava, S.K. Assessment of urban air pollution and it's probable health impact. *J. Environ. Biol.* **2010**, *31*, 913–920. [CrossRef]

63. The World Air Quality Project Anand Vihar, Delhi, Delhi Air Pollution: Real-Time Air Quality Index. Available online: https://aqicn.org/city/delhi/anand-vihar/ (accessed on 28 March 2021).

64. C40 Fossil Fuel Free Streets Declaration. Available online: https://www.c40.org/other/green-and-healthy-streets (accessed on 28 March 2021).

65. Burch, I.; Gilchrist, J.; Burch, B.I.; Gilchrist, J. *Survey of Global Activity to Phase Out Internal Combustion Engine Vehicles*; Center of Climate Protection: Santa Rosa, CA, USA, 2020.

66. Banister, D. Sustainable urban development and transport -a eurovision for 2020. *Transp. Rev.* **2000**, *20*, 113–130. [CrossRef]

67. Baggonkar, S.; Modi, A. Govt to leapfrog to BS-VI from 2020 | Business Standard News. Available online: https://www.business-standard.com/article/economy-policy/govt-to-leapfrog-to-bs-vi-from-2020-116010700054_1.html (accessed on 28 March 2021).

68. Jha, S. Govt. to Implement BS-VI Norms by 2020 - The Hindu. Available online: https://www.thehindu.com/business/Govt.-to-implement-BS-VI-norms-by-2020/article13984521.ece (accessed on 28 March 2021).

69. Roychowdhury, A.; Chattopadhyaya, V. *Bharat Stage VI (BS-VI) Readiness and Roadmap in India*; Round Table on BSVI Readiness: New Delhi, India, 2019.

70. Patil, A.A.; Joshi, R.R.; Dhavale, A.J.; Balwan, K.S. Review of Bharat Stage 6 Emission Norms. *Int. Res. J. Eng. Technol.* **2019**, *6*, 1359–1361.

71. Planko, J.; Cramer, J.; Hekkert, M.P.; Chappin, M.M.H.; Planko, J.; Cramer, J.; Hekkert, M.P.; Chappin, M.M.H. Technology Analysis & Strategic Management Combining the technological innovation systems framework with the entrepreneurs' perspective on innovation. *Technol. Anal. Strateg. Manag.* **2017**, *29*, 614–625. [CrossRef]

72. Kolleck, N.; Jörgens, H.; Well, M. Levels of Governance in Policy Innovation Cycles in Community Education: The Cases of Education for Sustainable Development and Climate Change Education. *Sustainability* **2017**, *9*, 1966. [CrossRef]

73. Henshilwood, E.; Swilling, M.; Naidoo, M.L. A transdisciplinary inquiry into sustainable automobility transitions: The case of an urban enclave in Cape Town. *Int. J. E -Plan. Res.* **2019**, *8*, 13–37. [CrossRef]

74. Kozinets, R.V. Marketing Netnography: Prom/ot(Ulgat)ing a New Research Method. *Methodol. Innov. Online* **2012**, *7*, 37–45. [CrossRef]

75. Ampofo, L.; Collister, S.; O'Loughlin, B.; Chadwick, A. Text Mining and Social Media: When Quantitative Meets Qualitative and Software Meets People. In *Innovations in Digital Research Methods*; Halfpenny, P., Procter, R., Eds.; SAGE Publications Limited: London, UK, 2015; pp. 161–192.

76. Ananiadou, S.; Rea, B.; Okazaki, N.; Procter, R.; Thomas, J. Supporting Systematic Reviews Using Text Mining. *Soc. Sci. Comput. Rev.* **2009**, *27*, 509–523. [CrossRef]
77. Rajput, H. Social Media and Politics in India: A Study on Twitter Usage among Indian Political Leaders. *Asian J. Multidiscip. Stud.* **2014**, *2*, 63–69.
78. Ahmed, S.; Cho, J.; Jaidka, K. Leveling the playing field: The use of Twitter by politicians during the 2014 Indian general election campaign. *Telemat. Inform.* **2017**, *34*, 1377–1386. [CrossRef]
79. Bouvier, G. What is a discourse approach to Twitter, Facebook, YouTube and other social media: Connecting with other academic fields? *J. Multicult. Discourses* **2015**, *10*, 149–162. [CrossRef]
80. Livingstone, S. Taking risky opportunities in youthful content creation: Teenagers' use of social networking sites for intimacy, privacy and self-expression. *New Media Soc.* **2008**, *10*, 393–411. [CrossRef]
81. Suárez-Rico, Y.; Gómez-Villegas, M.; García-Benau, M. Exploring Twitter for CSR Disclosure: Influence of CEO and Firm Characteristics in Latin American Companies. *Sustainability* **2018**, *10*, 2617. [CrossRef]
82. Heale, R.; Forbes, D. Understanding triangulation in research. *Evid. Based. Nurs.* **2013**, *16*, 98. [CrossRef] [PubMed]
83. Graham, T.; Broersma, M.; Hazelhoff, K.; van 't Haar, G. Between broadcasting political messages and interacting with voters: The use of Twitter during the 2010 UK general election campaign. *Inf. Commun. Soc.* **2013**, *16*, 692–716. [CrossRef]
84. Gee, J.P. *How to do Discourse Analysis: A Toolkit*, 2nd ed.; Routledge: London, UK, 2014.
85. Mohammadi, M.; Javadi, J. A Critical Discourse Analysis of Donald Trump's Language Use in US Presidential Campaign, 2016. *Int. J. Appl. Linguist. English Lit.* **2017**, *6*, 1. [CrossRef]
86. Hamaguchi, T. Prospects for Self-Reliance and Indigenisation in Automobile Industry: Case of Maruti-Suzuki Project. *Econ. Polit. Wkly.* **1985**, *20*, M115–M122.
87. Kathuria, S. Commercial Vehicles Industry in India-A Case History, 1928-1987. *Econ. Polit. Wkly.* **1987**, *22*, 1809–1823.
88. Bhave, P.; Kulkarni, N. Air Pollution and Control Legislation in India. *J. Inst. Eng. Ser. A* **2015**, *96*, 259–265. [CrossRef]
89. Beckerman, W. *Economic Development and the Environment: Conflict of Complementarity?* The World Bank Policy Research Working Paper Series: Washington, DC, USA, 1992.
90. Ranawat, M.; Tiwari, R. *Influence of Government Policies on Industry Development: The Case of India's Automotive Industry*; Hamburg University of Technology (TUHH), Institute for Technology and Innovation Management: Hamburg, Germany, 2009.
91. Thakurta, P.G. Ideological Contradictions in an Era of Coalitions: Economic Policy Confusion in the Vajpayee Government. *Glob. Bus. Rev.* **2002**, *3*, 201–223. [CrossRef]
92. Sen, G. *Million Cars for Billion People*; Vemuri, S., Ed.; Platinum Press: Mumbai, India, 2014.
93. George, S.; Jha, R.; Nagarajan, H. The Evolution and Structure of the Two-wheeler Industry in India. *Int. J. Transp. Econ.* **2002**, *29*, 63–89. [CrossRef]
94. ACMA The Automotive Component Manufacturers Association of India-ACMA. Available online: https://www.acma.in/ (accessed on 28 March 2021).
95. SIAM Society of Indian Automobile Manufactures. Available online: https://www.siam.in/members.aspx?mpgid=1&pgidtrail=4 (accessed on 28 March 2021).
96. FADA About FADA-FADA India. Available online: http://www.fadaindia.org/about-fada.html (accessed on 28 March 2021).
97. ASDC Automotive Skills Development Council: Automotive Training Certification Centre-ASDC. Available online: https://www.asdc.org.in/ (accessed on 28 March 2021).
98. Supreme Court of India M.C. Mehta vs Union Of India on 24 October, 2018. Available online: https://indiankanoon.org/doc/73307198/?__cf_chl_jschl_tk__=99ff6f2ccfb0126a2b2fa4446979491edc28fb2b-1616939067-0-Ab8pAzQ12CuaplwxLpnMPGWs4SHPFjoOPfKVBFn0c2uWXPQwRIG5-XhigfKPKasEfyBERMkfcTRpHbk2B-Zy33UF2P13ATSZoAAS5UAPXL7aV72CBJiGrg-FeUmeW6-SMC137yNZhN_2YAwHfOdTwTyfbCd_VQzBxJauz9sUCuD7Iq_PGMZDH3gVaYV1VV73EZVA0qo4G8OF5o5qSTe58HUT9C83-7oCMOn4Us80jZylJ1-pTSCSPFVjBnnxRRcClX2jojDVirmtV5jLGdgK2LZ-VnftGuWd-SLpkiLje9DYn-5RuNOeQUOxgUzoBl3jvbGm2I5HBd8LGRX9YC5Q6kGWqWi9kgVskCeZSuWdWMhSMV_bWYm_3SEi9bEf2tmxpDo_ZTyR0oLj0l6DAZ1J-Ks (accessed on 28 March 2021).
99. The Times Of India Passenger Vehicle Sales Fall for Ninth Straight Month, Dive 31% in July. Available online: https://timesofindia.indiatimes.com/business/india-business/passenger-vehicle-sales-fall-for-ninth-straight-month-dive-31-in-july/articleshow/70654709.cms?utm_source=contentofinterest&utm_medium=text&utm_campaign=cppst (accessed on 28 March 2021).
100. The Economic Times Budget 2019: Pollution Control in Focus, Environment Ministry Gets Rs 2,954 Crore. Available online: https://economictimes.indiatimes.com/news/economy/policy/budget-2019-pollution-control-in-focus-environment-ministry-gets-rs-2954-crore/articleshow/70095085.cms (accessed on 28 March 2021).
101. Mukherjee, S. India's New Passenger Vehicle Financing Market to Double to Rs 1600 Billion by 2020-The Economic Times. Available online: https://economictimes.indiatimes.com/industry/auto/indias-new-passenger-vehicle-financing-market-to-double-to-rs-1600-billion-by-2020/articleshow/57269910.cms?from=mdr (accessed on 28 March 2021).
102. Hadkar, C. Supreme Court Allow BS4 Registrations – MotorOctane. Available online: https://motoroctane.com/news/207604-bs4-registrations (accessed on 28 March 2021).
103. Hudson, B.; Hunter, D.; Peckham, S. Policy failure and the policy-implementation gap: Can policy support programs help? *Policy Des. Pract.* **2019**, *2*, 1–14. [CrossRef]

104. Berkhout, F. Technological regimes, path dependency and the environment Technological regimes, path dependency and the environment. *Glob. Environ. Chang.* **2002**, *12*, 1–4. [CrossRef]
105. Berggren, C.; Magnusson, T.; Sushandoyo, D. Transition pathways revisited: Established firms as multi-level actors in the heavy vehicle industry. *Res. Policy* **2015**, *44*, 1017–1028. [CrossRef]
106. Shiroyama, H.; Kajiki, S. Case Study of Eco-town Project in Kitakyushu: Tension Among Incumbents and the Transition from Industrial City to Green City. In *Governance of Urban Sustainability Transitions*; Springer Japan: Tokyo, Japan, 2016; pp. 113–132.
107. Bajwa, N. Automobile Industry in India-Auto Sector Growth Analysis. Available online: https://www.investindia.gov.in/sector/automobile (accessed on 28 March 2021).
108. Markandya, A.; Sampedro, J.; Smith, S.J.; Van Dingenen, R.; Pizarro-Irizar, C.; Arto, I.; González-Eguino, M. Health co-benefits from air pollution and mitigation costs of the Paris Agreement: A modelling study. *Lancet Planet. Heal.* **2018**, *2*, e126–e133. [CrossRef]
109. Backhaus, J. Intermediaries as Innovating Actors in the Transition to a Sustainable Energy System. *Cent. Eur. J. Public Policy* **2010**, *4*, 86–109.
110. Flanagan, K.; Uyarra, E.; Laranja, M. Reconceptualising the "policy mix" for innovation. *Res. Policy* **2011**, *40*, 702–713. [CrossRef]

 *sustainability* 

Article

What Affects Chinese Households' Behavior in Sorting Solid Waste? A Case Study from Shanghai, Shenyang, and Chengdu

Yanmin He [1,*], Hideki Kitagawa [2], YeeKeong Choy [3], Xin Kou [4] and Peii Tsai [5]

1 Faculty of Economics, Otemon Gakuin University, Osaka 567-8502, Japan
2 Faculty of Policy Science, Ryukoku University, Kyoto 612-8577, Japan; kitagawa@policy.ryukoku.ac.jp
3 Faculty of Economics, Keio University, Tokyo 108-8345, Japan; choy3293@gmail.com
4 School of Management, Shenyang Jianzhu University, Shenyang 110015, China; cat.1639@hotmail.com
5 GCI, Yokohama City University, Yokohama 236-0027, Japan; peii.tsai555@gmail.com
* Correspondence: kaenmin@hotmail.com; Tel.: +81-72-665-5371

Received: 22 September 2020; Accepted: 22 October 2020; Published: 23 October 2020

Abstract: The main aim of this study was to examine residents' environmental behavior in sorting solid household waste, and to identify the integrative factors that contribute to their waste-separation cooperation and other related pro-environmental behaviors. This was achieved based on a questionnaire survey in Shenyang, Chengdu, and Shanghai. Methodologically, we applied a discrete choice model to examine whether individuals' garbage sorting behaviors differ based on their characteristics, social attributes, residential circumstances, and environmental awareness, and whether these factors are correlated with individuals' receptiveness to a refuse charge system, or to policies requiring garbage sorting. We further examined whether individuals' garbage sorting behavior, their receptiveness to fee-based waste collection, and their receptiveness to policies requiring garbage sorting differ across areas. In this particular survey, we introduced a 16 item scale of pro-environmental behavior and a nine item scale of altruism to ascertain the ways in which internal motivational factors affect people's environmentally conscious voluntary behavior. Overall, the present work is expected to contribute to an important understanding of the motivational forces and incentives behind human pro-environmental behavior and action. It also brings relevance to the analysis of moral solidarity in relation to the household waste disposal problems currently confronting us today.

Keywords: municipal solid waste; garbage sorting behavior; environmental awareness; pro-environmental behavior; altruism

1. Introduction

Rapid economic growth, urbanization, and the steep growth in the global population and consumption rate have resulted in increased waste production at an unprecedented rate. In 2016, the worlds' cities generated 2.01 billion tonnes of municipal solid waste (MSW). East Asia and Pacific regions currently generate most of the world's waste, at 23% (468 million tonnes). However, the fastest growing regions in waste generation are Sub-Saharan Africa and South Asia, where the total waste generation is expected to more than double by 2050, making up 35% of the world's waste. The Middle East and North Africa regions are also expected to double their waste generation by 2050. It is also noteworthy that at least 33% of the global MSW is not managed in an environmentally safe manner. Ineffective waste management will cause serious air, soil, and groundwater pollution. This will not only hamper sustainable urban environment but will also threaten the health of residents [1]. Chen et al. shows that, without stringent policy directives, due to both the strong continued growth of the total waste generation and the slow increase of sustainable treatment shares, there is no absolute decoupling effect for the waste observed as countries become richer (with the exception of Japan, where other institutional and cultural factors may play a role), and there is thus little evidence for a waste-related environmental Kuznets curve [2]. It thus follows that sustainable waste management will be a major challenge for many countries, especially developing countries, in the coming decade.

However, effective waste management is costly, often taking up 20% to 50% of municipal budgets [3]. With limited funding, citizen engagement involving behavioral change and public participation has become one of the most cost-effective means in promoting sustainable waste management [4]. Many countries—such as Korea, the Philippines, and Thailand—use behavior-changing variable fees to motivate waste reduction, source-separation, and reuse. In Nepal, the government uses results-based financing to create a sustainable behavioral change in waste disposal, while in Jamaica, the government employs various educational strategies to induce behavioral change in safe and environmentally-friendly waste disposal practices [1].

In view of the far-reaching environmental and health repercussions arising from continued and increased waste production, sustainable solid waste management is every country's business. This is particularly true for China, which has the largest population in the world. Demonstrably, China generates large volumes of MSW annually due to its huge population and high consumption rates. In 2016, about 47 percent (220 million tonnes) of waste in the East Asia and Pacific regions was generated by China, although its daily per capita waste generation rate of 0.43 kg is below the regional average of 0.56 kg [1]. In addressing its increased waste production problems, China has also adopted various environmental behavioral control strategies to induce sustainable household waste generation, separation, and disposal practices. The present article aims to examine the empirical evidence of the determinants of household environmental behavior in MSW management in China.

To begin with, the impressive economic growth in China for the past few decades has accompanied rapid and massive urbanization across the country. Inevitably, this has given rise to one of the most formidable challenges facing the country today: the MSW problem. Viewed from China's perspective, MSW includes solid wastes defined under national laws and administrative regulations, as well as solid wastes discharged from residents' daily activities or from the necessary services provided for such activities. In general, MSW is categorized into household solid waste, commercial solid waste, market solid waste, solid waste from street cleaning, solid waste from public facilities, and business-related solid waste. In this paper, we focus our research on the city's daily household solid waste (HSW).

In 2017, the amount of HSW collected and transported to HSW disposal sites and final disposal facilities located nationwide reached approximately 200 million metric tonnes [5]. The massive amount of HSW has intuitively given rise to an urgent need for the government to establish a proper solid waste management system in order to contain its environmentally destructive effects. It may well be that the improper handling of HSW can cause soil contamination, foul odors, water pollution, and other environmental problems, which impair human health and hamper urban sustainable development.

In light of the above, it may be remarked that rapid urbanization and the improvement of people's living standards have significantly raised human environmental awareness and interest in environmental issues, especially in relation to HSW disposal. However, the local communities where 'Not In My Back Yard' (NIMBY) facilities are located, especially in developing countries with high population densities, have quite strongly opposed land acquisitions for the development of solid waste disposal sites. In China, annually, and with more than five anti-incinerator demonstrations, local residents have claimed the relocation of municipal solid waste incineration facilities from 2007 to 2016, and the facility that would benefit the public the most was aborted. This caused a drastic lack of disposal sites or land resources for the safe disposal of HSW [6,7].

In order to address the above issues, the Chinese government launched its HSW sorting policy within eight cities in the early 2000s as a sustainable waste management pilot project under the auspices of the Ministry of Construction, in order to encourage waste minimization in all of the sectors of the community. Nonetheless, the project was poorly implemented, and failed to achieve its aim as expected.

Indisputably, if the HSW problem is left unmitigated, it would exert great impacts in upsetting the government's vision of an 'Ecological Civilization' (EC).

The EC is a political vision introduced in 2007 by the former president, Hu Jintao. It aims to uphold harmony between humanity and nature as one of the basic strategies to promote sustainable development, especially in relation to pollution reduction, a circular economy, a low-carbon economy, and green development. As regards MSW, the green concept of ecological civilization was formally endorsed under China's Circular Economy Promotion Law in 2009, as discussed below. The EC concept was progressively elevated to the rank of a paramount objective of the Chinese Communist Party (CCP) during the 18th National People's Congress of the Communist Party of China, held in 2012. It was further enshrined as a constitutional principle in the Constitution of the People's Republic of China (PRC) in 2018 [8].

It is worth reiterating, in light of the foregoing, that the EC vision is distinctly concerned with the orientation of a low-carbon, eco-friendly, and resource-efficient society that underpins the promotion of a green economy. Here, it may be remarked that proper waste disposal is related to not only the reuse/recycling of resources, but also to the creation of a sustainable society through the optimal utilization of waste as an urban energy source. A case in point is the transformation of household kitchen garbage into biogas. Increasingly, efforts to promote the separate collection of HSW based on the guiding principles of garbage reduction, recycling, and detoxification have attracted the attention of many individuals in various parts of China in recent years [9].

Generally speaking, solid waste management comprises two methods of waste collection, namely, a fixed-price system and a quantity-based pricing system. A municipality with a fixed-price system charges a fixed fee per household or household member for waste disposal, regardless of the amount of garbage originating from the household. The fee is not linked to the amount of garbage discarded, so the system is not very effective in reducing the amount of garbage disposed. In contrast, for the quantity-based pricing system, the fee levied for waste disposal changes in accordance with the amount of garbage discarded. One prominent example is the paid garbage bag system, which requires residents to use garbage bags that meet certain standards (designated bags).

Manifestly, there have been various legislations put in place to promote sustainable HSW management in China. One of the most prominent legal instruments is China's Circular Economy Promotion Law, which was implemented in 2009, as briefly noted above. It legislates the principles of waste reduction, reuse, and recycling. Moreover, the government controlling bodies—such as the Ministry of Environmental Protection (MEP), the Ministry of Housing and Urban-Rural Development (MOHURD), and the National Development and Reform Commission (NDRC)—also play an important role to ensure effective law enforcement and observation. The local People's Governments also enact various regulations and measures to strengthen the sustainable disposal and recycling of HSW practices. For example, in 2011, the State Council issued a Notice on the Opinions on Further Strengthening the Work of Municipal Solid Waste Disposal (State Council, 2011, document No. 9) to reinforce sustainable HSW practices [10]. It further introduced new HSW disposal fees based on a 'discarder pays principle', which is similar to the 'polluter pays principle' (those, irrespective of being consumers or producers, who generate it are the ones who pay). Subsequently, efforts to promote waste sorting and waste reduction via the implementation of the fee-based waste disposal principle have become active at the local government level.

Despite these novel efforts, casual observation on the ground seems to indicate that they have not been able to contribute effectively to foster proper waste-sorting habit-formation among the residents. Furthermore, the separation of HSW was also not properly maintained after collection in conformity with the stipulated waste segregation guidelines. Experience shows that waste still remains practically unsorted. Worse yet, the residents were indifferent to the adoption of the green principles of waste sorting and separation in line with the EC concept. In order to resolve these problems, the State Council issued the Proposed Method of Implementing the Sorting System for Municipal Solid Waste [11] in

March 2017, with the aim to mandate the separate collection of HSW in 46 designated cities by the end of 2020.

Against this backdrop, the present study aims to assess the effects of the new mandate on the behavioral change of waste sorting and separation in three selected regions out of the 46 designated pilot cities, namely, Shanghai, Shenyang (Liaoning Province), and Chengdu (Sichuan Province). This is achieved based on questionnaire surveys. In particular, the questionnaire research aims to examine residents' behavior in sorting HSW. This covers the assessment of the factors that contribute to residents' receptiveness of waste separation policies, as well as other variables in relation to sustainable HSW management. The latter include personal demographics, personal attitudes, external moderators such as circumstances and economic incentives, and internal moderators such as environmental concerns and human altruism.

2. Literature Review

The separate collection of HSW makes it possible to sustainably curb the amount of garbage generated at its sources and mitigate its adverse environmental impacts at the final disposal sites. Requiring households to sort garbage, however, burdens the government with an enormous cost of monitoring compliance. Therefore, the effectiveness of waste management requirements strongly depends on whether the household residents comply with them without being monitored or coerced. However, the reality is that the waste sorting programs, as implemented by the government, often fail to persuasively induce residents to comply with the guidelines as stipulated [12].

That said, to enhance the effectiveness of garbage sorting compliance, it is necessary to influence household residents' waste disposal behavior. Linde'N and Carlsson-Kanyama [13] and Antonides and van Raaij [14] divide the factors that affect residents' waste disposal behavior into external motivational factors and internal motivational factors. External motivational factors include administrative measures, such as laws and regulations, economic measures, information measures, and physical measures. The authors concerned posit that a policy package that combines an assortment of these four measures can affect residents' garbage sorting behavior. Insofar as the internal motivational factors are concerned, environmental knowledge, environmental concern, environmental values and attitudes, behavioral preferences, lifestyle preferences, and social influence are often considered to be some of the important motivating means of behavioral change [8].

On the other hand, Lindhqvist unveiled three factors that promote the separate collection of household garbage, namely, economic incentives, the level of convenience associated with discarding garbage, and information [15]. However, Dahlén and Lagerkvist argue that access to opportunities and places for separate disposal is an important factor affecting the rate of separate collection [16]. In addition, Chappells et al. assert that the introduction of a system that monitors the separate disposal of garbage encourages proper garbage sorting behavior [17]. Others, such as Judge and Becker [18], Linde'N and Carlsson-Kanyama [13], and Ando and Gosselin [19] emphasize the effects of the placement of bins, the ease of sorting garbage, and increased convenience in terms of the timing of garbage collection. Houtven and Morris, however, argue that residential ownership has a noticeable effect on the household's waste generation and disposal behavior [20].

From China's perspective, Ghorbani et al. argue that members of the Communist Party of China (CPC) and the All-China Federation of Trade Unions not only have a stronger environmental awareness, but also a greater willingness to pay for environmental protection activities than the general public [21]. Similarly, Clark et al. claim that being altruistic and having environmentally friendly attitudes affects people's environmentally-conscious voluntary behavior, and that altruistically inclined individuals are more likely to participate in a green electricity program [22]. People with a strong sense of place or belonging to their city or community tend to actively participate in the city's government activities. A sense of place may be defined as the meaning ascribed, and the attachment formed, to a place by an individual or communities. This sense of place may be grounded on the emotional values that individuals ascribe to the surrounding environment [23–25]. Another factor that strongly influences an individual's sense of belonging to their city or community is the sense of one's own status. Here, it is appropriate to remark that, in China, people's household registration cannot be freely changed. Thus, even within a city, there is a stark difference between those who have an urban household registration and those who have a rural household registration. In view of this, an individual's sense of their own status may impact considerably on their degree of their sense of belonging to their city or community. For this reason, this study takes into account the survey respondents' household registration status, with a view to ascertain the relationship between a sense of place, a sense of belonging, and environmental behavior in relation to HSW disposal.

The above studies provide sufficient indication that there are many factors that influence one's environmental behavior and action. However, as regards China, it is unclear as to what categories of factors or behavioral determinants contribute to drive an individual's environmental behavioral change. Against this premise, what follows is an attempt to examine the motivational forces and other integrative factors that underpin household residents' garbage sorting habits.

3. Materials and Methods

3.1. Research Hypotheses and Model

Environmental behavior is not only related to social and economic factors. We must never lose sight of the fact that it is also unalterably affected by the levels of environmental knowledge, environmental awareness, and environmental concern. Following logically from this line of thought, and inspired by the work of Rylander and Allen, we introduce an internal variable of altruistic attitudes into an integrative environmental behavioral framework in order to assess the motivational forces that influence an individual's pro-environmental behavior (Figure 1) [26]. The framework considers demographic variables and knowledge to be the integrative factors that shape individual attitudes towards environmentally friendly behavior. Here, it must be admitted that attitudes do not necessarily influence receptiveness directly, as their effect is moderated by multifaceted internal and external variables, such as environmental concern, altruism, information, and economic incentives. The close link between attitudes and receptiveness may be the most tenuous aspect of the model. This study attempts to examine whether individuals' garbage sorting behavior is contingent on their characteristics, social attributes, residential circumstances, or environmental awareness. We also consider whether these factors are correlated with individuals' receptiveness to garbage sorting behavior, to fee-based waste collection, or to policies requiring garbage sorting. Here, it may be noted that our research is premised on self-reported receptiveness, which may be over-stated by the respondents concerned.

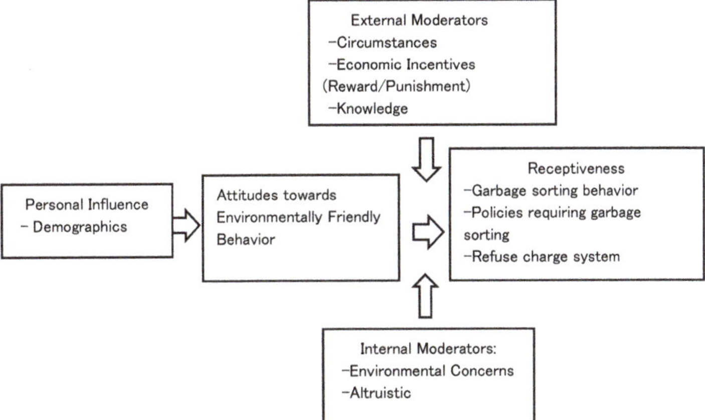

Figure 1. The research framework.

3.2. Survey Area

The survey was conducted in Shanghai, Shenyang, and Chengdu, which are representative cities of the eastern region, northern region, and western region of China, respectively (Figure 2). In 2017, the permanent registered population was 14.55 million in Shanghai, 7.4 million in Shenyang, and 14.35 million in Chengdu. The annual GDP was 3063.3 billion yuan in Shanghai, 586.5 billion yuan in Shenyang, and 1388.9 billion yuan in Chengdu in 2017 [27].

Figure 2. Location map of Shanghai, Shenyang, and Chengdu. Source: Ministry of Natural Resources of the People's Republic of China (PRC).

Generally, these three cities categorize household garbage into four types: recyclable, hazardous, biodegradable (specifically known as 'wet waste' in Shanghai), and other waste (categorically called 'dry waste' in Shanghai). Both Shanghai and Chengdu stipulate local rewards for residents' sorting behaviors. Shanghai was selected in June 2000 for the implementation of China's first nationwide pilot program for garbage separation and collection, along with Beijing, Nanjing, Hangzhou, Guilin, Guangzhou, Shenzhen, and Xiamen. As past experience reveals, a high monitoring cost makes it hard for the government to constantly detect the separation of solid waste from each household. It is therefore necessary to introduce some kind of incentive to induce social conscience for sustainable waste disposal behavior and practices.

In 2014, the Shanghai Huizhong Green Corporate Social Responsibility was established, with the aim to encourage citizens to sort kitchen garbage and other waste based on an economic incentive system, called the Green Account program. Under this system, one Green Account card is issued to one household upon request. If the household discards sorted kitchen garbage during the designated hours (7:00 to 9:00 in the morning, and 17:00 to 19:00 in the evening), it receives ten points each time (up to a maximum of 20 points per day). The points received are valid for two years, and can be exchanged for everyday items, movie tickets, or park tickets, among other things [28,29].

In Chengdu, Sichuan Province, the Chengdu Green Earth Environment Technology Co., a pioneer in the waste classification field in China, introduced the Green Earth program in 2008, with the aim to encourage residents to sort recyclable waste. Green Earth provides each family with a unique barcode sticker that can be put on the trash bag for identification. Households can receive reward points based on the weight of the trash bags placed in the designated recycling bins. For example, every 100 g of normal recyclable garbage (paper, plastic, and metal), every 200 g of glass, or every 500 g of clothes will receive one green point. The points received are valid for two years, and can be exchanged for detergents, toothpastes, and other everyday items [30].

3.3. Questionnaire

The questionnaire comprised eight sections: social demographic information, circumstances, behavior, knowledge, receptiveness, a pro-environmental behavior scale, and an altruism scale. Table 1 shows the question items for all of the respondents.

Table 1. Question items.

Demographic Information	Gender, age, occupation, education level, marital status, residence type and ownership, income, household registration status, years residing, political affiliation
Circumstances of discarding sorted household garbage	Rules on garbage sorting, equipment at garbage collection spot, time to collection spot, elevator exists or not, frequency of garbage disposal
Knowledge on sorting household garbage	Knowledgeable or not knowledgeable about garbage sorting, have or do not have the correct knowledge on garbage sorting
Behavior related to sorting household garbage	Pre-disposal sorting at home or not, end-point sorting at collection spot or not
Economic incentives	On the Green Account program in Shanghai: - Knowledgeable or not knowledgeable about the Green Account - Knowledgeable or not knowledgeable about the purpose of the Green Account - Knowledgeable or not knowledgeable about the use of the Green Account On the Green Earth program in Chengdu: - Knowledgeable or not knowledgeable about the Green Earth program - Knowledgeable or not knowledgeable about the purpose of the Green Earth program - Knowledgeable or not knowledgeable about the use of the Green Earth program In Shenyang, we used the following hypothetical question, because there is no economic incentive yet: - Do you agree or not agree with the "Pay-As-You-Throw" program?
Receptiveness to policies requiring garbage sorting	Support or oppose garbage sorting
Receptiveness to a refuse charge system	Support or oppose quantity-based charging for garbage collection, amount willing to pay for a refuse charge system, charging method (charging according to water use fee, charging designated bags, charging according to the number of household members, or charging according to household income)
Other scales	Pro-environmental behavior scale Altruism scale

3.4. Sample Characteristics

The empirical research was conducted in two stages. Firstly, data were collected by a preliminary survey (300 people) in September 2017 as a pretest; secondly, a main survey (2100 people) was conducted through an online platform (Wen Juan Xing). The present research was analyzed based on the data accumulated from the main survey. The main survey, which involved various groups of adults aged between 18 and 70 years, was conducted between December 2017 and the end of January 2018. A sample of 2100 adults between 18 and 69 years old was randomly selected from the local population. After the elimination of responses with outliers and missing values, the number of valid responses was 612 for Shanghai, 484 for Shenyang, and 525 for Chengdu (1621 in total).

The demographic composition of the sample is shown in Table 2. In terms of the respondents' gender, the sample contains slightly more females than males (819 women vs. 802 men). As for the age category, the number of respondents in their 40s is the largest, and the number of respondents in their 60s is significantly smaller (505 respondents aged 18 to 29 years, 439 respondents in their 30s, 535 respondents in their 40s, 126 respondents in their 50s, and 16 respondents aged 60 years and over). The smaller number could be attributed to the low rate of internet use among seniors. The sample consists of 124 respondents who have achieved a middle school diploma or obtained a lower educational achievement, 456 respondents with a high school or specialized school diploma, 944 respondents with a university or advanced specialized school diploma, and 97 respondents with a graduate school diploma. It is thus clear that 54% of the respondents have completed at least a university-level education. The sample is more or less evenly split between those who have political affiliation and vice versa: 308 Communist Party of China members, 348 Communist Youth League members, 21 members of other non-communist or minor parties, 879 respondents with no political affiliation, and 65 respondents identified as 'other.' The number of respondents who had a household registration for their area of residence (referred to as 'natives') is 916, which is higher than the number of respondents who did not have a household registration for their area of residence (referred to as 'outsiders'), which is 705. The gender ratio, education, and income level are all consistent with the Statistical Yearbook. Hence, the survey reflects the real situation in the three cities under consideration.

Table 2. The profile of subjects (N = 1621).

	N (number)	Percentage (%)
Gender		
Male	802	49.5
Female	819	50.5
Age		
18–29	505	31.2
30–39	439	27
40–49	535	33
50–59	126	7.8
60 or above	16	1
Education level		
Junior high and below	124	7.7
Senior high or senior secondary	456	28.1
Undergraduate or junior college	944	58.2
Postgraduate or above	97	6
Occupation		
Administrative	33	2
Public-sector organizations	162	10
State-owned enterprise	265	16.3
Private enterprise	411	25.4
Social organization	17	1.1
Self-employed/liberal profession	176	10.9
Foreign capital enterprise	133	8.2
Farming, forestry, fishery workers, or others	123	7.6
Housewife	66	4.1
Retired	36	2.2
Student	88	5.4
Unemployed	21	1.3
Others	90	5.6
Income (RMB, thousand yuan)		
1–3	184	11.2
3.001–6	435	26.5
6.001–9	335	20.7
9.001–12	215	13.3
12.001–15	167	13.3
15.001–20	125	7.71
20.001–30	81	5
30.001 or above	79	4.9

Table 2. *Cont.*

	N (number)	Percentage (%)
Political affiliation		
Member of the Communist Party of China (CPC)	308	19
Member of the Communist Youth League of China	348	21.5
Member of other non-communist or minor parties	21	1.3
Commoner	879	54.2
Other	65	4
Marital status		
Single	396	24.4
Married	1184	73
Other	41	2.5
Household registration status		
Native	916	56.5
Outsiders	705	43.5

Source: authors' own calculations, 2020.

4. Data Analysis and Results

4.1. Descriptive Statistics

It was uncovered that approximately 80% of the targeted individuals responded that they usually require less than three minutes (300 m) to arrive at the closest garbage collection spot. It was also revealed that more than half of the communities adhere to the rules on garbage sorting, and had installed garbage bins for separate disposal. Nonetheless, the proportion of respondents who consistently or almost frequently sort their garbage at home before discarding is less than 40%. Roughly 40% of the respondents revealed that they separately discard garbage at the collection spot. Among them, 55.6% confirmed that they are intellectually equipped with the correct knowledge of garbage sorting. In contrast, about 40% of the respondents were found to sort garbage based on incorrect knowledge (see Figure 3).

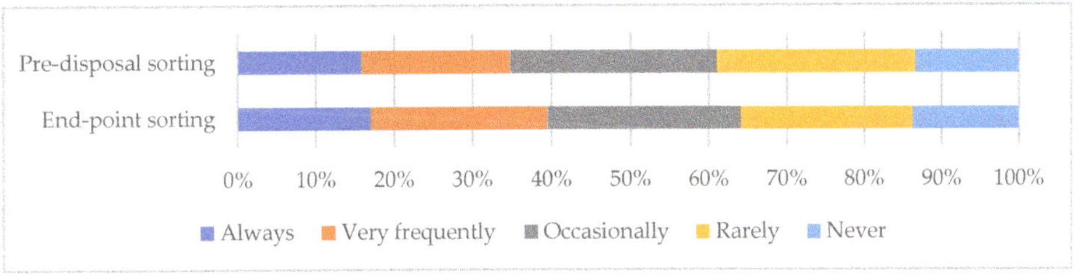

Figure 3. Behavior related to sorting household garbage. Source: authors' own calculations, 2020.

With respect to the garbage sorting program, 60% of the respondents asserted that they support it, 20% asserted that they somewhat support it, and only 1% opposed it. However, the proportion of respondents who agree or slightly agree to comply with the refuse charge system is 40% (see Figure 4). In other words, although 60% of the respondents are highly supportive of the garbage sorting program, an equal number (60%) of them are indifferent to the refuse charge system.

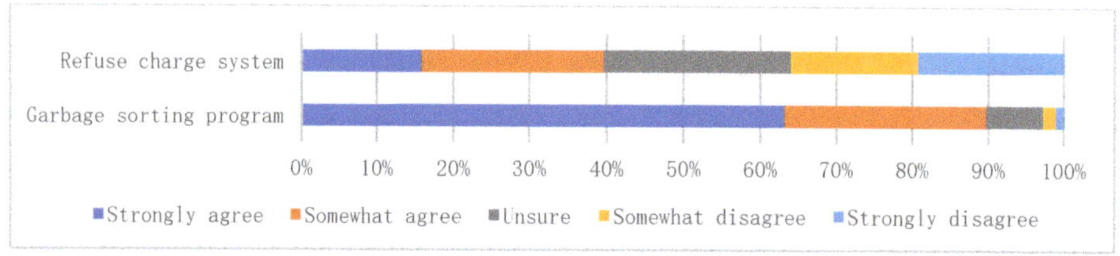

Figure 4. Receptiveness to the refuse charge system. Source: authors' own calculations, 2020.

The survey also inquired into the environmental and moral value judgements of the respondents in relation to a 16-item scale of pro-environmental behavior and a nine-item scale of altruism (see Table 3). In creating the pro-environmental behavior scale, this study took its clues from a study by Zhao et al., and used a new approach that produces an accurate and reliable measurement based on the separate analysis of three aspects of pro-environmental behavior: purchase, use, and reuse [31].

Table 3. Pro-environmental behavior scale items.

	SA	SWA	U	SWD	SD	$r_{i\text{-}t}$	α	FL
(1) If I can avoid using disposable items, for example, plastic bags and disposable dishware, the environmental quality will be improved.	37.26	39.36	14.99	5.37	3.02	0.336	0.748	0.423
(2) If no one is protecting the environment together with me, I will not do it myself.	30.66	21.84	21.28	17.27	8.95	0.113	0.771	0.064
(3) I always purchase home appliances with 'China Environmental Labelling' or 'Energy Conservation Certificate' stickers.	56.57	27.51	12.09	2.65	1.17	0.517	0.736	0.634
(4) I always take a taxi or use my own car when I go out. I seldom take public transportations such as subways and buses.	27.21	23.75	16.1	18.57	14.37	0.011	0.783	−0.037
(5) I use my own mug cup at work instead of disposable paper cups.	72.61	16.59	5.55	2.47	2.78	0.459	0.740	0.587
(6) I print double-sided and use both sides of paper when I write too.	55.52	26.77	11.54	3.89	2.28	0.550	0.732	0.664
(7) I use food storage containers instead of cling wrap or aluminium foil.	43.49	26.96	17.95	7.9	3.7	0.462	0.737	0.568
(8) I always turn off the computer when I get off or when I think that it will not be used for a long time.	64.71	22.52	7.9	3.02	1.85	0.561	0.733	0.664
(9) I always purchase and use rechargeable batteries.	28.19	25.17	20.85	14.37	11.41	0.267	0.756	0.371
(10) I always reuse wastewater, for example, wastewater from the washing machine and washing rice.	31.15	31.4	19.19	11.97	6.29	0.476	0.735	0.514
(11) I always purchase pens with refills or fountain pens.	44.66	30.6	14.62	6.48	3.64	0.564	0.728	0.632
(12) I usually throw my used clothes away directly into bins.	25.11	26.4	19.86	16.96	11.66	0.515	0.737	0.034
(13) I only use an air conditioner when it is necessary.	61.26	27.27	7.53	2.41	1.54	0.086	0.773	0.594
(14) I will set the air conditioner at 28 °C in the summer.	28.19	28.32	19.74	13.08	10.67	0.343	0.748	0.361
(15) I will set the air conditioner (heater) at 20 °C in the winter.	26.1	30.1	21.28	11.54	10.98	0.369	0.745	0.371
(16) I will try to run larger and fuller loads of laundry.	50.22	29.98	10.73	5.31	3.76	0.452	0.738	0.502

Source: Authors' own calculations, 2020. Notes: SA is 'strongly agree', SWA is 'somewhat agree', U is 'unsure', SWD is 'somewhat disagree', SD is 'strongly disagree', ri-t is the item-total correlations, α is Cronbach's coefficient alpha, and FL is factor loading. Percentages may not sum to 100 due to rounding.

The altruism scale was constructed as a new scale for this research. We applied the Schwartz norm-activation principles to measure altruistic attitudes (see Table 4). According to Schwartz's model, altruistic behavior arises from personal norms if two criteria are met: an individual must be aware that particular actions (or inactions) have consequences for the welfare of others (awareness of consequences,

AC), and an individual must ascribe responsibility for the consequences of those actions to himself or herself (ascription of responsibility, AR) [32,33]. The simultaneous presence of AC and AR in a specific situation enables pertinent personal norms to motivate behavior.

Table 4. Altruism scale items.

Item	SA	SWA	U	SWD	SD	r_{i-t}	α	FL
(1) I implement garbage sorting only when it helps to lower my own expenses.	12.95	19.68	21.84	28.93	16.59	0.188	0.403	−0.109
(2) Contributions to community organizations can greatly improve the lives of others.	1.54	2.28	11.54	36.21	48.43	0.293	0.370	0.755
(3) The individual alone is responsible for his or her satisfaction in life.	31.71	29.55	16.66	13.63	8.45	0.095	0.447	−0.262
(4) It is my duty to help other people when they are unable to help themselves.	0.99	1.85	11.66	36.71	48.8	0.306	0.369	0.762
(5) Many of society's problems result from selfish behavior.	1.91	3.64	16.59	32.63	45.22	0.134	0.422	0.580
(6) Households like mine should not be blamed for environmental problems caused by energy production and use.	27.82	26.9	24.98	14.37	5.92	0.109	0.437	−0.223
(7) My responsibility is to provide only for my family and myself.	22.33	31.65	18.26	18.88	8.88	0.253	0.371	−0.097
(8) Using renewable energy is the best way to combat global warming.	1.79	4.38	18.14	33.87	41.83	0.081	0.440	0.547
(9) It is possible that my personal actions can greatly improve the well-being of other people.	1.97	3.89	18.75	36.4	38.99	0.237	0.386	0.631

Source: Authors' own calculations, 2020. Notes: SA is 'strongly agree', SWA is 'somewhat agree', U is 'unsure', SWD is 'somewhat disagree', SD is 'strongly disagree', ri-t is item-total correlations, α is Cronbach's coefficient alpha, and FL is factor loading. Percentages may not sum to 100 due to rounding.

Durkheim provided an eloquent analysis of the importance of moral norms in influencing human collective behavior [34]. Personal norms are perceived moral norms that represent personal beliefs about what is right and wrong when acting in a particular manner in a specific situation. Moral norms are a shared set of beliefs, values, and ideas on what is presumed to be the right behavior. They regulate social life or human affairs by guiding and restraining individual behavior and action that produces adverse consequences for other members of society. Thøgerson [35] discussed the role of social norms in restricting individualism in favor of collectivism; in 2009, Thøgerson [36] further explored the strength of a person's norms—namely, subjective social norms and personal norms—in guiding environmentally responsible behavior. Personal norms or moral norms inform our sense of identity and behavior. They represent the will of individuals to altruistically prioritize collective interest over self-interest.

The research presented here applies Schwartz's principles in the form of a general altruism scale based on Clark et al. [22]. The scale contains a total of nine items that test for the presence of individual personal norms, AC, and AR. Specific items are listed in Table 4. Items 1, 3, and 4 refer to personal norms; Items 2, 5, and 8 represent AC; and Items 6, 7, and 9 represent AR.

For the altruism scale, Cronbach's coefficient alpha was used to evaluate internal consistency. When all of the nine items were used, Cronbach's coefficient alpha was found to be 0.434. Given this, the items with higher values (namely, Items 3, 6, and 8) were eliminated from the analysis.

This study used a discrete choice model for the analysis. The objective variables considered are end-point garbage sorting behavior ('sorting behavior'), receptiveness to a refuse charge system ('receptiveness to fees'), and receptiveness to policies requiring garbage sorting ('receptiveness to policies'). The values of the garbage sorting behavior variable, an objective variable, were computed based on the responses derived from the questionnaires. These include, for example, "Do you sort garbage when you discard it at a collection spot in your community?" In answering the question, the respondents are required to choose one of the five response options: always, very frequently, occasionally, rarely, and never. A five-point Likert scale was used to assign values to the responses (5 = always, ... , 1 = never). The variable 'receptiveness to fees' corresponds to the question, "Do you support the idea that people discarding more garbage should be charged a higher fee?" Likewise, the variable 'receptiveness to policies' is related to the question: "Do you support the garbage sorting system?" For these questions, the respondents were required to select one of the five response options: strongly agree, somewhat agree, unsure, somewhat disagree, or strongly disagree. Again, a five-point Likert scale was used to assign values to the responses as stated (5 = strongly agree, ... , 1 = strongly disagree).

As shown in Table 5, the explanatory variables include: (1) *EI* (Knowledgeable or not knowledgeable ('Agree or not agree' in Shenyang) of the local economic incentive program: yes = 1, no = 0), (2) *PEB* (the score of the pro-environmental behavior scale), (3) *ALT* (the score of the altruism scale), (4) *OWNERSHIP* (have or do not have a residential ownership: yes = 1, no = 0), (5) *MANAGEMENT* (the communities have or do not have a property management: yes = 1, no = 0), (6) *RULE* (the communities have or do not have the rules on garbage sorting: yes = 1, no = 0), (7) *INFRA* (the communities have or do not have the infrastructure to support garbage sorting: yes = 1, no = 0), (8) *DIST* (distance to the garbage collection spot: 5 = more than or equal to 10 min, 4 = 7–9 min, 3 = 4–6 min, 2 = 1–3 min, 1 = less than 1 min), (9) *KNOWLEDGE* (knowledge on the city's household garbage sorting regulations, derived from the answers to the question: 5 = very familiar, 4 = familiar, 3 = have heard about, but not familiar, 2 = never heard about the regulations, but have heard about the sorting instructions, 1 = never heard about the regulations or the sorting instructions), and (10) demographic variables including *GENDER* (male = 1, female = 2), *AGE* (above 60 = 5, 50–59 = 4, 40–49 = 3, 30–39 = 2, 18–29 = 1), *MARRIAGE* (marital status, yes = 1, no = 0), *REGIS* (household registration status, native = 1, outsiders = 0), *EDUCATION* (education level: postgraduate or above = 4, undergraduate or junior college = 3, senior high or senior secondary = 2, junior high and below = 1), *OCCUPATION* (yes = 1, no = 0), *INCOME* (above 4001\$ = 8, 3001–4000\$ = 7, 2501–3000\$ = 6, 2001–2500\$ = 5, 1501–2000\$ = 4, 1001–1500\$ = 3, 501–1000\$ = 2, less than 500\$ = 1), *POLITICAL* (political affiliation, member of the CPC = 4, the Communist Youth League = 3, non-communist or minor parties = 2, commoner or others = 1), and *CITY* (Shanghai = 1, Shenyang = 2, Chengdu = 3).

Table 5. Description of the explanatory variables.

Explanatory Variables	Description	Mean	SD (%)
EI	Knowledgeable or not knowledgeable ('Agree or not agree' in Shenyang) of the local economic incentive program	0.27	0.45
yes = 1			27.45%
no = 0			72.55%
PEB	The score of the pro-environmental behavior scale	62.61	8.49
ALT	The score of the altruism scale	22.57	3.21
OWNERSHIP	Have or do not have a residential ownership	5.97	0.49
yes = 1			59.72%
no = 0			40.28%
MANAGEMENT	The communities have or do not have a property management	0.78	0.41
yes = 1			78.16%
no = 0			21.84%
RULE	The communities have or do not have the rules on garbage sorting	0.5	0.5
yes = 1			50.34%
no = 0			49.66%
INFRA	The communities have or do not have the infrastructure to support garbage sorting	0.59	0.49
yes = 1			58.54%
no = 0			41.46%

Table 5. *Cont.*

Explanatory Variables	Description	Mean	SD (%)
DIST	Distance to the garbage collection spot	1.82	0.91
	more than or equal to 10 min = 5		2.47%
7–9 min = 4			2.65%
4–6 min = 3			11.54%
1–3 min = 2			41.52%
	less than 1 min = 1		41.83%
KNOWLEDGE	Knowledge on the city's household garbage sorting regulations	3.85	1.02
very familiar = 5			5.19%
familiar = 4			7.88%
	have heard about, but not familiar = 3		35.38%
	never heard about the regulations, but have heard about the sorting instructions = 2		35.10%
	never heard about the regulations or the sorting instructions = 1		16.44%
GENDER		0.49	0.5
male = 1			49.48%
female = 0			50.52%
AGE		2.2	1
above 60 = 5			0.99%
50–59 = 4			7.77%
40–49 = 3			33%
30–39 = 2			27.08%
18–29 = 1			31.15%
MARRIAGE	Marital status	0.73	0.44
yes = 1			73.04%
no = 0			26.96%
REGIS	Household registration status	0.57	0.50
native = 1			56.51%
outsiders = 0			43.49%
EDUCATION	Education level	2.63	0.71
	postgraduate or above = 4		7.65%
	undergraduate or junior college = 3		28.13%
	senior high or senior secondary = 2		58.24%
	junior high and below = 1		5.98%
OCCUPATION	Occupation status	0.81	0.39
yes = 1			81.43%
no = 0			18.57%
INCOME	Income level	3.52	1.93
above 4001\$ = 8			4.87%
3001–4000\$ = 7			5.00%
2501–3000\$ = 6			7.71%
2001–2500\$ = 5			10.30%
1501–2000\$ = 4			13.26%
1001–1500\$ = 3			20.67%
501–1000\$ = 2			26.84%
less than 500\$ = 1			11.35%
POLITICAL	Political affiliation	2.01	1.25
member of:			19.00%
the CPC = 4			
	the Communist Youth League = 3		21.47%
	non-communist or minor parties = 2		1.30%
	commoner or others = 1		58.24%
CITY		1.95	0.84
Shanghai = 1			37.75%
Shenyang = 2			28.86%
Chengdu = 3			32.29%

Source: authors' own calculations, 2020. Note: N = 1621, SD: Standard deviation, SD/%: this column refers to the standard deviation (SD) unless otherwise noted (%).

4.2. Estimation Results

The results of the analysis are shown in Tables 6 and 7. Table 6 shows the estimation results for all of the respondents. The estimated coefficients on *PEB* and *EI* are statistically significant in the expected direction. More specifically, the positive signs on both variables indicate that the stronger the environmentally friendly activities and knowledge are, the higher the probabilities of participating in the garbage sorting program, the receptiveness to fees, and the receptiveness to policies are. The result supports the idea that economic measures exert a positive impact on inducing and promoting residents' garbage sorting behavior. They also serve to increase their receptiveness to the introduction of a refuse charge program, or to policies mandating garbage sorting. Nonetheless, the altruism scale (ALT) only has a significant positive correlation with receptiveness to policies. This indicates that altruistic individuals will engage in pro-environmental behaviors when there are environmental benefits. However, when it comes to private benefits, various responses were noted. For instance, in response to the question of "What charging method do you think is appropriate if the government introduces quantity-based charging for garbage collection?", 416 respondents (the largest number) chose a designated bag system (Figure 5).

Table 6. Estimation results for all of the respondents.

Variable	Ordinal Logit Models			Ordinal Probit Models		
	Sorting Behavior	Receptiveness to Fees	Receptiveness to Policies	Sorting Behavior	Receptiveness to Fees	Receptiveness to Policies
EI	1.874 ***	0.706 ***	0.534 ***	1.127 ***	0.401 ***	0.300 ***
PEB	0.033 ***	0.020 ***	0.055 ***	0.020 ***	0.012 ***	0.030 ***
ALT	0.013	0.002	0.141 ***	0.008	0.003	0.082 ***
OWNERSHIP	0.022	0.106	−0.201	0.029	0.067	−0.11
MANAGEMENT	−0.1	0.148	0.365 **	−0.069	0.071	0.215 ***
RULE	1.464 ***	0.185	0.339 **	0.828 ***	0.105	0.188 **
INFRA	0.709 ***	0.052	−0.013	0.406 ***	0.032	0.006
DIST	0.02	0.063	−0.098 *	0.01	0.045	−0.057 *
GENDER	0.154 *	0.313 ***	0.117	0.075	0.186 ***	0.058
AGE	0.108 **	0.144 ***	0.056	0.061 *	0.086 ***	0.035
MARRIAGE	−0.159	0.079	0.08	−0.088	0.067	0.055
REGIS	−0.254 **	−0.166	−0.360 ***	−0.159 **	−0.111 *	−0.218 ***
EDUCATION	−0.221 ***	−0.007	−0.038	−0.133 ***	−0.013	0.002
OCCUPATION	0.261 **	0.146	0.206	0.145 **	0.091	0.115
INCOME	0.026	0.045 *	0.021	0.015	0.030 *	0.011
POLITICAL	0.059	0.05	0.082 *	0.039 *	0.033	0.047 *
CITY	0.029	0.139 **	0.253 ***	0.01	0.081 **	0.130 ***
chi2	911.259	145.688	303.707	910.965	143.944	299.89
R-squared	0.177	0.028	0.098	0.177	0.028	0.097
N	1621	1621	1621	1621	1621	1621

Source: authors' own calculations, 2020. Note: the symbols *, **, and *** indicate significance at the 10%, 5%, and 1% levels, respectively.

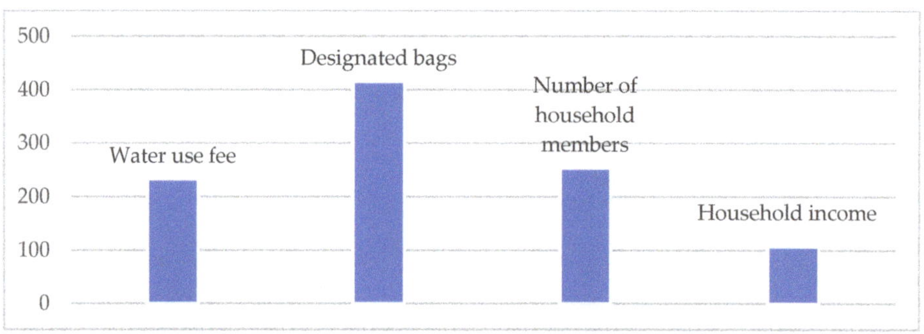

Figure 5. The refuse charge system. Source: authors' own calculations, 2020.

However, in response to the further question of "How much are you willing to pay if a designated bag system is introduced?", approximately 60% of the respondents indicated 0.01 to 0.02 yuan, which is, by any standard, surprisingly low (Figure 6). Implicitly, even though some respondents revealed that they support a refuse charge system, they are egoistically unwilling to bear any financial burden. This contradicts Schwartz's principles of altruism, in that the awareness of consequences (AC) did not lead to the ascription of responsibility. In other words, these respondents displayed a high self-interest, and are unwilling to engage in pro-environmental behavior or action in the presence of a perceived cost of sacrifice, as expressed in terms of a financial burden.

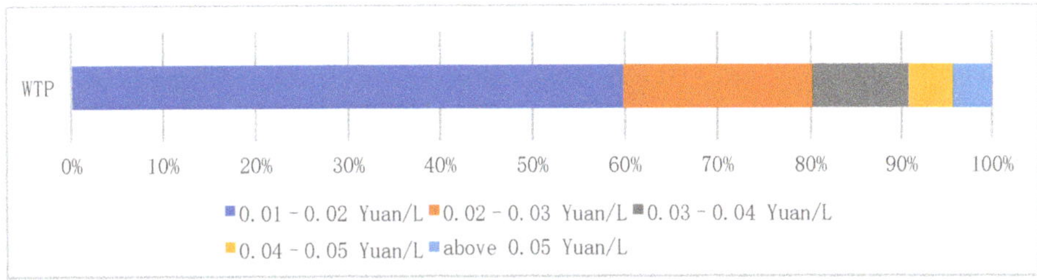

Figure 6. WTP (Willingness to Pay) for waste collection. Source: authors' own calculations, 2020.

The signs and coefficients for *AGE* indicate that the elderly are more likely to participate in the garbage sorting program, and display a higher receptiveness to fees. One reason for these tendencies could be that the young workers are unable to freely join sorting efforts that are scheduled during specific hours in the morning and evening in areas where the separate collection and disposal of HSW is implemented. There is also an asymmetric information issue. It may well be that campaigns to advertise the Green Account, the Green Earth program, the refuse charge system, and policies related to garbage sorting are mostly held during the daytime, with the consequence that the information fails to reach young workers. This may be one of the main factors that impede the wider recognition and adoption of sustainable HSW programs and policies or refuse charge systems among the younger population.

GENDER positively correlates with receptiveness to fees. That is, males were shown to be more likely to support a refuse charge system. The signs and significance of *OCCUPATION* indicate that people who have a stable job are more likely to become 'willing' participants. This may be attributed to the fact that businesses are more effective and quicker at educating employees on policies requiring garbage sorting and actual waste sorting. However, *OCCUPATION* does not have an effect on the receptiveness to fees and the receptiveness to policies. The signs and significance of *INCOME* show that the higher an income is, the higher the willingness to pay the refuse charge is.

Furthermore, the coefficients for *MARRIAGE* and *OWNERSHIP* are not significantly deviated from zero, indicating that neither affects the probability of participation, receptiveness to fees, or receptiveness to policies. The estimated coefficients on *EDUCATION* were found to be statistically significant in the negative direction of the garbage sorting behavior; that is, the higher a person's education level is, the less likely he or she will be willing to engage in garbage sorting. This result contradicts existing studies.

One the main reasons behind the above contradicting trend is that people with higher education levels tend to work longer. Hence, if sorted garbage is scheduled to be collected during specific hours in the morning and evening, as mentioned above, these people, in most cases, are not able to participate in HSW separation and collection. The signs and significance of *REGIS* indicate that the 'outsiders', who do not have a household registration for their area of residence, are more likely to be willing participants. They also displayed a higher receptiveness to policies. This result is in contrast to what was expected before the estimation. One of the main contributing factors behind this deviation is probably due to the 'sense of place' psychological factor of the 'outsiders', which leads to an increase of their feeling of identification, and to a positive change in their strength of relationship with the environment within which they exist. Choy [23,24] empirically established a strong relationship between the indigenous peoples' close attachment to the natural environment (sense of place) and their strong altruistic inclination for environmental protection, based on field research conducted in the tropical forest in Borneo, Malaysia. Choy [25] examined the values that the forest-dwelling indigenous people placed on the forested environment that they called home. He classified this as a 'sense of place' value. Semken [37] provides a good discussion of the strong sense of place of the American Indian and Alaskan Native people.

POLITICAL has a significantly positive correlation with behavior and receptiveness to policies, although its magnitude is small. It seems that members of the CPC display not only a stronger environmental awareness, but also a higher receptiveness to policies than the general public at large. One reason behind this pro-environmental inclination could be that, in the garbage separation pilot area, Primary Party organizations also play a key role in garbage sorting, and members of the CPC have acquired a green habit of actively taking part in garbage sorting [38,39]. This spontaneously sets an example to inspire and induce their neighbors and relatives to adopt environmentally friendly practices by introspection. This result also supports existing studies, such as Ghorbani et al. [21].

For matters governing the disposal of the sorted household garbage, the estimated coefficients on *RULE* imply that communities which establish garbage sorting rules are more likely to participate in a garbage sorting program. They also revealed a higher receptiveness to waste management policies. However, as regards *INFRA*, the existence of proper equipment at the garbage collection spot significantly and positively impacts garbage sorting behavior. The signs and coefficients for *MANAGEMENT* indicate that communities with a property management company show a higher level of receptiveness to policies. More specifically, in a community with a property management company, the company undertakes the task of sorting the garbage on behalf of the residents. This reduces the burdens of the residents. Inexorably, this tends to induce a moral sense of obligation incumbent upon the residents to react positively to policies requiring garbage sorting. The coefficients for *DIST* show that the more time required to arrive at the collection spot, the less receptive the individuals are to policies requiring garbage sorting.

Additionally, the estimated coefficients on *CITY* statistically, significantly, and positively correlate with the receptiveness to fees and to policies. This implies that there are differences in receptiveness to fee and policy support among cities.

The estimation results for the respondents who perform end-point sorting are shown in Table 7. The results of the estimation are similar to those of the case with all of the respondents. However, the coefficients for *MANAGEMENT*, *DIST*, *MARRIAGE*, and *POLITICAL* are not significantly correlated with the objective variables, indicating that none of them affect the probability of participation, receptiveness to fees, or receptiveness to policies.

The signs and significance of *INCOME* show that people with a higher income are more likely to take part in garbage sorting. *OCCUPATION* is also significantly and positively correlated with the receptiveness to policies. It seems that because businesses are quick to educate employees on policies requiring garbage sorting and actual sorting wastes, they have contributed to the enhancement of individuals' receptiveness to such policies. The estimated coefficients on *AGE* are not statistically significant with respect to sorting behavior and the receptiveness to policies. On the other hand, *RULE* and *INFRA* have a significantly positive correlation with garbage sorting behavior. The estimated coefficients on *KNOWLEDGE* were found to be statistically, significantly, and positively correlated with garbage sorting behavior, the receptiveness to fees, and the receptiveness to policies. It seems that the knowledge of the city's household garbage sorting regulations stimulates people to participate in the garbage sorting program. This has the effect of boosting their receptiveness to fees.

The estimation results show that garbage sorting behavior, the receptiveness to fees, and the receptiveness to policies significantly vary across the three cities. An analysis with propensity score matching was also performed. The receptiveness to fees and the receptiveness to policies were set as outcomes, the existence of economic incentives was set as an assignment variable, and the respondents' attributes were set as covariates. All of the coefficients were negative. In other words, compared with the residents in Shanghai and Chengdu, where economic incentive measures such as the Green Account and the Green Earth program have been introduced, the residents of Shenyang, where such measures are non-existent, were found to engage more actively in garbage sorting. They are also more receptive to a refuse charge system and policies requiring garbage sorting.

Table 7. Estimation results for respondents who perform end-point sorting.

	Ordinal Logit Models			Ordinal Probit Models		
Variable	Sorting Behavior	Receptiveness to Fees	Receptiveness to Policies	Sorting Behavior	Receptiveness to Fees	Receptiveness to Policies
EI	0.632 ***	0.489 ***	0.421 ***	0.380 ***	0.272 ***	0.221 **
PEB	0.026 ***	0.019 **	0.069 ***	0.016 ***	0.011 **	0.037 ***
ALT	0.008	−0.025	0.131 ***	0.004	−0.014	0.076 ***
OWNERSHIP	−0.071	0.083	−0.350 *	−0.037	0.038	−0.177
MANAGEMENT	−0.115	0.111	0.22	−0.072	0.052	0.116
RULE	0.777 ***	0.058	0.470 **	0.482 ***	0.036	0.262 **
INFRA	0.568 ***	−0.032	−0.138	0.319 ***	−0.024	−0.077
DIST	−0.092	0.085	−0.067	−0.055	0.055	−0.04
GENDER	0.134	0.312 ***	0.092	0.074	0.183 ***	0.035
AGE	−0.001	0.130 **	0.086	−0.004	0.080 **	0.057
MARRIAGE	−0.151	0.109	−0.108	−0.079	0.079	−0.065
REGIS	−0.287 **	−0.173	−0.359 **	−0.170 **	−0.115	−0.219 **
EDUCATION	−0.284 ***	−0.013	−0.087	−0.171 ***	−0.015	−0.004
OCCUPATION	0.171	0.212	0.482 **	0.097	0.13	0.288 ***
INCOME	0.083 **	0.048	0.084 **	0.048 **	0.032	0.045 *
POLITICAL	0.066	0.025	0.04	0.038	0.02	0.021
CITY	0.199 **	0.178 **	0.383 ***	0.109 **	0.095 **	0.199 ***
KNOWLEDGE	0.725 ***	0.221 ***	0.083	0.401 ***	0.125 ***	0.048
chi2	344.905	93.957	205.554	338.65	89.985	195.363
R-squared	0.152	0.029	0.116	0.15	0.027	0.11
N	1040	1040	1040	1040	1040	1040

Source: authors' own calculations, 2020. Note: the symbols *, **, and *** indicate significance at the 10%, 5%, and 1% levels, respectively.

4.3. Overall Estimate Outcomes and Policy Implications

All of the study areas clearly show that many factors influence residents' sustainable waste disposal behavior and practice. These may be widely divided into the following categories:

(i). Personal characteristics, such as gender, marital status, education, age, and a sense of belonging or sense of place, among others.

(ii). Economic characteristics, such as a reward system or government incentives (for example, movie tickets or park tickets, as mentioned above).

(iii). Attitudinal variables, such as environmental beliefs in relation to the change in attitude towards waste problems.

(iv). In relation to (iii), attitudinal variables are highly dependent on the promotion of environmental and moral education, such as the promotion of the concepts of the Ecological Civilization (ecocentric environmental beliefs and the ethic of altruism).

(v). Other related variables, such as political influence in relation to the spread of environmentally friendly practices, or the time required for HSW management and to arrive at a garbage collection spot or recycling center (here, it may be remarked that all of the study areas showed that the shorter the distance to the garbage collection spot is, the larger the impact factor is on environmental behavioral change).

Furthermore, the analysis based on the external variables indicates that the participants of the garbage sorting program tend to live in communities that have clear rules on garbage sorting and have installed the proper equipment at garbage collection spots. Additionally, the development of the solid waste management infrastructure—such as smart garbage collecting stations that are designed to recognize different types of waste automatically—can encourage citizens to cooperate with the local government in waste sorting.

The analysis also uncovered the interesting fact that participants of the garbage sorting program tend to be elderly and employed. In addition, the study revealed a lesser known, let alone well-analyzed, issue that the 'sense of place' can serve as a crucially important intrinsic impact factor in inducing an individual's green behavioral practices.

Our investigation also uncovered the important fact that the power of the external influence arising from politically influential and environmentally inclined elites, such as CPC members, is positively related to the state of green environmental consciousness or the green mentality of the individuals surrounding them. These findings reflect the combined significance of external and internal moderations, and the importance of advertising and educational activities with respect to the garbage sorting policies in each community. In the future, the local government could optimize the use of metro or subway advertising media or social media such as WeChat or Tiktok in order to disseminate information concerning sustainable waste disposal practices. These critical measures also serve to induce communities to strengthen their waste sorting rules and set up their waste collection spots properly.

Our study further revealed that residents' receptiveness to a refuse charge system varies across cities, and many respondents tend to oppose the implementation of a refuse charge program. It was also found that, if a fee-based system were introduced, a designated bag system would be the most effective to draw support from the residents. In addition, compared with the residents in Shanghai and Chengdu, residents from Shenyang—where economic incentive measures have not been introduced are found to be more actively engage in garbage sorting. They are also more receptive to a refuse charge system and policies requiring garbage sorting. This implies that mandatory garbage sorting would be more effective than economic measures.

5. Conclusions

The present study examined residents' garbage sorting behavior based on a questionnaire survey, and clarified the factors that contribute to their green cooperation and other environmental morality issues in three selected regions, namely, Shanghai, Shenyang (Liaoning Province), and Chengdu (Sichuan Province). The accumulated data, which was analyzed using ordinal logit models and ordinal probit models, indicated that pro-environmental behavior arising from environmental awareness is a significant explanatory variable in promoting personal norms in HSW sorting habits and the social endorsement of refuse charge systems and policies, in line with the concept of an ecological civilization. Altruism appears to influence the receptiveness to policies only. However, both altruism and environmental awareness are preconditions for the enhancement of pro-environmental behavior and regulative environmental observance.

Overall, our findings ineluctably signal the sovereign importance of environmental and moral education in inducing and promoting personal norms in sustainable HSW management and practices. This is all the more necessary because it is infeasible for a government to strictly monitor daily household residents' HSW sorting behavioral practices due to the massive monitoring cost involved. That said, it is through further publicity, and environmental and moral education, that household residents can determine what must be valued in green development, and what actions must be taken to be in line with the Ecological Civilization philosophy that is rigorously promoted by the government.

Our study has far-reaching implications that the repertory of regulative control or legislative constraint alone is far from adequate to effectively hold society accountable for sustainable HSW management practices. What is important is to inculcate a collective moral interest through various educational activities and a moral sense that can bind a society towards embracing an Ecological Civilization, as discussed above. Working incisively in tandem with regulative control, environmental and moral education can serve as an effective means of promoting an 'ecologically civilized' society par excellence.

Author Contributions: Conceptualization, Y.H. and H.K.; methodology—formal analysis, Y.H.; writing—original draft preparation, Y.H.; supervision, H.K.; funding acquisition, H.K.; writing—review and editing, Y.C.; data curation, X.K.; validation, P.T. All authors have read and agreed to the published version of the manuscript.

Funding: This work was supported by JSPS KAKENHI Grant Number 16K12664. Any opinions, findings, and conclusions or recommendations expressed in this material are those of the author(s) and do not necessarily reflect the views of the author(s)' organization, JSPS, or MEXT.

Conflicts of Interest: The authors declare that there is no conflict of interest.

References

1. Kaza, S.; Yao, L.C.; Bhada-Tata, P.; Van Woerden, F. What a Waste 2.0: A Global Snapshot of Solid Waste Management to 2050. In *Urban Development*; World Bank: Washington, DC, USA, 2018. Available online: https://openknowledge.worldbank.org/handle/10986/30317 (accessed on 10 October 2020).
2. Chen, D.M.-C.; Bodirsky, B.L.; Krueger, T.; Mishra, A.; Popp, A. The world's growing municipal solid waste: Trends and impacts. *Environ. Res. Lett.* **2020**, *15*, 074021. [CrossRef]
3. World Bank. Solid Waste Management. 2019. Available online: https://www.worldbank.org/en/topic/urbandevelopment/brief/solid-waste-management (accessed on 10 October 2020).
4. Ferrara, I. Waste Generation and Recycling. In *Household Behaviour and the Environment. Reviewing the Evidence*; OECD: Paris, France, 2008; pp. 19–58. Available online: https://www.oecd.org/environment/consumption-innovation/42183878.pdf (accessed on 10 October 2020).
5. The Ministry of Housing and Urban-Rural Development. *China Urban-Rural Construction Statistical Yearbook 2017*; China Statics Press: Beijing, China, 2018. Available online: http://www.stats.gov.cn/tjsj/ (accessed on 20 March 2020). (In Chinese)
6. Yang, Q.; Zhu, Y.; Liu, X.; Fu, L.; Guo, Q. Bayesian-Based NIMBY Crisis Transformation Path Discovery for Municipal Solid Waste Incineration in China. *Sustainability* **2019**, *11*, 2364. [CrossRef]
7. Huang, X.; Yang, D.L. NIMBYism, waste incineration, and environmental governance in China. *China Inf.* **2020**, *34*, 342–360. [CrossRef]
8. Goron, C. Ecological Civilisation and the Political Limits of a Chinese Concept of Sustainability. *China Perspect.* **2018**, 39–52. [CrossRef]
9. Wang, H.; Jiang, C. Local Nuances of Authoritarian Environmentalism: A Legislative Study on Household Solid Waste Sorting in China. *Sustainability* **2020**, *12*, 2522. [CrossRef]
10. The State Council. Notice on the Opinions on Further Strengthening the Work of Municipal Solid Waste Disposal. 2011. Available online: http://www.gov.cn/zwgk/2011-04/25/content_1851821.htm (accessed on 20 March 2020). (In Chinese)
11. The State Council General Office. The Proposed Method of Implementing the Sorting System for Municipal Solid Waste. 2017. Available online: http://www.gov.cn/zhengce/content/2017-03/30/content_5182124.htm (accessed on 20 March 2020). (In Chinese)
12. Yokoo, H.; Wada, H.; Yamada, M. 家庭ごみ分別制度と社会的規範: —日本とシンガポールにおけるアンケート調査の比較—. Review of Environmental Economics and Policy Studies. *Soc. Environ. Econ. Policy Stud.* **2014**, *8*, 85–88. [CrossRef]
13. Linde'N, A.-L.; Carlsson-Kanyama, A. Environmentally Friendly Disposal Behaviour and Local Support Systems: Lessons from a metropolitan area. *Local Environ.* **2003**, *8*, 291–301. [CrossRef]
14. Antonides, G.; Fred, W. *Consumer Behaviour. A European Perspective*; Wiley: Hoboken, NJ, USA, 1999; pp. 514–515.
15. Lindhqvist, T. *Extended Producer Responsibility in Cleaner Production: Policy Principle to Promote Environmental Improvements of Product Systems*; IIIEE; Lund University: Lund, Sweden, 2000. Available online: https://lup.lub.lu.se/search/ws/files/4433708/1002025.pdf (accessed on 20 March 2020).

16. Dahlén, L.; Lagerkvist, A. Evaluation of recycling programmes in household waste collection systems. *Waste Manag. Res.* **2009**, *28*, 577–586. [CrossRef]

17. Chappells, H.; Klintman, M.; Lindèn, A.L.; Shove, E.; Spaargaren, G.; van Vliet, B. *Domestic Consumption, Utility Services and the Environment*; Final Domus Report; Wageningen University: Wageningen, The Netherlands, 2000.

18. Judge, R.; Becker, A. Motivating Recycling: A Marginal Cost Analysis. *Contemp. Econ. Policy* **1993**, *11*, 58–68. [CrossRef]

19. Ando, A.W.; Gosselin, A.Y. Recycling in multifamily dwellings: Does convenience matter? *Econ. Inq.* **2005**, *43*, 426–438. [CrossRef]

20. Van Houtven, G.L.; Morris, G.E. Household Behavior under Alternative Pay-as-You-Throw Systems for Solid Waste Disposal. *Land Econ.* **1999**, *75*, 515–537. [CrossRef]

21. Ghorbani, M.; Gunderson, M.; Lee, Y.S.B. Union and Communist Party Influences on the Environment in China. *Ind. Relat.* **2019**, *74*, 552–576. [CrossRef]

22. Clark, C.F.; Matthew, J.K.; Michael, R.M. Internal and external influences on pro-environmental behavior: Participation in a green electricity program. *J. Environ. Psychol.* **2003**, *23*, 237–246. [CrossRef]

23. Choy, Y.K. Land Ethic from the Borneo Tropical Rainforests in Sarawak, Malaysia: An Empirical and Conceptual Analysis. *Environ. Ethics* **2014**, *36*, 421–441. [CrossRef]

24. Choy, Y.K. Cost-benefit Analysis, Values, Wellbeing and Ethics: An Indigenous Worldview Analysis. *Ecol. Econ.* **2018**, *145*, 1–9. [CrossRef]

25. Choy, Y.K. *Global Environmental Sustainability: Case Studies and Analysis of the United Nations' Journey toward Sustainable Development*; Elsevier: Amsterdam, The Netherlands; Oxford, UK; Cambridge, CA, USA, 2020.

26. Rylander, D.; Allen, C. Understanding Green Consumption Behavior: Toward an Integrative Framework. *Am. Mark. Assoc. Winter Educators' Conf. Proc.* **2001**, *12*, 386–387.

27. National Bureau of Statistics of China. Available online: http://data.stats.gov.cn/ (accessed on 20 March 2020). (In Chinese)

28. The Information Is Based On Interviews (Conducted On 16–17 August 2017) with a Representative from Shanghai Laogang Solid Waste Utilization Co., Ltd., Ms. Qi Yumei (Deputy Chief) and Mr. Yuehua Zhang of the Environmental Sanitation Management Office of the Shanghai Landscaping and City Appearance Administrative Bureau, and Ms. Liqiong Hao (Director) of the Shanghai IFINE Environmental Protection Organization, and Information from Green Account Website. Available online: https://www.greenfortune.sh.cn/dist/index.html (accessed on 20 March 2020). (In Chinese).

29. Zhou, M.H.; Shen, S.L.; Xu, Y.S.; Zhou, A.N. New Policy and Implementation of Municipal Solid Waste Classification in Shanghai, China. *Int. J. Environ. Res. Public Health* **2019**, *16*, 3099. [CrossRef] [PubMed]

30. Chengdu Green Earth Environment Technology Co. Available online: http://www.lvsediqiu.com (accessed on 16 August 2020). (In Chinese).

31. Zhao, H.H.; Gao, Q.; Wu, Y.P.; Wang, Y.; Zhu, X.D. What affects green consumer behavior in China? A case study from Qingdao. *J. Clean. Prod.* **2014**, *63*, 143–151. [CrossRef]

32. Schwartz, S.H. Elicitation of Moral Obligation and Self-Sacrificing Behavior: An Experimental Study of Volunteering to Be a Bone Marrow Donor. *J. Personal. Soc. Psychol.* **1970**, *15*, 283–293. [CrossRef]

33. Schwartz, S.H. Normative Influences on Altruism. *Adv. Exp. Soc. Psychol.* **1977**, *10*, 221–279.

34. Durkheim, E. *Emile Durkheim on Morality and Society. Selected Writings*; Robert, N.B., Ed.; The University of Chicago Press: Chicago, IL, USA; London, UK, 1973.

35. Thøgerson, J.; Biel, A. Activation of social norms in social dilemmas: A review of the evidence and reflections on the implications for environmental behaviour. *J. Econ. Psychol.* **2007**, *28*, 93–112. [CrossRef]

36. Thøgersen, J. The Motivational roots of norms for environmentally responsible behaviour. *Basic Appl. Soc. Psychol.* **2009**, *31*, 348–362. [CrossRef]

37. Semken, S. Sense of place and place-based introductory geoscience teaching for American Indian and Alaska native undergraduates. *J. Geosci. Educ.* **2018**, *53*, 149–157. [CrossRef]

38. Beijin Review. A Use for Refuse -Garbage Sorting Becomes Mandatory for a Cleaner Environment and a More Civilized Society. 2019. Available online: http://www.bjreview.com/Current_Issue/Editor_Choice/201908/t20190819_800176310.html (accessed on 11 October 2020).

39. Shanghai Municipal People's Government. Press Release for 28 June 2019. Available online: http://sh.eastday.com/m/20190628/u1ai12629076.html (accessed on 11 October 2020). (In Chinese).

sustainability

Article

Impact of COVID-19 on Food and Plastic Waste Generated by Consumers in Bangkok

Chen Liu [1,*], Pongsun Bunditsakulchai [2,*] and Qiannan Zhuo [3]

[1] Sustainable Consumption and Production Area, Institute for Global Environmental Strategies, 2108-11 Kamiyamaguchi, Hayama 240-0115, Japan
[2] Department of Civil Engineering, Faculty of Engineering, Chulalongkorn University, 254 Phayathai Road, Phathum Wan, Bangkok 10330, Thailand
[3] Graduate School of Media and Governance, Keio University, 5322 Endo, Fujisawa 252-0882, Japan; s12381ss@gmail.com
* Correspondence: c-liu@iges.or.jp (C.L.); pongsun.b@chula.ac.th (P.B.)

Abstract: The crisis ignited by COVID-19 has transformed the volume and composition of waste generation and requires a dynamic response from policy makers. This study selected Bangkok as a case study to semi-quantitatively examine the impact of the COVID-19 outbreak on consumer-generated food and plastic waste by examining changes in lifestyles and consumption behaviour through a face-to-face questionnaire survey. Travel bans and diminished economic activity due to COVID-19 have led to a dramatic reduction in waste from the business sector and in the total amount of municipal waste generated. However, the results of the survey showed that both food and plastic waste generated by households in Bangkok increased during COVID-19. The shift from eating out to online food delivery services led to an increase in plastic bags, hot-and-cold food bags, plastic food containers, and food waste. Reasons for the increase in household food waste during COVID-19 varied, with respondents citing excessive amounts of food and unappetising taste, followed by exceeding the expiration date and rotting/foul odours. These reasons may be the result of the inability to predict quantity and quality when ordering online, and inadequate food planning and management by consumers. To achieve more effective food and plastic waste management, home delivery services, consumer food planning and management, and the formation of a circular economy based on localised supply chains may be considered as important intervention points.

Keywords: COVID-19; food waste; plastic waste; household; lifestyle; Bangkok

Citation: Liu, C.; Bunditsakulchai, P.; Zhuo, Q. Impact of COVID-19 on Food and Plastic Waste Generated by Consumers in Bangkok. *Sustainability* **2021**, *13*, 8988. https://doi.org/10.3390/su13168988

Academic Editors: Dimitrios Komilis, Yasuhiko Hotta, Tomohiro Tasaki and Shunsuke Managi

Received: 11 July 2021
Accepted: 6 August 2021
Published: 11 August 2021

Publisher's Note: MDPI stays neutral with regard to jurisdictional claims in published maps and institutional affiliations.

1. Introduction

With shifts in lifestyles and consumption habits, supply chain interruptions, changes in material flows, waste sorting and recycling logistics, falling oil prices, and reduced demand for recycled waste, the COVID-19 outbreak has posed significant challenges for waste management, waste recycling, and the circular economy around the world [1–3]. This waste is not only medical, infectious, and healthcare waste but also general waste such as food waste (FW) and plastic waste (PW).

Within the past decade, food waste and plastic pollution have become key sustainability issues of international concern to policymakers, corporations, local communities, and researchers who are searching for solutions to the resulting environmental impacts across a range of academic disciplines [4–6]. There have been a number of literature review papers on FW and PW issues published in recent years. For example, Muriana [7] assessed the use of mathematical models in food waste and loss, and clarified food waste's dependency on supply chain strategies, while Amicarelli and Bux [8] outlined global approaches, characteristics, limitations, opportunities, and results of food waste measurement methodologies through a systematic and configurative literature review. Bernstad Saraiva Schott et al. [9] reviewed existing life cycle assessment studies on food waste management to clarify the impacts of each treatment method on global warming and decisive factors in setting system boundaries. De Menna et al. [10] systematically reviewed different aspects and approaches for life cycle costing methodologies to evaluate FW management and valorisation routes.

The number of review studies on consumer-generated food waste in particular has risen since private households were first identified as key actors in food waste generation in developed economies [11]. For example, Reynolds et al. [12] reviewed literature on FW prevention at the consumption stage. Roodhuyzen et al. [13] developed a framework that conceptualised the generation of consumer food waste in relation to stages of the household supply chain and categorised 116 potential factors of consumer food waste into four groups (behavioural, personal, product, and societal factors). Schanes et al. [14] reviewed the rising number of empirical studies on consumer food waste practices and the factors that foster and impede the generation of food waste at the household level. These studies reveal food waste to be a complex and multi-faceted issue that cannot be attributed to a single variable. Given its complex nature, the growing body of literature sheds light on food-related practices and routines, ranging from planning and shopping, to storing, cooking, eating, and managing uneaten food within the context of food waste generation by adopting practical theories and other conceptual approaches.

With respect to PW studies, Heidbreder et al. [15] provided an overview of the existing social-scientific literature on plastic, ranging from awareness and consumer preference to political and psychological intervention strategies through a review of 187 studies. The review concludes that future studies should further investigate plastic-specific behaviour and implement behaviour-based solutions.

Meanwhile, many FW and PW reduction targets had been set before the onset of this global crisis, including Sustainable Development Goal 12.3 which aims to 'halve per capita global food waste at the retail and consumer levels' by 2030, as well as manage and control waste emissions and reduce marine pollution, with specific reference to targets 12.4, 12.5, and 14.1 [16]. In particular, several countries had issued bans on specific plastic products, optimistic in the hope of reducing serious environmental pollution [17]. For example, the Thai government released a 'Plastic Waste Management Road Map' to phase out the use of plastic by 2030 and issued a ban on single-use plastics in January 2020. However, the advent of the COVID-19 pandemic has enhanced the complexities of FW and PW management. Single-use plastic usage is expected to snap back due to growing concerns with hygiene (such as gloves, masks, packaging, etc.) and increased demand for online shopping during the pandemic. However, household food waste generation may abate along with the trends of more conscious food management during lockdowns due to fear and anxiety associated with logistic systems amidst concerns about food shortages [2,18,19].

Although there have been a number of studies conducted on the impacts of COVID-19 on household food waste, there has been no research on developing Asian countries or cities, and little is known about the conditions and determinants of consumer food waste during the pandemic. Lockdowns may affect consumer behaviour and attitudes towards FW and PW due to changes in lifestyle habits. It is especially urgent to view this as an opportunity to promote studies that examine the implications of food waste reduction policies in the cities of Asian developing countries where levels of FW and PW are spiking, but where both existing data and the capacity to tackle this issue are limited.

In an earlier study conducted by the authors in 2018 in Bangkok that investigated FW generation trends in Bangkok and the relationship between daily lifestyles and FW [20], it was found that FW issues in Bangkok have quite distinctive features when compared with existing studies (although most are case studies from developed countries). For example, although the proportion of organic waste and FW normally decreases in the context of growing urbanisation, this type of waste has increased in Bangkok since 2015 due to the growth of tourism and changes in food consumption lifestyles. Furthermore, it has been reported that the largest single contribution to FW in developed countries is at the consumption stage (mostly at the household level), while in developing countries, greater food losses occurred at the production and post-harvest stages [21]. However, FW generation in Bangkok is still high compared to developed cities. Moreover, consumers in Thailand eat out frequently and consume ready-made food, which has resulted in the broad distribution of FW throughout the entire supply chain. However, since the advent of the pandemic, people have isolated themselves at home and avoided eating out, giving rise to the research question: What impact, if any, does the pandemic have on behaviour in relation to FW and PW? For these reasons, Bangkok was selected as a case study to investigate the effects of COVID-19 lockdown conditions on household FW, PW, and correlating behaviour. To the authors' best knowledge, this is the first paper to report changes in household FW and PW in Bangkok due to COVID-19 lockdowns and is the only face-to-face questionnaire survey conducted during the outbreak in 2020.

Specifically, the objectives of this study are: (1) to examine the impact of COVID-19 on food and food-related plastic waste generated by consumers, and (2) to evaluate options for preventing and reducing FW and PW even after the crisis to assist the Bangkok Metropolitan Administration (BMA) in implementing medium- to long-term improvements. For this purpose, a cross-sectional questionnaire survey was conducted to capture shifts in respondents' lifestyles during the pandemic, including a focus on behavioural changes in working on-site versus remotely, eating styles, cooking and shopping practices, as well as waste generation before and during the COVID-19 pandemic. Survey results provided insights into policy implications for addressing issues.

Following this introduction, the paper first outlines the state of COVID-19 and FW and PW generation trends in the Bangkok metropolitan area based on existing data in Section 2. The methods employed in the study are presented in Section 2, and results are discussed in Section 3. Section 4 delves further into policy recommendations. Finally, Section 5 outlines the main conclusions and identifies both limitations to the study and recommendations for further research.

2. Review of the State of COVID-19 and Food Waste and Plastic Waste Generation

2.1. COVID-19 in Thailand

The national government published a notice declaring COVID-19 to be a dangerous infectious disease on 29 February, about six weeks after the first case of the virus was found in Thailand on 12 January 2020. An emergency decree and travel ban were issued on 26 March in response to the rising number of cases following a super-spreader event at a boxing stadium on 6 March and additional cases of local transmission. As the number of cases rose, a national curfew was imposed on 3 April, which was lifted in stages in May and June. Of the 3162 cases found between 12 January and 27 June 2020, a total of 3053 people recovered, 51 patients were hospitalised, and 58 deaths were recorded. This survey was conducted between 16 and 19 June, just after the first national curfew was lifted, which means the respondents of this survey had been under lockdown for more than two months. The main timeline of the COVID-19 outbreak in Thailand around this survey is shown in Figure 1.

The Thai government instituted a number of preventive measures for COVID-19. The emergency decree on 26 March restricted domestic and international travel, banned entry into and closed high-risk areas, encouraged masks to be worn and promoted hand washing and social distancing. Restaurants and food stalls were allowed to remain open, but only for take-out. The first national curfew that started on 3 April restricted people to their homes between 10 p.m. and 4 a.m.

To curb the rise in infections, the government distributed masks, offered subsidies for healthcare services, provided free COVID-19 screenings, subsidised the costs of testing, and instituted programmes to assist persons with disabilities. The government also implemented a number of relevant measures to support individuals and companies, including deferrals and exemptions for personal income tax payments, extensions for filing tax returns, and lower taxes for small- and medium-sized enterprises (SMEs) and corporations, as well as subsidies for electricity charges to support people working from home. The subsidy period for compressed natural gas (CNG) was extended for entrepreneurs, and the withholding tax rate was reduced, while cash subsidies were offered for employees at SMEs and value-added tax (VAT) refunds were expedited.

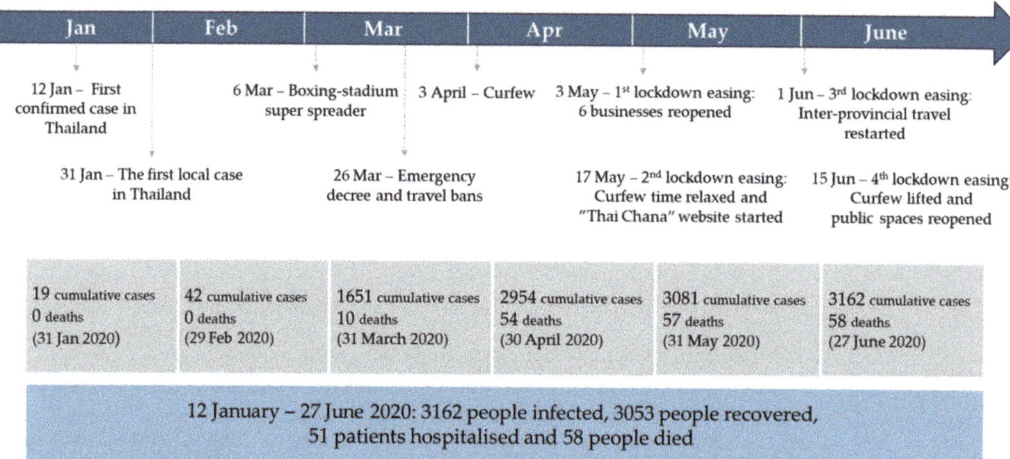

Figure 1. Timeline of COVID-19 outbreak in Thailand.

2.2. Review of Food Waste and Plastic Waste Generation in Bangkok

Based on the data collected in the authors' previous study [20] and officially reported data by the BMA's Department of Environment, time-series changes in FW, PW, and municipal solid waste (MSW) generation between 2003 and 2020 in Bangkok are shown in Figure 2. MSW is solid waste generated by municipal activities (including by residences, supermarkets, retail shops, businesses, service providers, marketplaces, and institutions) that is collected and treated by BMA. The amount of MSW generated fell by about 1000 tonnes per day in early 2020 after the COVID-19 outbreak due to the closure of hotels and restaurants, following a steady increase in the decade prior to the pandemic. Food waste accounts for 50–60% and plastic waste for 20–30% of the total MSW, but COVID-19 has prompted a reduction in food waste and an increase in plastic waste at the city level. Food waste contains unavoidable items such as peels, stems, and bones, as well as leftovers and other avoidable items, but excludes surplus food from the commercial sector and reused and recycled food such as animal feed, which increased rapidly in the late 2010s, mainly due to growing tourism and lifestyle changes. Since the COVID-19 outbreak and the resulting lockdowns, the total weight of MSW has fallen significantly due to a significant reduction in food waste from hotels and restaurants. On the other hand, the amount of plastic waste generated rose and fell along a gentle curve before the COVID-19 outbreak, averaging 2115 tonnes per day in 2019, but increased sharply by 62% in 2020, reaching an average of 3432 tonnes per day between January and April. In addition, contaminated plastic items from food delivery services, such as takeaway bags, containers, bottles, and cups, that are difficult to reuse and recycle comprised more than 80% of MSW in 2020.

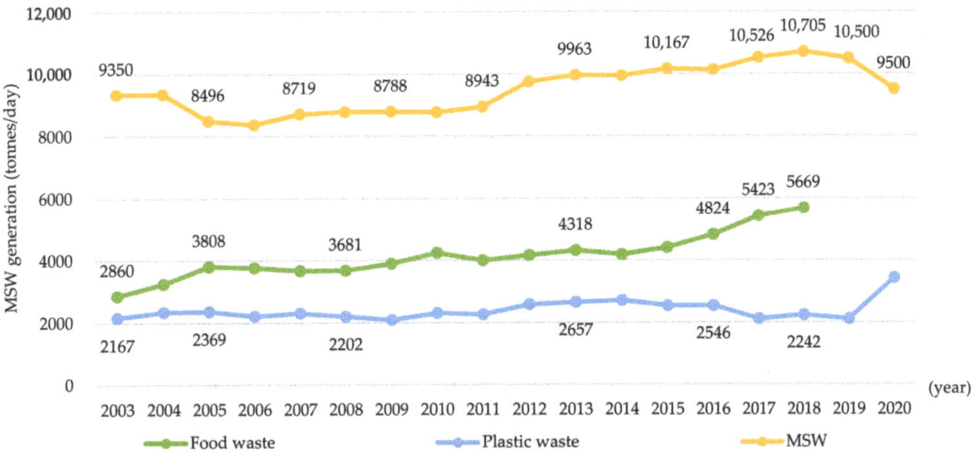

Figure 2. Food waste, plastic waste, and municipal solid waste generated in the Bangkok metropolitan area. Note: Data on PW and MSW generation in 2020 are the average between January and April 2020 as reported by the BMA's Department of Environment to BBC Thai. No data on food waste generation are available in 2019 and 2020.

3. Materials and Methods

3.1. Sampling Size and Analytical Approach

An extensive face-to-face questionnaire survey of residents in the Bangkok metropolitan area conducted between 16 and 19 June 2020 (just after the first national curfew was lifted) presented a snapshot of changes in respondents' lifestyles during the COVID-19 pandemic. This study applied the calculation formula developed by Yamane [22] to determine sample sizes. Considering the budget and labour required to conduct the survey, the confidence level was set at 93% (or at a precision level of ±7%); accordingly, the appropriate sample size was 222. In this survey, passers-by were randomly sampled on the streets [23] of the Bangkok metropolitan area. The questionnaire included queries about working days, eating habits, and purchasing routes for food both before and after the preliminary outbreak of COVID-19 in January 2020, and responses were expected to reflect the ways in which food and plastic waste has been generated by consumers. Statistical tests (*t*-test, Kruskal–Wallis test, and Dunn's multiple comparisons test) were used to detect behavioural differences before and during the preliminary outbreak. Food delivery services were also evaluated as a potential key component in the COVID-19 success story. Related environmental impacts on food waste, plastic waste, and other problems caused by new food consumption paradigms were also discussed and statistically tested.

3.2. Content of Questionnaire

The questionnaire survey (see Supplementary Material) for consumers on food waste mainly consisted of four sections and covered a range of daily activities. The first section included questions designed to elicit basic information about the respondents, such as gender, age, occupation, educational level, and household income, as well as working days in the office prior to and during the pandemic. The second section posed questions about changes in respondents' food-related habits in their daily routines, including purchasing, cooking, eating, and disposal. The third section highlighted trends in the food delivery service sector prior to and during the pandemic, including primary reasons cited by respondents for the use of food delivery services, main factors considered when selecting specific food delivery services from several alternatives, and frequency of ordering different types of food (Thai, Chinese/Japanese/Korean or western cuisine, fast food, street food, desserts, and beverages) using online food services prior to and during the pandemic. The fourth section focused on behavioural changes in relation to the generation of household waste, as well as the respondents' attitudes towards and intentions in reducing food waste.

4. Results and Discussion

4.1. Respondent Attributes

The attributes of respondents are shown in Table 1. Primary data were collected from 238 individuals (50% male and 50% female). The sample showed a broad range of employment conditions, with the majority of respondents employed at companies (41%), followed by students (22%), the self-employed (16%), and government officials (12%), with the remainder comprising full-time homemakers (5%), the unemployed (3%), and others. The highest percentage of respondents (31%) take home a monthly income of THB 50,001–100,000, with 29% earning more than THB 100,000. The remainder earn between THB 15,001 and 50,000 (34%), while 6% earn less than THB 15,000. Fifty-one percent of respondents live with other adults, 17% with elderly family members, 15% live alone, 10% reside with children, and 7% live with both children and older family members.

Table 1. Attributes of respondents.

Characteristics	Number of Respondents (N = 238)	Percentage
Gender:		
Female	120	50%
Male	118	50%
Occupation:		
Company employee	98	41%
Student	53	22%
Self-employed	38	16%
Government official	27	12%
Full-time homemaker	12	5%
Unemployed	7	3%
Other	3	1%
Education:		
Undergraduate	173	73%
Master's degree or higher	36	15%
High school degree or lower	17	7%
Vocational or technical university	12	5%
Household type:		
Only adults	121	51%
Family with older adults	40	17%
Living alone	37	15%
Family with children	23	10%
Family with children and older adults	16	7%
Other	1	0%
Income:		
>THB 100,000	68	29%
THB 50,001–100,000	75	31%
THB 30,001–50,000	41	17%
THB 15,001–30,000	40	17%
<THB 15,000	14	6%
Residence type:		
Detached house	89	7%
Apartment/Condominium	41	13%
Town house	62	20%
Shop house (Shop is on the first floor)	23	27%
Dormitory/Share house	23	33%

4.2. Changes in Work–Life Balance

Changes in the number of days respondents worked or attended classes outside the home are shown in Figure 3. Before COVID-19, almost half of all respondents (49.58%) commuted to their workplace/school five days a week, with 13.5% of respondents working/studying outside the home more than 5 days a week, and 18.5% of respondents either working/studying on a flexible schedule or travelling to their workplace/school less than five days a week. After the start of the COVID-19 pandemic, respondents either switched over to teleworking full-time (33%), at least five days/week (21%), or three days/week (19%), respectively. These figures show a substantial shift in work–life balance due to the COVID-19 outbreak.

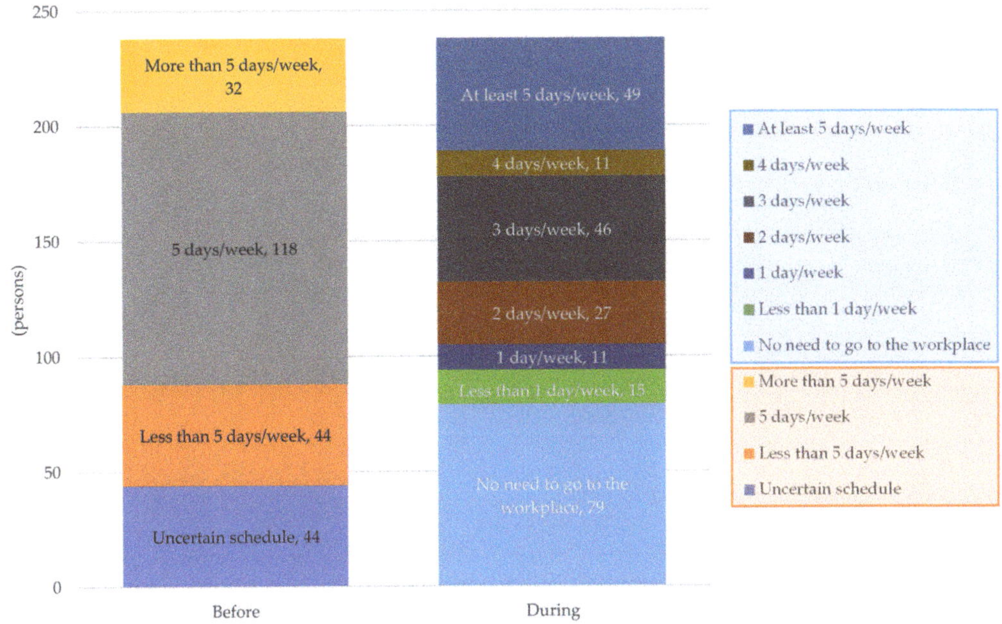

Figure 3. Working days at offices before and during the COVID-19 pandemic.

4.3. Changes in Eating Styles and Food Consumption Behaviour

4.3.1. Eating Styles

Changes in eating styles are shown in Figure 4, and the results of the *t*-test are shown in Table 2. Before COVID-19, respondents ate out on average 6.31 times a week. However, the number of times respondents dined out fell to an average of 2.42 meals a week after the outbreak began and were replaced by other styles, including the consumption of ready-made meals (an increase of 1.1 times from 5.14 to 5.80 meals/week), use of food delivery services (an increase of 1.6 times from 2.42 to 3.90 meals/week), and eating at home (an increase of 1.3 times from 6.12 to 8.26 meals/week). This may be attributed to the government's social distancing and 'stay-at-home' policies to prevent the spread of the virus. The study also found a slight increase in the number of people cooking for themselves or with meals prepared by other family members.

Table 2. *p*-value of eating styles.

Eating Style	Alternative Hypothesis	*p*-Value
Eating out	after–before < 0	$<2.2 \times 10^{-16}$
Ready-made meals	after–before > 0	0.00052
Food delivery services	after–before > 0	4.09×10^{-15}
Eating at home	after–before > 0	2.66×10^{-14}

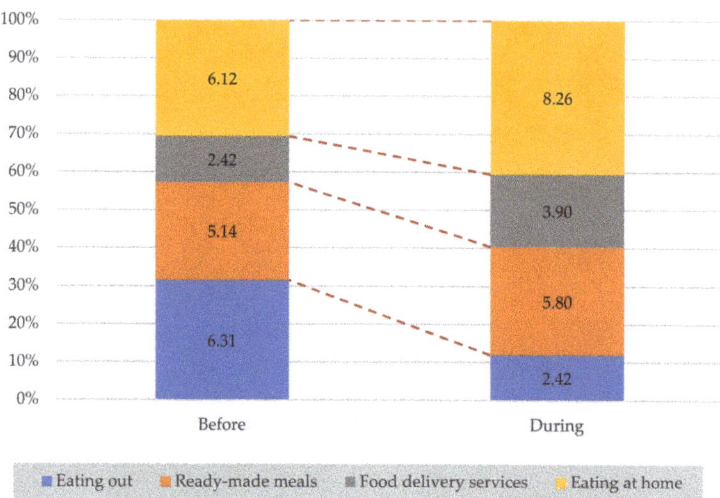

Figure 4. Eating styles before and during the COVID-19 pandemic (meals/week).

4.3.2. Food Consumption

Changes in food consumption in each category are shown in Figure 5, and the results of the *t*-test are shown in Table 3. The amount of food consumed by people in Thailand changed as they complied with the government's 'stay-at-home' orders. Because more food was consumed at home, they needed to purchase and stock up on greater amounts of rice and other ingredients than usual. The survey also found that respondents increased their consumption of meat, vegetables and fruit, eggs and dairy products, and ready-to-eat food. In contrast, there was a significant reduction in the amount of seafood and alcoholic beverages consumed (Table 3). The decreased consumption of seafood may indicate respondents' strong health concerns during the COVID-19 pandemic, while the reduced consumption of alcoholic beverages may be due to temporary bans imposed by the government during lockdowns.

Table 3. *p*-value of change in consumption.

Food and Ingredients	Alternative Hypothesis	*p*-Value
Rice, powder, bread, noodles	Greater	$<2.2 \times 10^{-16}$
Meat	Greater	$<2.2 \times 10^{-16}$
Seafood	Less	0.007796
Vegetables and fruit	Greater	$<2.2 \times 10^{-16}$
Eggs and dairy products	Greater	$<2.2 \times 10^{-16}$
Oil for cooking	Greater	0.008256
Semi-processed food	Greater	8.06×10^{-16}
Instant processed food	Greater	0.001344
Ready-to-eat food	Greater	1.10×10^{-16}
Frozen food	Greater	0.006507
Snacks, desserts, soft drinks	Less/Greater	0.784
Alcoholic beverages	Less	$<2.2 \times 10^{-16}$

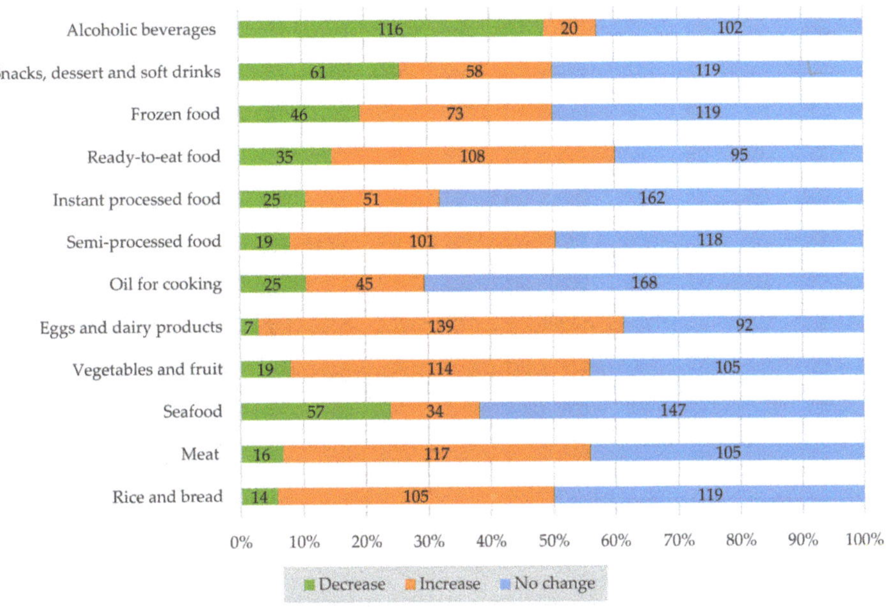

Figure 5. Change in consumption (number of respondents).

4.4. Changes in Shopping Behaviour

4.4.1. Purchasing Routes

The types of routes used to purchase food and other ingredients and the frequency in which they were used before and during the COVID-19 pandemic are shown in Figure 6, and the results of the *t*-test are shown in Table 4. Responses demonstrated that since the outbreak, there has been a considerable rise in the frequency of online shopping. Respondents also indicated that they have significantly reduced the number of times they visit temporary markets, mom-and-pop stores, street stalls, fresh markets, and supermarkets, although there has not been much change in the frequency of shopping at convenience stores and co-ops.

Table 4. *p*-value of frequency of purchases at different types of markets.

Market	Alternative Hypothesis	*p*-Value
Fresh market	Less	1.871×10^{-8}
Temporary market	Less	$<2.2 \times 10^{-16}$
Supermarket	Less	0.000157
Convenience store	Less/Greater	1
Mom-and-pop store	Less	4.50×10^{-16}
Co-op	Less/Greater	0.5078
Street stall	Less	$<2.2 \times 10^{-16}$
Online store	Greater	1.77×10^{-14}

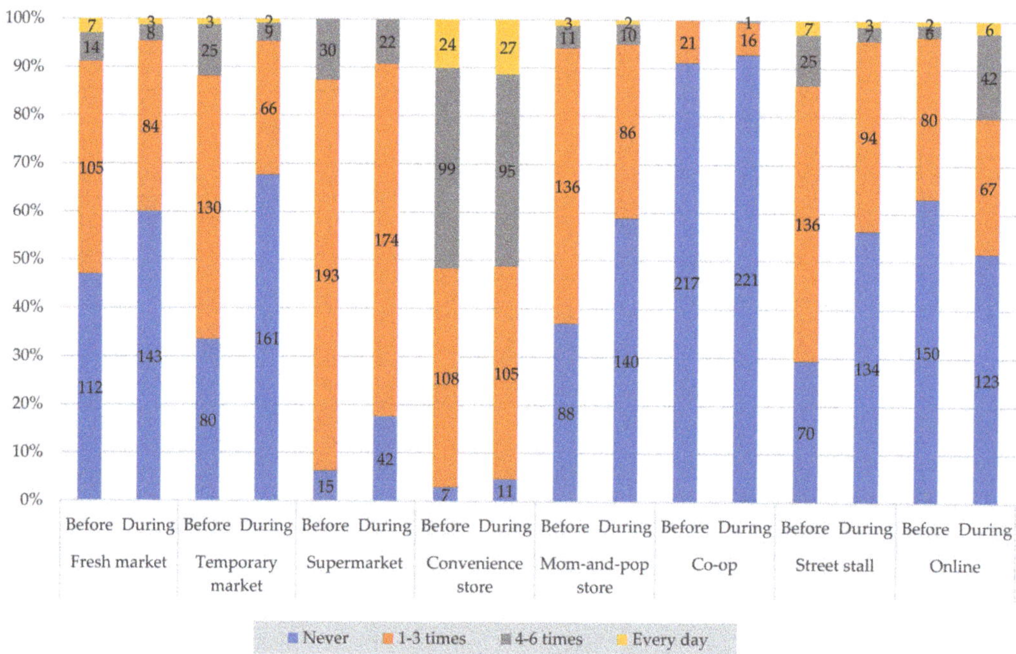

Figure 6. Purchasing routes and frequency before and during the COVID-19 pandemic (unit: number of respondents and percentages).

4.4.2. Food Delivery Service Trends

Similar to people around the world who are apprehensive about COVID-19, residents in Thailand also refrained from leaving home to shop for food. According to the results of the survey, respondents used food delivery services because this option allowed them to stay at home or in the office. A second factor cited was that respondents did not want to wait in long queues, while the third factor driving the increased use of food delivery services was the prevalence of discount coupons or promotions. Furthermore, respondents cited discount coupons and promotions, reasonable delivery costs, and user friendliness as the primary reasons for choosing online applications (Grab Food, Foodpanda, and LINEman). Moreover, according to the *t*-test results, respondents increasingly used applications for food delivery services, official restaurant websites, and phone calls when ordering food after the outbreak started. In addition, data from the survey showed variations in the types of frequently ordered foods, including an increase in the consumption of Thai, Chinese, Japanese, and Korean food, as well as fast food. The frequency of orders for desserts and beverages also rose slightly, although the frequency of orders for street food stayed flat, while that for Western cuisine fell.

4.5. Changes in Food and Food-Related Plastic Waste by Household

4.5.1. Changes in Food Waste Generation

Changes in food waste generated in households are shown in Figure 7. Seventy-six percent of respondents indicated that they felt the amount of waste generated had increased and that most of this could be attributed to a rise in the use of online food delivery services and other ready-made meals.

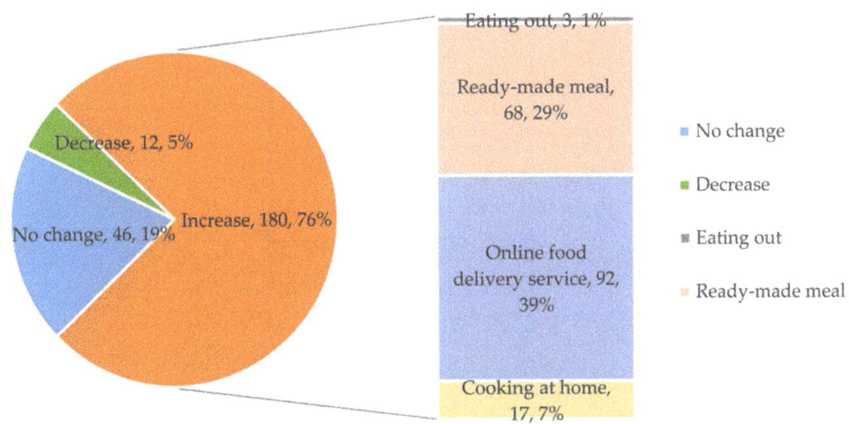

Figure 7. Change in food waste generation.

4.5.2. Changes in Causes of Increased Food Waste Generation

Respondents were queried about the primary reasons for increased food waste generated during the COVID-19 pandemic and changes before and during the COVID-19 pandemic. The top five reasons for food waste, as indicated by respondents, included products that had exceeded their expiration date, rotting/foul odours, excessive amounts of food, unappetising taste and deteriorated quality (Figure 8). Meanwhile, the results of the *t*-test indicated an increase in every cause of food waste since the outbreak (deteriorated quality, rotten/foul odours, exceeding expiration date, excessive amounts of food, taste, and no plans to consume further).

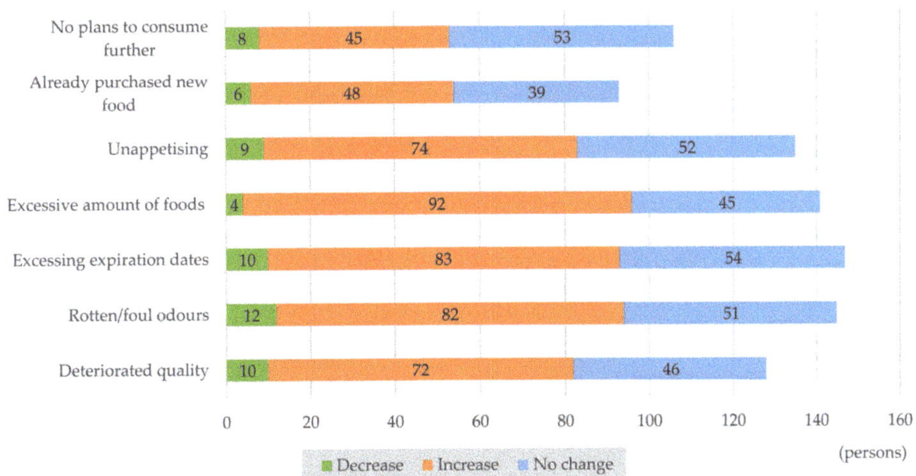

Figure 8. Trend of changes in main reasons for food waste.

The same question was investigated in an earlier study by the authors in 2018 [20,24], which identified the two dominant reasons for increased FW at home as exceeding expiration dates and deteriorated quality, while other reasons, such as excessive amounts, were not cited as a primary cause for FW. In contrast, the reasons mentioned in this survey during the

COVID-19 pandemic are more varied, with the top five listed as exceeding expiration dates, rotten food or foul odours, excessive amounts, unappetising taste, and deteriorated quality (Figure 7). Changes in the primary reasons for FW in households (decrease, increase, and no change) were also queried in this survey, with 'increase' higher than 'decrease' for all reasons (Figure 7), which may be related to online food delivery. For example, reasons selected by respondents (exceeding expiration dates and rotten/foul odours) may be due to a shorter shelf life for prepared food, while other reasons (such as that the food is unappetising) may be related to the inability to predict taste when using online food delivery services. Furthermore, excessive amounts of food became one of the primary reasons for increased FW in households during the pandemic. The inability to predict quantities when ordering online may have caused respondents to over-order, which in turn led to food waste. On the other hand, the results also indicated inadequate food planning and management tendencies by consumers at home during the pandemic.

4.5.3. Waste Generated from the Use of Online Food Delivery Services

Respondents considered plastic bags (E), hot-and-cold food bags (B), plastic food containers (K), and food waste (A) to be the top four types of waste generated from the use of online food delivery services (Table 5). The results of Dunn's multiple comparisons test also showed that the top three types of waste (E, B, A) have significant statistical differences (Table 6).

Table 5. Average score for each type of waste.

No.	Type of Waste	Average Score
A	Food waste	0.70
B	Hot-and-cold food bags	1.27
C	Plastic spoons/forks	0.57
D	Seasoning packages	0.54
E	Plastic bags	1.44
F	Rubber bands	0.17
G	Paper napkins	0.01
H	Toothpicks	0.02
I	Staples	0.05
J	Chopsticks	0.01
K	Plastic food containers	0.86
L	Other food packages such as paper or foam	0.40

Table 6. Dunn's multiple comparisons test for various waste types.

	A	B	C	D	E	K	L
A							
B	3.31×10^{-7}						
C	0.178479	7.72×10^{-11}					
D	0.20438	1.62×10^{-10}	0.894884				
E	1.33×10^{-13}	0.027823	1.97×10^{-18}	4.35×10^{-18}			
K	0.177766	0.000312	0.006085	0.008516	2.81×10^{-9}		
L	0.008463	5.1×10^{-15}	0.19965	0.17456	7.99×10^{-24}	4.28×10^{-5}	

Note: The darker the background color, the lower the p-value indicating the significant differences among various waste types as the cause of environment problems due to the online food delivery service.

4.6. Environmental and Social Concerns and Efforts to Reduce Waste

4.6.1. Environmental and Social Problems Caused by Food Delivery Services

In terms of the environmental and social problems caused by food delivery services, most respondents expressed concern about the increased amount of both food and plastic waste. Air pollution caused by the increased volume of traffic from the use of food delivery services was cited as a secondary concern, followed by higher food prices.

4.6.2. Concerns about Food Due to COVID-19

The top three concerns about COVID-19 and food include an upward swing in food prices (A), uncertainty about the safety of food and ingredients (G), and government epidemic control measures that caused many restaurants to temporarily suspend or permanently shutter their businesses (H) (Table 7).

Table 7. Average scores for concerns due to COVID-19.

No.	Concerns Due to COVID-19	Average Score
A	Upward swing in food prices	1.00
B	Deteriorating quality	0.59
C	Reduced variety of food/ingredients in the market because food manufacturers cannot operate as usual	0.75
D	Deteriorating freshness of food/ingredients due to logistical issues	0.78
E	Lack of time to cook	0.21
F	Uncertainty about the taste of home-cooked food in the family	0.11
G	Uncertainty about the safety of foods/ingredients	1.00
H	Government measures causing restaurants to temporarily suspend or permanently shutter their businesses	0.99
I	Increased amount of food and plastic waste leading to hygienic problems	0.56

4.6.3. Efforts to Reduce FW and PW

The survey showed that the top three actions taken by respondents to reduce food and plastic waste (Table 8) included advance meal planning, use of cloth bags or reuse of plastic bags when shopping, and regular checks of leftover food.

Table 8. Average score for actions to reduce food waste, plastic waste, and other waste.

No.	Specific Actions to Reduce Waste	Average Score
A	Avoid cooking excessive amounts of food	0.25
B	Regularly check leftover food in refrigerators or cupboards	1.03
C	Regularly check expiration dates to avoid throwing away food	0.41
D	Avoid over-shopping	0.51
E	Plan ingredients in advance	1.23
F	Consider how to make different meals from leftovers	0.25
G	Use cloth bags or reuse plastic bags when shopping	1.14
H	Try to consume all food prepared for meals to avoid leftovers	0.26
I	Avoid over-ordering when eating out or using online food delivery services	0.16
J	Request smaller amounts when ordering at restaurants	0.04
K	Request restaurants to avoid excessive packaging when using online food delivery services	0.32
L	Offer leftovers to other people or pets	0.08
M	Reuse tableware and food packaging that are still in good condition	0.35
N	Other	0.01

4.7. Practical Implications of this Study

There have been a few questionnaire-based academic studies on the impacts of COVID-19 on household FW, and Table 9 provides a summary of the survey sites, methods, and content, and the main outcomes that are relevant to this study. The resulting changes brought about by the pandemic have been confirmed by researchers at all stages, from purchasing, cooking and eating to disposal. Compared with the relevant outcomes of the existing literature, the results of this study mainly presented the following practical insights:

(1) The pandemic has had an impact on people's awareness towards health, as they have shown greater concern about nutritional balance. In looking at the categories of food that have been purchased and disposed, it is clear that people have consumed more fruits, fresh vegetables, and meat than usual [25–27]. The total amount of

food purchased, especially canned goods and frozen foods, also increased, as people may have experienced fear or anxiety about logistical systems as a result of food shortages [19,25,28]. Similar trends have been confirmed in Bangkok as well (see Section 4.3.2).

(2) Numerous studies have shown that the COVID-19 pandemic has streamlined people's attitudes toward food waste reduction and more sustainable consumption models [18,29], and subsequent effects, such as stockpiling due to the fear of difficulty in finding food in the medium to long term and staying at home, have had a positive influence on reducing FW. In comparison with Thailand, bulk purchases have not caused significant food waste generation in Italy, Portugal, Spain, and many other countries (see Table 9). Pires et al. [30] revealed that people in Portugal reduced the frequency of food purchases and turned to local shops due to restrictions on outdoor movement, becoming more circumspect in their choices and purchases. However, according to this survey, respondents in Bangkok reported an increase in the amount of household food waste generated during the pandemic, with a significant amount generated from online food delivery or other ready-made meals.

(3) Contrary to what has been reported in the literature, where it has been verified that people in the U.S., Portugal, and several other countries (see Table 9) started cooking at home after the outbreak, most respondents in Bangkok relied on food delivery services. With support from the government and local residents' familiarity with food delivery services such as Grab Food, Foodpanda, and LINEman, the business practices of these services grew aggressively during the pandemic. While many studies indicated a reduction in the amount of food waste generated with more people cooking at home and trying to reduce leftovers, the use of online food delivery services in Bangkok actually resulted in an increase in the amount of food waste generated. This situation differs significantly from that in Brazil where nearly half of the respondents never purchased food online. The convenience of online food delivery services and excessive food supplies might overshadow the chance for people in Bangkok to improve their skills in food planning and management.

(4) Several studies [28,31–33] have clarified that socio-economic and demographic factors such as age, household size and composition, income, attitudes, subjective norms, perceptions of behaviour control, and personal values might impact food management behaviours. Everitt et al. [26] directly measured the quantity and composition of household FW disposed during the first wave of the COVID-19 pandemic and examined how household demographics, socio-economic conditions and local food environment characteristics may influence household FW in the city of London, Ontario, Canada. Further studies may combine waste component analyses and quantitative surveys with socio-economic and demographic components to provide a more in-depth understanding of the food waste situation in Bangkok.

(5) Due to social distancing and travel restrictions during the pandemic, almost all existing literature used online surveys. While it is possible to collect information from people who are interested in the research topic, it has proven difficult to obtain information from people with no access to social networks (e.g., low income, poorly educated, elderly, etc.). This may therefore affect how representative the sample of the population is. As this survey in Bangkok was conducted face-to-face, there is less sampling bias than may be found in an online survey.

Table 9. Relevant international studies and main outcomes on household food and plastic waste during the COVID-19 pandemic.

Country/City	Methods and Contents	Main Outputs
Japan [31]	Nationwide online survey ($n = 1959$) conducted on 2 July 2020 containing questions on thoughts and behaviours related to food purchase, management, consumption, and food waste during COVID-19.	Regions highly impacted by the pandemic appeared to be more careful about their food preparation, purchasing, and management, considering the amount, type, and cost of daily household food waste, while residents in low-impact regions appeared to buy more 'excessive' or 'unnecessary' food.
Tunisia [18]	An online self-administered questionnaire conducted from 24 March to 7 April 2020 ($n = 284$) asking about food purchase behaviour, knowledge of food labelling, attitudes toward food waste, information needs to reduce food waste, and sociodemographic characteristics.	A loss of income and the fear of food shortages led to well-planned shopping behaviours which effectively helped reduce food waste. According to the study, cooking excessive amounts and long-term storage were cited as the major reasons for food waste, indicating that further efforts are needed for food planning and management to reduce food waste and maintain positive changes in behaviour.
U.S. (New York State) [25]	Internet-based survey ($n = 300$) conducted in August 2020 containing 20 questions on household purchases and food waste between mid-March and mid-July 2020.	Food purchases, especially stockpiling food and cooking supplies, increased during the pandemic since more people started to cook at home. However, bulk purchases did not cause massive amounts of food waste to be generated; rather, the results of the study indicated a slight decrease. This may be due to the tendency of people to improve cooking and storage skills and to prepare plans before shopping during the pandemic.
U.S. [34]	Online survey conducted in the United States in October 2020 ($n = 946$) asking about individual demographic factors, household characteristics, COVID-19-related household changes, and changes in food-related behaviours due to the pandemic.	More people tended to cook at home since they spent more time in their houses, especially households with children, or as a result of lost income or a need to work from home. Thus, over 75% of respondents purchased more food during the pandemic. Stockpiling food was identified as a significant predictor of increased food waste. Of all food ingredients, fresh vegetables and frozen food accounted for the majority of food waste.
Italy [33]	Self-administered online survey ($n = 1078$) from 10 April to 3 May 2020 focusing on food management habits before and during lockdown.	The study showed that respondents spent more per week over lockdowns (an average of EUR 132 per week compared to EUR 110 pre-COVID), likely due to greater amounts of food consumed at home. Most households reported that they threw away less food during COVID-19 lockdowns. Fifty-nine percent of respondents prepared shopping lists for food purchases in regular times, compared to 86.5% during lockdowns. The spread of planning-related food management practices (compiling shopping lists, planning purchases and meals in advance, reuse of leftovers for other recipes) played a key role in reducing FW.
Italy [19]	Online survey ($n = 1188$) from 20 to 25 April 2020 that included a set of qualitative questions about changes in purchasing behaviour, food expenditures, waste production, and other food-related behaviours during the COVID-19 pandemic.	The increase in food purchases during the pandemic did not generally lead to a higher rate of food waste. About 33% and 16% of the sample reported that the amount of food waste decreased substantially or mildly, respectively, during lockdowns. About 45% reported no change, while only 5% and 1% indicated that food waste increased mildly or substantially, respectively. The decrease in food waste is related to the purchase of non-perishable foods.
26 Brazilian States and Federal District (27 states) [27]	Online self-administered questionnaire from 21 May to 30 May 2020 ($n = 458$) that included questions about food purchase behaviour, knowledge of food labelling, attitudes toward food waste, information needs to reduce food waste, and sociodemographic characteristics.	Empirical results confirmed that 'intentions to reduce wastage', 'management routines for leftover or uneaten food', and 'routines of purchasing food on sale' are positively related to the reduction of FW. However, 'planning purchases', 'knowledge about labels', and 'activities to avoid food waste' were not confirmed as having an effect on reducing FW. Additionally, the surveyed population preferred shopping in person, with 45.6% never having made purchases online, while in contrast, 33.0% of respondents reported an increased frequency of online purchases and 16.4% indicated no changes in their online purchasing habits. There was no substantial change in purchasing behaviours of Brazilian households in the specific context of the COVID-19 pandemic with in-person shopping and payment methods using cash.
Spain [28]	Online survey conducted from 14 May to 11 June 2020 ($n = 6293$) consisting of 36 questions on purchasing, storage, cooking habits, waste generation, and changes brought about by the pandemic.	Although most people reported that they did not generate more food waste than usual and some started to be more creative in cooking with leftovers, people who bought food due to fear or anxiety tended to waste more. Respondents who worked from home reported that were stressed since they needed to work more hours than usual and showed the same tendency as those who stored food to waste more due to fear or anxiety.

Table 9. *Cont.*

Country/City	Methods and Contents	Main Outputs
Portugal [30]	Online survey conducted from 22 May to 5 June ($n = 841$), which is the same 36-question survey used in Spain.	From the study, it appears that people in Portugal reduced the frequency of purchases and preferred local shops, but purchases online did not increase. Respondents also reported that they did not change their diet nor the type of waste. A reduction in the total amount of food waste was seen since people tended to buy food and be more circumspect in what and how they prepared food, although producers' associations reported that they had been forced to discard large quantities of perishable products due to the cancellation of purchases in food services/supermarkets.
Turkey [32]	Self-administered questionnaire conducted in January 2021 ($n = 511$) to investigate changes in food management behaviour during the pandemic.	This study divided people into three groups and provided suggestions to each. People who do not prepare detailed plans should improve both shopping and cooking skills. Resourceful planners and cooks have less problems in these areas so they can maintain their food management behaviours. Those who are poor at planning but are resourceful cooks with adequate food preparation skills only need to plan better to purchase and cook food.
Apulia Region, Italy [35]	Online survey conducted from 14 to 30 November 2020 ($n = 323$) that included questions on sociodemographic characteristics, shopping habits, time management, perceptions of food waste, and behaviours during the pandemic.	Based on the results of the survey, the respondents were divided into three groups according to food consumption and food waste habits. One group had a high level of environmental awareness but still generated a large amount of food waste. The second group has limited awareness on food waste but wastes less. Only the last group of responders had a sufficient level of knowledge on food waste and was able to put that knowledge into practice to reduce food waste. It is necessary to offer contrasting information and educational programmes to different group of people.
London, Ontario, Canada [26]	Collection and analysis of waste samples between 9 and 16 June 2020 ($n = 100$) to investigate the food waste situation during the pandemic.	Each week, 2.81 kg of food waste per household was disposed, with fruit and vegetables accounting for over half. Larger households generated more food waste than smaller households. People living closer to grocery stores generated less waste. This may be because the larger the family, the further away they may live and the larger the bulk purchases may be, which may lead to a larger amount of food waste being generated.

5. Policy Implications and Potential Intervention Actions

The authors' previous study on the FW situation in Bangkok before COVID-19 [20] found that the sources of FW are widely distributed throughout the supply chain due to the higher frequency of use of food services and ready-made products and diversification of diets and eating habits. However, due to inadequate management and insufficient detailed regulations and laws, the amount of FW generated is on the rise, with most mixed together with MSW and landfilled. In this study in Bangkok, we found that the COVID-19 pandemic shifted the main source of FW from businesses to households and that both food and plastic packaging waste from households rose due to the increased use of online food delivery services. Post-pandemic, FW and PW generated by households in Bangkok are expected to continue to rise due to hygiene concerns, infection prevention measures, and rebounds in economic activity. At the same time, online shopping is expected to grow even after the pandemic, as the Thai government is partnering with private financial institutions to develop a platform for online shopping to promote digitalisation and cashless shopping.

Food waste and plastic pollution are viewed as two key drivers for achieving the Paris Agreement and the SDGs, but the COVID-19 pandemic may intensify challenges for FW and PW. Therefore, we believe that the policy implications proposed in the previous study [20] must be further strengthened, aiming to: (1) develop comprehensive policies along the entire supply chain; (2) enhance concrete implementation plans with clear targets for reducing and recycling waste based on 3R strategies; (3) develop practical source separation and collection systems; (4) promote the application of appropriate waste management technologies together with 'recycling loop' business models; (5) promote platforms for stakeholder collaboration and community-based interventions; (6) create uniform standards and understanding of 'date labels'; (7) encourage the provision and consumption of smaller portions; (8) utilise health as a driving force to motivate public concern; and (9) develop a policy mix targeting consumers' daily lifestyles and social practices. In addition to these policies, more intense efforts will be needed to achieve the

SDGs. Based on the results of this study, consumer food planning and management, online platforms/delivery services, and the formation of local circular economy frameworks could be considered as three important intervention points for reducing waste and achieving more effective food and plastic waste management practices. Further discussion on these three intervention points is as follows.

5.1. Improve Consumers' Capabilities to Plan, Manage Food and Cook without Waste

The survey found that (1) the number of people cooking and eating at home rose during the pandemic (4.3.1); (2) the excessive amount of food was identified as one of the top reasons for food waste (4.5.2) as a result of poor food management; and (3) the top two actions that respondents would take to reduce food and plastic waste included meal preparation and regularly checking leftover food (4.6.3). These responses may present governments, the food industry, and businesses with an opportunity to support people's abilities to plan, manage, and cook food to reduce the amount of food that is leftover or wasted at home. Not much effort is being implemented by the Thai national government and Bangkok local government on these issues, which needs to be addressed. Qian et al. [31] show evidence to support this result, as people who prepare their own food demonstrate more concern for food management and food waste than those who do not cook. Meanwhile, routines related to planning food purchases and their preparation are highly influenced by the skills or confidence that consumers have in their ability to perform such activities. Cooking classes, refrigerator cameras, shopping lists, and information campaigns on reducing food waste have been widely proven through case studies worldwide to have positive effects, though credibility needs to be further verified [12]. Additionally, Hebrok and Heidenstrøm [36] identified decisive moments and contexts for food waste prevention and discussed examples of measures that could be further explored by applying a practice-oriented approach to food waste drivers through food management practices. Furthermore, preparing and ordering excessive amounts of food might be the result of difficulties in estimating the amount of food, so an important measure to avoid this could be to provide hints on enhancing consumers' food planning and management capabilities and cooking skills, thereby reducing the increased amount of food waste generated during COVID-19. For example, the food industry could indicate the number of servings on food packaging instead of weight, which may help consumers while purchasing. Similar suggestions will effectively help consumers manage food in households, such as by packaging smaller portions and showing consumers how to manage uneaten food and extend expiration dates. Social media platforms, including television programmes, recipe apps, and cooking videos will also play a role in improving the ability of people who may lack skills or have few ideas about what to cook. In a similar fashion, supermarkets may also be able to provide suggestions to consumers about food preparation by displaying the ingredients needed for certain meals.

5.2. Develop Eco-Friendly Online Platforms and Food Delivery Services

The rapid expansion of e-commerce and online food delivery services is a visible change that has occurred as a result of the COVID-19 pandemic. This study shows that during the pandemic, most food services shifted to online food delivery, resulting in an increase in both FW and PW. Although online food delivery services benefit society as a whole in terms of lowering the number of potential routes for infectious diseases, while simultaneously providing a certain level of comfort to consumers and stimulating economic activity, there is a risk that incentives may encourage a rise in the use of these services, leading to overconsumption and other adverse behaviours in terms of FW and PW. Therefore, the key to preventing and reducing both FW and PW is determining how to build eco-friendly online platforms/food delivery services and business models to encourage consumers to act in environmentally friendly ways.

The temporary relaxation of bans on the use of single-use plastics during the pandemic may indicate a breakdown in sustainable patterns of behaviour. To mitigate the problem of plastic waste, research and development in materials science to streamline plastic packaging should be emphasised for sustainable development [37]. Consideration should also be given to establishing more sustainable options, such as deposit refund schemes, or default options, such as delivering food in reusable containers. At the same time, governments must institute educational curriculums and communication campaigns to highlight and promote environmentally friendly behaviour.

Meanwhile, along with the expansion of green food delivery options and online shopping, access to food has become easier and more efficient, which could lead to lower GHG emissions and achieve low-carbon lifestyles. Besides changes in shopping habits, Galanakis [29] points out that digital technologies, including information and communication technologies (ICT), apps, the Internet of things (IoT) platforms, big data and artificial technology, will enable food to be delivered precisely when demand arises, potentially leading to a reduction in FW. Additionally, digital platforms such as food rescue apps can be used as a mechanism for mobilising the active participation of stakeholders along the entire supply chain. Therefore, the system should be designed in advance to maximise the numerous synergies between the promotion of online platforms and the prevention of FW and PW, as well as to minimise trade-offs.

In a positive development, the Ministry of Natural Resources and Environment has joined together with private entities, including six food delivery platforms such as Food Panda, Grab Food, Gojek, and LalaMove in a push to reduce the use of plastic under the concept of the 'New Normal Food Delivery with Environmental Care'. Food delivery services involved in this initiative will add an opt-in button to their applications that will allow customers to decline single-use plastics as well as work with their restaurants to incorporate more environmentally friendly packaging (glass jars, metal straws, non-plastic bags, etc.).

5.3. Promote a Circular Economy via Localised Supply Chains to Improve Food Safety and Well-Being

A system-level approach to address issues surrounding FW, PW, and MSW is needed. Circular economy strategies have opened up new avenues for potential measures to reduce FW [38]. The concept of the 'circular economy' is central to European environmental thinking and policy-making, and the transition to a more circular economy is a major goal towards developing a sustainable, low-carbon, resource-efficient, and competitive economy in the EU [39]. Food waste and plastics are two of five priority sectors in the EU Circular Economy Action Plan, which helps contain all materials within infinite loops through sustainable consumption and production and sound waste management, including greater recycling and re-use, and also by creating a market for secondary raw materials. The concept of a circular economy also encompasses waste prevention in the first place, which is positioned at the top of the waste hierarchy.

As transportation and logistics have been highly restricted during the pandemic, the use of local food supply chains to improve food safety and revitalise the local economy has been an effective measure to counter COVID-19. Based on a systematic literature review study on COVID-19, food systems, and the circular economy by Giudice et al. [1], the 'localisation' of food systems might present more resilient and sustainable solutions: localised food systems reduce waste and stress nutrition; combining local and seasonal elements in short supply chains reduces storage and transportation needs, provides a better supply–demand balance, creates more transparency, improves tracking capability, and contributes to waste reduction; and consumers seem to place higher value on food purchased in local markets. The localisation of food systems will also help reduce the amount of plastic packaging waste and provide fiscal security to fight similar pandemics in the future.

6. Conclusions

In this paper, Bangkok was selected as a case study to examine the impact of the COVID-19 outbreak on the generation of food and plastic waste by consumers by examining shifts in consumer lifestyles and consumption behaviours through a face-to-face questionnaire survey. The potential of food delivery services in the starring role of a COVID-19 success story and the related environmental consequences of food waste, plastic waste, and other problems caused by a new food consumption paradigm were also examined. This paper also provides policy implications and innovative actions for tackling the issues raised to achieve more effective food and plastic waste management.

Although travel bans and diminished economic activity due to COVID-19 have led to a dramatic reduction in MSW, both FW and PW generated by households in Bangkok were observed to have increased during COVID-19. Furthermore, the total amount of FW and PW as well as MSW in Bangkok is expected to rise post-pandemic in the absence of appropriate institutional frameworks and a lack of policy-level directions and effective measurements to address FW and PW issues. This increase may also likely affect our mid- and long-term goals for transitioning towards sustainability.

Although the data presented in this study are relatively uncertain due to the limited number of samples, this is the first study to contribute to a better understanding of how COVID-19 affects consumer behaviour and can help constitute a basis to further promote behaviour that prevents and reduces FW and PW in households, even outlasting the COVID-19 crisis in Asian developing countries.

Of course, there are many research questions left unanswered. For example, there is still a poor understanding of the impacts made by socio-economic and demographic factors on food management behaviour, and on the amount of FW and PW generated by households, as well as throughout the supply chain. Furthermore, to achieve SDG 12, ensuring harmonised data collection on FW and PW remains a challenge in Asia, and practical policies, strategies and actions on household prevention and reduction is an area that can be considered for future study. In addition, the upheaval caused by the COVID-19 crisis has created not only a major challenge, but also an opportunity for reshaping existing policy frameworks and production-consumption style socio-economic systems, as well as a chance to identify the underlying drivers of food waste and their links with plastic packaging. It may also present an opportunity to engage with relevant stakeholders including consumers to tackle the dual challenge of food waste and plastic waste in a systemic way, which is also a topic for further work.

Supplementary Materials: The supplementary materials are available online at https://doi.org/10.6084/m9.figshare.15131100.v1.

Author Contributions: Conceptualization, methodology, data analysis, and writing (original draft) by C.L. and P.B.; investigation and data curation by P.B.; funding acquisition and project administration by C.L.; literature review and writing (revision) by C.L. and Q.Z. All authors have read and agreed to the published version of the manuscript.

Funding: This research was funded by the Institute for Global Environmental Strategies (IGES) under the Strategic Research Fund 2020–2021 and supported by the Environment Research and Technology Development Fund (S-16) of the Environmental Restoration and Conservation Agency of Japan, 'Policy Design and Evaluation to Ensure Sustainable Consumption and Production Patterns in Asian Region' (2016–2020).

Institutional Review Board Statement: Not applicable.

Informed Consent Statement: Not applicable.

Data Availability Statement: Supplementary Material associated with this article are openly available in FigShare at https://doi.org/10.6084/m9.figshare.15131100.v1 (accessed on 8 August 2021).

Acknowledgments: The authors would like to thank Emma Fushimi and Christine Pearson Ishii for proofreading this paper and would also like to express gratitude to the referees for their useful comments and suggestions.

Conflicts of Interest: The authors declare no conflict of interest.

References

1. Giudice, F.; Caferra, R.; Morone, P. COVID-19, the food system and the circular economy: Challenges and opportunities. *Sustainability* **2020**, *12*, 7939. [CrossRef]
2. Sharma, H.B.; Vanapalli, K.R.; Cheela, V.S.; Ranjan, V.P.; Jaglan, A.K.; Dubey, B.; Goel, S.; Bhattacharya, J. Challenges, opportunities, and innovations for effective solid waste management during and post COVID-19 pandemic. *Resour. Conserv. Recycl.* **2020**, *162*, 105052. [CrossRef]
3. Vanapalli, K.R.; Sharma, H.B.; Ranjan, V.P.; Samal, B.; Bhattacharya, J.; Dubey, B.K.; Goel, S. Challenges and strategies for effective plastic waste management during and post COVID-19 pandemic. *Sci. Total Environ.* **2021**, *750*, 141514. [CrossRef] [PubMed]
4. Lemaire, A.; Limbourg, S. How can food loss and waste management achieve sustainable development goals? *J. Clean. Prod.* **2019**, *234*, 1221–1234. [CrossRef]
5. Löhr, A.; Savelli, H.; Beunen, R.; Kalz, M.; Ragas, A.; Van Belleghem, F. Solutions for global marine litter pollution. *Curr. Opin. Environ. Sustain.* **2017**, *28*, 90–99. [CrossRef]
6. Wang, J.; Zheng, L.; Li, J. A critical review on the sources and instruments of marine microplastics and prospects on the relevant management in China. *Waste Manag. Res.* **2018**, *36*, 898–911. [CrossRef]
7. Muriana, C. A focus on the state of the art of food waste/losses issue and suggestions for future researches. *Waste Manag.* **2017**, *68*, 557–570. [CrossRef]
8. Amicarelli, V.; Bux, C. Food waste measurement toward a fair, healthy and environmental-friendly food system: A critical review. *Br. Food J.* **2020**, *123*, 2907. [CrossRef]
9. Schott, A.B.S.; Wenzel, H.; La Cour Jansen, J. Identification of decisive factors for greenhouse gas emissions in comparative life cycle assessments of food waste management—An analytical review. *J. Clean. Prod.* **2016**, *119*, 13–24. [CrossRef]
10. De Menna, F.; Dietershagen, J.; Loubiere, M.; Vittuari, M. Life cycle costing of food waste: A review of methodological approaches. *Waste Manag.* **2018**, *73*, 1–13. [CrossRef]
11. Aschemann-Witzel, J.; de Hooge, I.; Amani, P.; Bech-Larsen, T.; Oostindjer, M. Consumer-related food waste: Causes and potential for action. *Sustainability* **2015**, *7*, 6457–6477. [CrossRef]
12. Reynolds, C.; Goucher, L.; Quested, T.; Bromley, S.; Gillick, S.; Wells, V.K.; Evans, D.; Koh, L.; Kanyama, A.C.; Katzeff, C.; et al. Review: Consumption-stage food waste reduction interventions—What works and how to design better interventions. *Food Policy* **2019**, *83*, 7–27. [CrossRef]
13. Roodhuyzen, D.M.A.; Luning, P.A.; Fogliano, V.; Steenbekkers, L.P.A. Putting together the puzzle of consumer food waste: Towards an integral perspective. *Trends Food Sci. Technol.* **2017**, *68*, 37–50. [CrossRef]
14. Schanes, K.; Dobernig, K.; Gözet, B. Food waste matters—A systematic review of household food waste practices and their policy implications. *J. Clean. Prod.* **2018**, *182*, 978–991. [CrossRef]
15. Heidbreder, L.M.; Bablok, I.; Drews, S.; Menzel, C. Tackling the plastic problem: A review on perceptions, behaviors, and interventions. *Sci. Total Environ.* **2019**, *668*, 1077–1093. [CrossRef] [PubMed]
16. United Nations. *Transforming Our World: The 2030 Agenda for Sustainable Development*; Department of Economic and Social Affairs, United Nations: New York, NY, USA, 2015.
17. Van Rensburg, M.L.; S'phumelele, L.N.; Dube, T. The 'plastic waste era'; social perceptions towards single-use plastic consumption and impacts on the marine environment in Durban, South Africa. *Appl. Geogr.* **2020**, *114*, 102132. [CrossRef]
18. Jribi, S.; Ismail, H.B.; Doggui, D.; Debbabi, H. COVID-19 virus outbreak lockdown: What impacts on household food wastage? *Environ. Dev. Sustain.* **2020**, *22*, 3939–3955. [CrossRef] [PubMed]
19. Pappalardo, G.; Cerroni, S.; Nayga, R.M.; Yang, W. Impact of Covid-19 on Household Food Waste: The Case of Italy. *Front. Nutr.* **2020**, *7*, 585090. [CrossRef]
20. Liu, C.; Mao, C.; Bunditsakulchai, P.; Sasaki, S.; Hotta, Y. Food waste in Bangkok: Current situation, trends and key challenges. *Resour. Conserv. Recycl.* **2020**, *157*, 104779. [CrossRef]
21. Food and Agriculture Organization of the United Nations. *Food Wastage Footprint: Full-Cost Accounting*; FAO: Rome, Italy, 2014; Available online: http://www.fao.org/3/i3991e/i3991e.pdf (accessed on 8 August 2021).
22. Yamane, T. *Statistics, An Introductory Analysis*; Harper and Row: New York, NY, USA, 1967.
23. Richardson, A.J.; Ampt, E.S.; Meyburg, A.H. *Survey Methods for Transport Planning*; Eucalyptus Press: Melbourne, Australia, 1995.
24. Bunditsakulchai, P.; Liu, C. Integrated Strategies for Household Food Waste Reduction in Bangkok. *Sustainability* **2021**, *13*, 7651. [CrossRef]
25. Babbitt, C.W.; Babbitt, G.A.; Oehman, J.M. Behavioral impacts on residential food provisioning, use, and waste during the COVID-19 pandemic. *Sustain. Prod. Consum.* **2021**, *28*, 315–325. [CrossRef]
26. Everitt, H.; van der Werf, P.; Seabrook, J.A.; Wray, A.; Gilliland, J.A. The quantity and composition of household food waste during the COVID-19 pandemic: A direct measurement study in Canada. *Socio Econ. Plan. Sci.* **2021**, 101110. Available online: https://www.sciencedirect.com/science/article/pii/S0038012120307904 (accessed on 8 August 2021). [CrossRef]
27. Schmitt, V.G.H.; Cequea, M.M.; Neyra, J.M.V.; Ferasso, M. Consumption Behavior and Residential Food Waste during the COVID-19 Pandemic Outbreak in Brazil. *Sustainability* **2021**, *13*, 3702. [CrossRef]
28. Vidal-Mones, B.; Barco, H.; Diaz-Ruiz, R.; Fernandez-Zamudio, M.-A. Citizens' Food Habit Behavior and Food Waste Consequences during the First COVID-19 Lockdown in Spain. *Sustainability* **2021**, *13*, 3381. [CrossRef]

29. Galanakis, C.M. The Food Systems in the Era of the Coronavirus (COVID-19) Pandemic Crisis. *Food* **2020**, *9*, 523. [CrossRef] [PubMed]
30. Pires, I.M.; Fernández-Zamudio, M.Á.; Vidal-Mones, B.; Martins, R.B. The impact of covid-19 lockdown on portuguese households' food waste behaviors. *Hum. Ecol. Rev.* **2020**, *26*, 59–69. [CrossRef]
31. Qian, K.; Javadi, F.; Hiramatsu, M. Influence of the COVID-19 pandemic on household food waste behavior in Japan. *Sustainability* **2020**, *12*, 9942. [CrossRef]
32. Özbük, R.M.Y.; Coşkun, A.; Filimonau, V. The impact of COVID-19 on food management in households of an emerging economy. *Socio Econ. Plan. Sci.* **2021**, 101094. [CrossRef]
33. Principato, L.; Secondi, L.; Cicatiello, C.; Mattia, G. Caring more about food: The unexpected positive effect of the Covid-19 lockdown on household food management and waste. *Socio Econ. Plan. Sci.* **2020**, 100953. [CrossRef]
34. Cosgrove, K.; Vizcaino, M.; Wharton, C. COVID-19-related changes in perceived household food waste in the united states: A cross-sectional descriptive study. *Int. J. Environ. Res. Public Health* **2021**, *18*, 1104. [CrossRef]
35. Amicarelli, V.; Tricase, C.; Spada, A.; Bux, C. Households' Food Waste Behavior at Local Scale: A Cluster Analysis After the COVID-19 Lockdown. *Sustainability* **2021**, *13*, 3283. [CrossRef]
36. Hebrok, M.; Heidenstrøm, N. Contextualising food waste prevention-decisive moments within everyday practices. *J. Clean. Prod.* **2018**, *210*, 1435–1448. [CrossRef]
37. United Nations Environment program. *Exploring the Potential for Adopting Alternative Materials to Reduce Marine Plastic Litter*; UNEP: Nairobi, Kenya, 2018.
38. Borrello, M.; Caracciolo, F.; Lombardi, A.; Pascucci, S.; Cembalo, L. Consumers' perspective on circular economy strategy for reducing food waste. *Sustainability* **2017**, *9*, 141. [CrossRef]
39. European Commission. Communication from the Commission to the European Parliament, the Council, the European Economic and Social Committee and the Committee of the Regions. *Closing the loop—An EU Action Plan for the Circular Economy.* 2015. Available online: https://eur-lex.europa.eu/legal-content/EN/TXT/?uri=CELEX:52015DC0614 (accessed on 8 August 2021).

 *sustainability*

Article

Evaluation of Household Food Waste Generation in Hanoi and Policy Implications towards SDGs Target 12.3

Chen Liu [1,*] and Trung Thang Nguyen [2]

[1] Institute for Global Environmental Strategies (IGES), 2108-11 Kamiyamaguchi, Hayama, Kanagawa 240-0115, Japan

[2] Institute of Strategy and Policy on Natural Resources and Environment (ISPONRE), 479 Hoang Quoc Viet str., Bac Tu Liem dist., Hanoi 100000, Vietnam; ntthang@isponre.gov.vn

* Correspondence: c-liu@iges.or.jp

Received: 14 July 2020; Accepted: 10 August 2020; Published: 13 August 2020

Abstract: The issue of food waste, especially in developing economies, is a puzzle. Hanoi was selected as a case study to examine the current situation of food waste generated by consumers through daily habits/practices and to evaluate options for preventing and reducing food waste at the policy level through a literature/policy review and interview-style survey. An analysis of responses found that the self-reported food waste generation rate in Hanoi averaged 1192 g/day/household in urban areas and 1694 g/day/household in rural areas; cooking waste generated during meal processing/preparation accounts for more than 70% of the total; less than 20% of respondents separated out kitchen waste for reuse/recycling before disposal; expiration dates and deteriorating quality were cited as primary reasons for food waste at home in contrast with larger portions and over-ordering outside the home; leftover food is used indirectly as animal feed in urban areas and directly in rural areas; and most respondents indicate a willingness to reduce, reuse, and recycle food waste. To achieve SDG target 12.3, policymakers and practitioners must develop comprehensive food waste policies and actions targeting the entire supply chain, implement practical food waste management systems, and promote sufficiency strategies for saving food, reducing food waste, and maintaining health and well-being.

Keywords: food waste; lifestyle; SDGs; households; Hanoi

---

1. Introduction

A third of all food produced worldwide for human consumption is lost along the food management chain [1]. This lost food is a source of enormous waste in the form of valuable resources, and also contributes to the degradation of the environment [2], as well as adding to the increasing volume of greenhouse gas (GHG) emissions [3] and skyrocketing social and economic costs [4]. Moreover, the generation of food waste is also a moral issue when considering that there are 795 million undernourished people around the world [5]. In light of this, over the course of implementing the Paris Agreement and the UN's 2030 Sustainable Development Goals (SDGs), policymakers, practitioners, and academics have increasingly acknowledged the urgency of addressing the issue of food waste. The SDGs adopted by the United Nations Member States set a target to "by 2030, halve per capita global food waste at the retail and consumer levels and reduce food losses along production and supply chains, including post-harvest losses" [6]. To achieve this target, policies must encourage widespread adoption of certain practices along the food supply chain [7].

Food loss and waste occur at all stages of the food supply chain in different countries for a variety of reasons and can vary by culture [3]. There is no universally-agreed upon definition for food loss and waste [8,9], which has resulted in the publication of inconsistent data on food loss and waste in existing literature. To ensure that the results from this study are comparable to data in other studies, this paper defines "food waste" as "any food, and inedible parts of food, removed from the food supply chain to be recovered or disposed," as proposed by FUSIONS [9]. The term "food loss" in this paper is used to describe a reduction in edible food at the beginning of the supply chain, while "food waste" refers to losses that occur at the end of the supply value chain, or consumption stage, as proposed by FAO in 2011 [8].

Research on food loss and waste began in the late 1980s, and since 2005, data on this issue has become more widely available [10]. However, there is little data to be found in the food waste research landscape in Asia [11], and there is even less literature on food waste in the consumption phase (see 2.1 below). It is generally recognised that on the global level, the consumption phase in developed countries is the largest single contributor to the rising generation of food waste, whereas larger food losses occur during the production and post-harvest phases in developing countries [4]. However, it has been argued that the boundary of consumption habits between developing and developed countries is more blurred in urban settings, meaning that food waste during the consumption phase in developing countries may be as extensive as in developed countries. For example, the volume of food waste generated in Bangkok is even higher than the average in many developed cities and countries [12]. Factors such as the rapidly growing urban middle class [13] and increased income [14], greater spending power [15], dietary transitions towards westernised consumption patterns [16], modern retail diffusion [17], and time scarcity [18] impact the generation of food waste in rapidly urbanising localities. Where developed countries have created adequate policies and systems to manage waste, developing countries have not yet reached this stage; even when waste management and 3R policies are in place, developing countries may falter over implementation. Research analysing food waste in the consumption phase in developing countries has remained static at the preliminary stage and little is known about the determinants of consumer food waste and the underlying factors that encourage, drive, or impede food waste prevention behaviours. It is especially urgent to promote food waste studies to link waste to its producers and examine food waste reduction policy implications in the cities of developing countries where the amount of food waste is dramatically increasing, but where both existing data and the capacity to tackle this issue are limited.

Today, Vietnam is one of the most dynamic emerging economies in the East Asia region, with substantial population growth (jumping from 61.1 million in 1986 to 94.7 million in 2018 and forecast to rise to 120 million by 2050), accelerated economic growth (GDP per capita more than doubled by 2.7 times between 2002 and 2018, reaching 7% GDP growth since 2018), rapid urbanisation (rising from 29% in 2008 to 38.4% in 2018), a swiftly emerging urban middle class (rising from 7.7% in 2014 to 13.3% in 2016) and fast-growing development of consumer goods and services, as well as tourism (total tourism revenue increased more than three times from 200 trillion VND (Vietnamese Dong) in 2013 to 637 trillion VND in 2018) in recent years [19–22]. These developments will lead to the consumption of more food and other goods, causing food waste generation rates to skyrocket. With farmland making up 34.7% of the total land area and the agriculture, forestry, and fishing sector accounting for 37.7% of total employment in the country, Vietnam faces challenges with food loss and food waste along the entire food supply chain in both production and consumption.

Hanoi, the capital of Vietnam, is one of the most vibrant cities in the Asia and Pacific region with relatively high economic growth that is expected to continue rising rapidly in the coming years. The city is facing a number of challenges in solid waste management as it sees an increase in the generation of solid waste in terms of quantity, type, and toxicity. The city produces 8629 tonnes of domestic solid waste per day (3,149,723 tonnes/year), and the average amount of municipal solid waste (MSW) generated daily per capita is 1.1 kg [23]. This waste is discharged from municipal sources, including households, restaurants, markets, and businesses. The rate of waste generated in Hanoi averagely increases by 5% each year due to the growing population and rising economy [23]. The quantity and quality of MSW generated in Hanoi have changed dramatically due to the large concentrated population and the effects of lifestyle changes as a result of economic development [24,25].

It is estimated that by 2030, this figure will have reached 1.72 kg/person/day [23]. If considering food waste as the organic fraction of MSW, food waste accounted for 53.81% of MSW in 2012 (excluding paper residue), and with a growing population, it is estimated that the volume of domestic solid waste in Hanoi, including food waste, will continue to increase. Meanwhile, more than 80% of MSW is landfilled, which contributes to rising greenhouse gas emissions (GHGs). The city faces a series of challenges related to economic growth, environmental protection, and sustainability in agriculture and food systems, including food security, food safety, and food waste, as well as waste management and improving the quality of life.

In this paper, we selected the city of Hanoi as a case study to examine the current situation of food waste generated by consumers through daily habits and practices, and evaluated options for preventing and reducing food waste at the policy level. For this purpose, we

(1) review relevant food waste policies, strategies, and plans through a literature/policy review, and provide insights into policy implications to address issues, and

(2) conduct a questionnaire survey that includes a range of questions on the daily food-related practices of respondents at the household level, including eating habits (eating out, consuming ready-made food at home, and eating in), waste disposal practices, and other relevant points. This glimpse into daily practices is expected to provide an insight into the ways in which consumers generate food waste.

2. Policy/Literature Review

2.1. Household Food Waste Studies in Asian Developing Countries

The number of studies on consumer-generated food waste has grown since private households were identified as key actors in food waste generation in developed economies. Aschemann-Witzel et al. [26] published a study on factors behind the generation of food waste by consumers in households and along supply chains, demonstrating that motivation to avoid food waste, management skills in providing and handling food, and trade-offs between priorities have an extensive influence on the food waste behaviour of consumers. Roodhuyzen et al. [27] developed a framework that conceptualised the generation of consumer food waste in relation to stages of the household supply chain and identified and categorised potential factors of consumer food waste as behavioural, personal, product, and societal factors. Schanes et al. [28] systematically reviewed the rising number of empirical studies on consumer food waste practices and the factors that foster and impede the generation of food waste at the household level. This study reveals food waste to be a complex and multi-faceted issue that cannot be attributed to a single variable. Given the complex nature of food waste, a growing body of literature sheds light on food-related practices and routines, ranging from planning and shopping, to storing, cooking, eating, and managing leftovers, within the context of food waste generation by adopting practice theories and other conceptual approaches [28]. Due to the increased level of interest in research on the topic of food waste over the past decade, academic studies dedicated to this topic have emerged in Asian developing economies. For example, a material flow analysis among middle-class households in Bengalaru, India was combined with a social practice approach to understand how and in what way food is wasted in the post-consumption stage [29]. Taste preference has also been examined in the context of food provision and wastage in Bengaluru and Metro Manila [30]. Soma explores the transformation of household food consumption and food waste practices with the rise of supermarkets [31,32]. In yet another example, Liu et al. [12] analysed a case study on Bangkok using a questionnaire similar to the one in this study to examine the ways in which food waste is generated by consumers. Although Asia offers rich socio-economic dimensions and cultures where food waste issues can be examined from a number of different angles, this study has been limited in scope to present readers with an opportunity to gain a deeper understanding of food waste and to improve the governance of food waste, especially in Asian developing economies.

2.2. Relevant Policies, Strategies and Initiatives on the Issue of Food Waste at the National and Local Government Level

A review of documentation on food, food security, waste management, and 3R (reduce, reuse, recycle) policies, as well as on related food loss and food waste issues, was conducted to consider food waste solutions along the entire supply chain. The overall structure of waste management policies in Vietnam, along with related food loss and food waste policies is summarised in Figure 1. This structure shows that (1) Vietnam has no specific laws, policies, and strategies addressing the issue of food waste, and (2) there are two separate dimensions to food loss and food waste: the first is a policy on reducing post-harvest loss (generated during handling, storage, processing, distribution, and at the market), and the second is a policy on food waste managed as regular municipal solid waste. At the policy level, food waste is considered to be similar to organic waste in that source segregation, waste reduction, and community composting or integrated waste treatment facilities at disposal sites are embedded in MSW plans. The primary guidance on this topic is incorporated into the following policies and strategic plans.

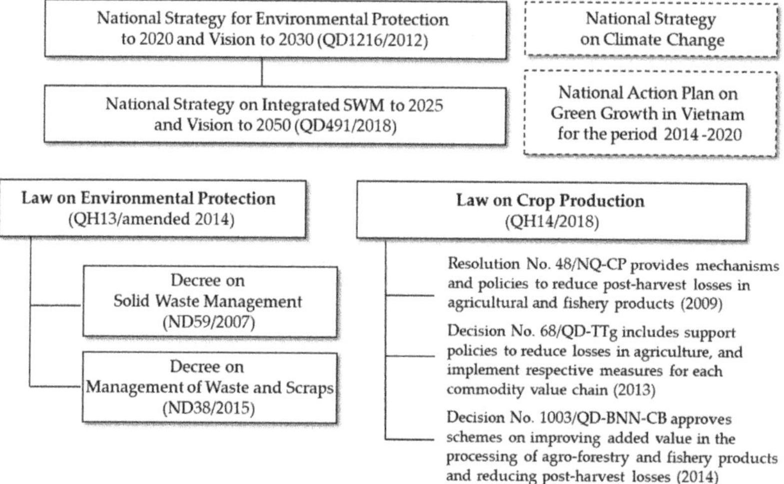

Figure 1. Overall structure of food loss and food waste-related policies in Vietnam.

In order to reduce post-harvest losses and facilitate the transformation and restructuring of agriculture, a number of policies have been put into place, including Resolution No. 48/NQ-CP in 2009 (providing mechanisms and policies to reduce post-harvest losses in agricultural and fishery products (2009)), Decision No. 68/2013/QD-TTg on 14 November 2013 by the Prime Minister on support policies to reduce losses in agriculture, and Decision No. 1003/QD-BNN-CB on 14 May 2014 by the Ministry of Agriculture and Rural Development (MARD) approving schemes to increase value-add in processed agricultural, forestry, and fishery products and setting 2020 as the target year for reducing post-harvest losses. In 2018, the Law on Crop Production (Law No. 31/2018/QH14) was issued with regulations on harvest activities to limit food loss and ensure quality and cost efficiency. The government has also introduced a number of policies to attract the private sector, including investments in technological innovation, increasing the rate of intensive processing products, ensuring food safety, setting competitive prices, and meeting market requirements, as well as investments in advanced technologies to produce high value-add products from agricultural waste.

Vietnam has a relatively comprehensive legal framework on solid waste already in place under the Law on Environmental Protection. This Law was first issued in 1993 and revised in 2005 and 2014, and contains a separate chapter on waste management (Chapter IX). Article 95 of this Law stipulates that waste generators have a "responsibility to classify ordinary solid waste at source to facilitate reuse, recycling, energy recovery, and processing." This was further clarified in Decree No. 38/2015/ND-CP on waste management and scraps issued on 24 April 2015 which expands on policies on waste prevention, reuse, recycling, energy recovery, and disposal. The Decree also mentions separation at source and fees in relation to household waste and contains a section on technology for the disposal of household waste that specifies the use of organic fertilisers.

Following this approach, Decision No. 491/QD-TTg was issued on 7 May 2018 approving revisions to the National Strategy on Integrated Solid Waste Management (SWM) to 2025, with a vision to 2050. The revised strategy addresses the high rate of food and organic waste in domestic waste and encourages environmentally friendly recycling, reuse, and disposal solutions. A number of targets to 2025 have been set up including (i) collecting and treating 90% of the total domestic solid waste generated in urban areas; (ii) enhancing the capability to reuse and recycle; (iii) striving to achieve a rate of less than 30% of collected domestic solid waste directly transported to landfill; (iv) collecting, storing, transporting, and treating 80% of domestic solid waste generated in concentrated rural residential areas on-site or through centralised treatment systems to meet environmental protection requirements and; (v) reusing, recycling, composting, or self-treating organic waste in households in rural areas to produce compost for local use.

To date, food waste has not been specifically mentioned in the Law on Environmental Protection, and waste management as a whole is referred to from the perspective of prevention through reuse, recycling, waste-to-energy, and disposal. The Law specifies the responsibilities of households in minimising and separating waste and partially covering costs for waste collection. The Law, which addresses the management of food waste generated by households, is currently under revision; the bill has been submitted to the National Assembly for debate and is slated for adoption in November 2020. Specifically, the draft Law specifies that household waste should be separated into (i) recyclables; (ii) food waste and organic waste; (iii) hazardous waste; (iv) bulky waste; and (v) other waste. The draft Law also states that collection and transportation fees for food waste should be lower than other household waste and that provincial and municipal authorities shall regulate waste management in households and set applicable fees. However, the draft Law does not mention the management of food loss in crop harvesting and post-harvest, logistics/distribution processes, or in restaurants (MONRE, 2020. Draft version of Law on Environmental Protection, which was submitted to the National Assembly on 23 March 2020).

In addition, the Law on Food Safety, issued in 2010 and revised in 2018, which acts as the guiding principle for food safety related incidents, states that "food should not cause harm to people's health and lives." This Law specifies requirements on hygienic levels at commercial sites and prohibits the use of expired materials and additives in food preparation, both of which could have a trade-off effect on food saving and food waste reducing.

On the ground, there are also a number of initiatives and good practices by local governments, the private sector, NPOs/NGOs, civil society, and other stakeholders on reducing food waste. For example, volunteer-based Hanoi Food Rescue was established in 2013 with sponsorship from REACH, a non-profit organisation operating in the field of training and job support for students in Vietnam who find themselves in difficult circumstances. Tet Donation is an annual event organised to collect quality leftover food after the Lunar New Year for the poor and homeless. Furthermore, many private enterprises have also invested in the processing and recycling of food waste and organic waste, such as the Vietnam Food Joint Stock Company (VNF), which has produced animal feed from shrimp shell by-products using enzymatic hydrolysis technology.

3. Methodology of Household Survey

3.1. Content of the Questionnaire

In order to clarify the impact of consumer behaviour on food waste generation and intentions to reduce food waste, a questionnaire survey was conducted in both urban and rural areas of Hanoi between 15 January and 28 February 2019 in collaboration with the Department of Environment and Sustainable Development under the Institute of Strategy and Policy on Natural Resources and Environment (ISPONRE) in Vietnam.

The questionnaire survey (see Supplementary Materials) for consumers on food waste mainly consisted of four sections and covered a range of daily activities. The first part included basic information about the respondents, such as sex, age, occupation, educational level, and household income. The second section asked respondents to self-report on daily food waste generation in their households, which included cooking waste and leftovers. The third part posed questions about the respondents' daily food related practices, such as their eating, shopping, cooking, and food management habits. The fourth section focussed on respondents' waste separation habits and food waste disposal in households, and the last part contained questions about the respondents' attitudes and intentions to reduce food waste. The questionnaire also included questions about single-use plastics, waste collection and time use, although that has been excluded from this paper for brevity.

3.2. Study Site, Sampling Size and Analytical Approach

Statistics from 2018 indicate that Hanoi [19] has an area of 3358.59km^2 and a population of 7,852,600, including an urban population of 3,847,300 and suburban population of 3,978,300. Population density is 2338 persons/km^2, but is particularly high in 12 urban districts where the density is 11,468 persons/km^2. In addition, there are hundreds of thousands of non-residents and labourers from other provinces earning their living in Hanoi.

To select the survey sites, the research team used a random stratified sampling method. Firstly, the team divided Hanoi into two major areas — an urban area (12 districts, 1 town) and a rural area (17 districts) — and conducted surveys in both areas. Following the sample size calculation formula developed by Yamane [33] to determine sample size and considering the budget and availability of human resources for the survey, the survey's confidence level was set at 95% with a precision level at 7% and an appropriate sample size of 204. After accounting for missing data, a total of 252 responses were received. Secondly, the team chose six out of the 13 districts and towns in the inner city area of Hanoi as representative of the urban area, namely the districts of Long Bien, Nam Tu Liem, Thanh Xuan, Ha Dong, Cau Giay, and Ba Dinh. In the rural area, the team selected six out of 17 rural districts in the suburban area, namely Dong Anh, Thanh Tri, Thach That, Quoc Oai, Gia Lam, and Ung Hoa. Thirdly, the team selected one ward or commune in each district to survey in consideration of how representative it was in providing a picture of the entire city and its accessibility. Lastly, the team randomly selected 20 to 21 households in each ward or commune to survey.

Twelve trained investigators visited the wards or communes and carried out personal interviews with the residents there, rather than requesting that respondents complete the surveys on their own. This style of conducting the survey resulted in highly accurate responses with excellent completion rates.

Using the collected data, the team developed a database in Microsoft Excel on daily food and food waste-related practices in households. A non-parametric t-test (Mann–Whitney U test) was conducted using SPSS (Statistical Package for the Social Sciences) software to clarify the differences between urban and rural areas. The results of the survey are described in detail below.

4. Results of the Survey and Discussion

4.1. Attributes of Respondents

Table 1 provides a detailed composite of the respondents, including gender, age, occupation, education and household income. Since the questionnaire survey was conducted through personal interviews, collection and response rates are nearly 100%. In total, the team collected primary data from 252 individuals (118 in the urban area and 130 in the rural area with four non-responses) comprising 92 males and 156 females. Ages are distributed along a range of 20 to over 60 years of age, with the largest group falling between 30 to 39 years of age (41%). The educational level of the sample varies from primary school to post-graduate degrees, and out of these, 91 respondents have graduated university and 51 have a master's degree or higher. Monthly household income ranges from below 2 million to over 300 million VND. Among them, 165 respondents (66%) earn 10–50 million VND. The sample provides a diverse picture of the residents in Hanoi; the majority of respondents are employed as government officers (31%) and company employees (25%), followed by those engaged in self-employment (20%), farming (6%), and day labour (4%), with the remainder comprising full-time housewives, students, part-timers, the unemployed (including pensioners), and others.

Table 1. Characteristics of respondents ($n = 252$).

		Number of Respondents
Location	Urban area	118
	Rural area	130
Gender	Male	92
	Female	156
Age	≤19	0
	20–29	39
	30–39	104
	40–49	56
	50–59	28
	≥60	18
Occupation	Company employee	64
	Self-employed	50
	Day labourer	11
	Part-timer	4
	Full-time housewife	8
	Farmer/Agriculturist	16
	Government officer	77
	Student (university, junior college, etc.)	7
	Unemployed (including pensioners)	7
	Others	8
Education	No schooling	0
	Primary school	6
	Lower secondary school	38
	Upper secondary school (High school)	27
	Vocational or technical university	35
	University	91
	Master's degree or higher	51
Household income	<2 million VND/month	4
	2–5 million VND/month	11
	5–10 million VND/month	63
	10–50 million VND/month	165
	50–100 million VND/month	7
	≥100 million VND/month	2

4.2. Self-Reported Generation of Food Waste in Households

The total amount of food waste generated in households averages 1192 g/day/household in urban areas and 1694 g/day/household in rural areas. However, at a 5% significance level, there was no significant difference between urban and rural areas. When considering the average number of members in a household (4.2 persons/household in urban areas and 4.0 persons/household in rural areas), food waste generation is adjusted to 285 g/day/person in urban areas and 423 g/day/person in rural areas. Cooking waste generated during the processing and preparation of meals, such as peels, scraps, corn stub, and bones, accounts for more than 70% of the total food waste generated (Table 2). While the amount of edible food wasted (leftovers and untouched food) is relatively low in both urban and rural areas, a report by Natural Resources Defence Council found that an average of 0.23 kg of food per person was wasted per day in households in three U.S. cities (Denver, Nashville, and New York), which contained more than two-thirds (68%) of edible food [34]. WRAP [35] reports that the average amount of food wasted per person per day was 0.42 kg in the United Kingdom in 2015, which included 0.30 kg per person per day from household waste and 0.13 kg per person per day from supply chain waste. According to official data by the Tokyo Metropolitan Government in 2012, the average amount of food wasted per person per day was around 0.39 kg, which included 0.20 kg from household waste and 0.19 kg from supply chain waste [12]. Although data is not directly comparable because of differing definitions of food waste and the methods employed in estimating, it is clear that the total amount of food waste generated in Hanoi is on the same high level as the average in many developed cities and countries.

Table 2. Amount of food waste generated in households.

Average Amount (g/Day/Household)	Urban Area	Rural Area	Total
Cooking waste (generated during processing and preparation of meals)	864	1187	1043
Leftovers from cooked staple foods (rice, bread, etc.)	163	224	199
Leftovers from cooked dishes	87	134	113
Untouched food such as food that has passed the "sell-by date" or "use-by date"	38	41	40
Tea leaves and coffee grounds	38	60	51
Others	2	48	26
Total	1192	1694	1443

4.3. Current and Future Eating Habits

This survey offers a portrait of the balanced diet of a segment of Hanoi's population and provides a glimpse into the types of ingredients consumed by respondents, including meat (pork, chicken), eggs, fish and seafood, dairy products, vegetables, and fruits. This paper focuses on eating habits in terms of the respondents' frequency of consumption and settings in which they partake of meals, such as dining out, consuming store-bought, ready-to-eat meals at home, and eating home-cooked meals at their place of residence.

Table 3 shows the number of times respondents eat each day. It shows that staple foods can be found on the plates of respondents in rural areas about three times a day on average, compared to two times a day in urban areas. People in both urban and rural areas consumed rice, noodles, bread, vegetables, and fruits more than once a day. There was a significant difference between urban and rural areas in regard to staple foods, other grains/cereals, pork, other meats, milk/dairy products, desserts, and dietary supplements at a 5% level of significance in the Mann–Whitney U-test.

The average number of times respondents eat in, consume ready-made meals, and eat out each day is shown in Table 4, among which showing a significant difference between urban and rural areas at the 5% level of significance by Mann–Whitney U-test. People in urban areas ate out 0.5 times per day, a figure that is higher than those in rural areas (0.3 times per day), while people in rural areas ate in 2.3 times per day, which is more frequent than respondents in urban areas (1.7 times per day).

Both residents in rural and urban areas consumed ready-made meals relatively infrequently. There is a strong indication that people eat at home more often than dining outside or consuming ready-made meals, in comparison with cities in other countries, such as Bangkok [12].

Table 3. Number of times respondents eat each day.

	Average Times a Day		
	Urban	Rural	Total
Staple foods (rice, noodle, bread) *	1.9	2.7	2.3
Other grains/cereals *	0.4	0.4	0.4
Beans and pulses/processed beans	0.4	0.4	0.4
Pork *	0.7	1.1	0.9
Chicken	0.4	0.4	0.4
Beef	0.4	0.3	0.4
Other meats *	0.4	0.1	0.2
Eggs	0.5	0.5	0.5
Fish and seafood	0.4	0.5	0.5
Milk/dairy products *	0.9	0.7	0.8
Vegetables *	1.7	2.1	1.9
Fruits	1.4	1.3	1.3
Sweets and desserts *	0.7	0.4	0.5
Dietary supplements *	0.4	0.3	0.3

* Mann–Whitney U-test: $p < 0.05$.

Table 4. Average number of times respondents eat in, consume ready-made meals and eat out each day.

	Average Times a Day		
	Urban	Rural	Total
Eat out *	0.5	0.3	0.4
Consume ready-made meals *	0.3	0.2	0.2
Eat in *	1.7	2.3	2.0

* Mann–Whitney U-test: $p < 0.05$.

Dietary changes over the past five years and intentions in terms of eating habits are shown in Figure 2. More than half of the respondents plan to maintain their current eating habits in the future. In particular, a greater number of people in the urban area plan to cook and eat in more often. This indicates that most people intend to continue eating in more often in both urban and rural areas. Responses to the survey also showed that health and food safety is a key factor in decision-making on diet amongst consumers in Hanoi, which may be a springboard for identifying ways that consumer behaviour can be adapted to decisively reduce food waste [28].

4.4. Frequency and Reasons for Food Waste in Diverse Settings

Figures 3 and 4 illustrate the frequency with which food waste is generated when eating at home and dining out. More than half of the respondents in urban and rural areas indicated that they rarely or never waste food when eating at home. This is in contrast to a shift in habits when dining out, with almost 50% of urban residents responding that they often waste food in these settings. While rural residents still waste less food on average than their counterparts in urban areas, the percentage of people in rural areas who indicated that they sometimes waste food when they eat out is almost twice as high as when they eat at home.

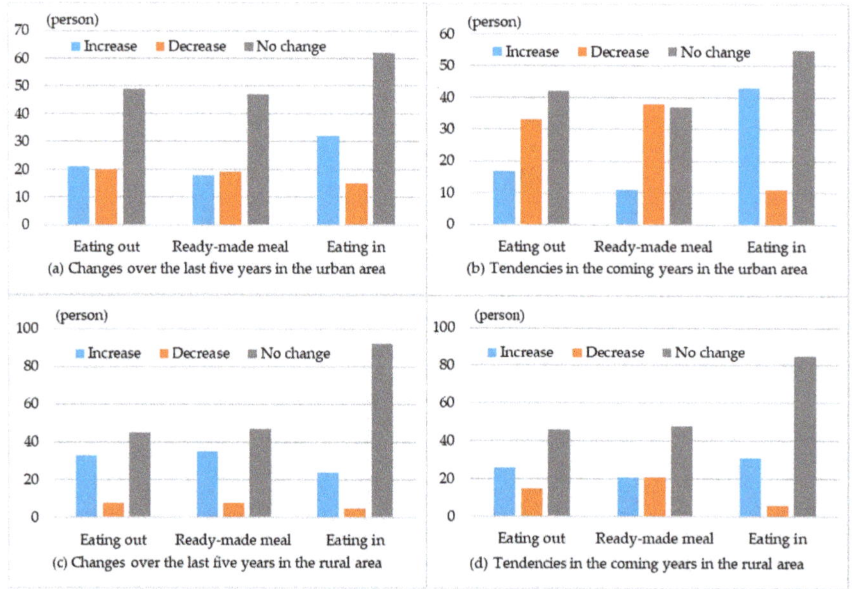

Figure 2. Dietary changes and intended future eating habits in urban and rural areas.

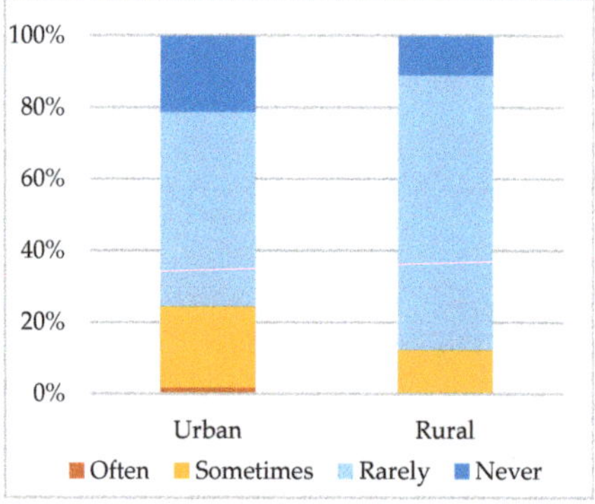

Figure 3. Food wastage when eating at home in urban and rural areas.

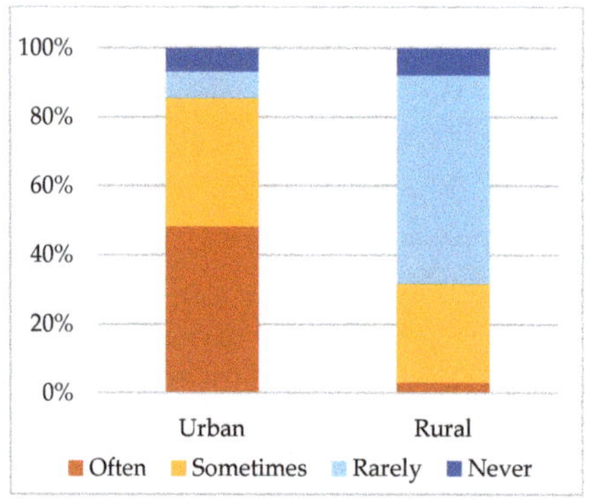

Figure 4. Food wastage when eating out.

At home, the prevailing reason for the generation of food waste in both urban and rural areas is that food had passed its "use-by" date, followed by deteriorating quality (Figure 5). Although labels that indicate expiration dates generally reflect the estimated date that food is at peak quality or taste, it rarely indicates the actual safety of a food product, and may result in the unnecessary disposal of large amounts of food that can still be consumed.

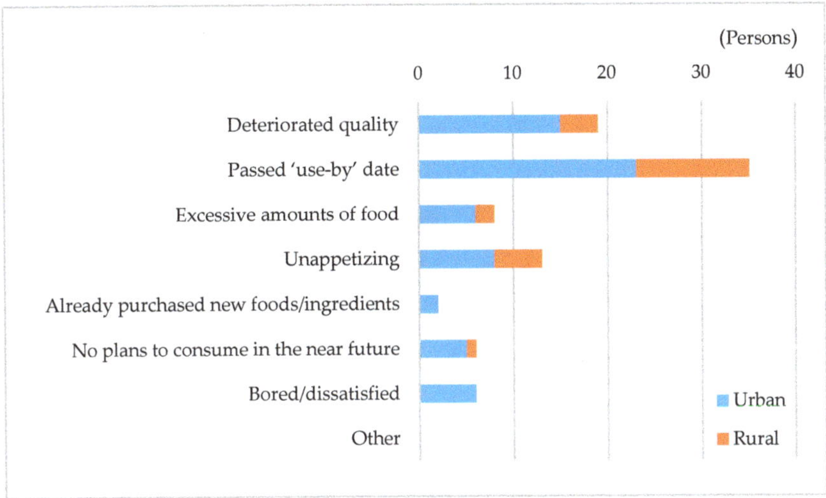

Figure 5. Reasons for food waste generated at home.

The top reasons for food waste when eating out were that people ordered too much and servings were too large (Figure 6). Respondents also cited dislike or dissatisfaction with the taste or that they preferred more variety as reasons for leftover food.

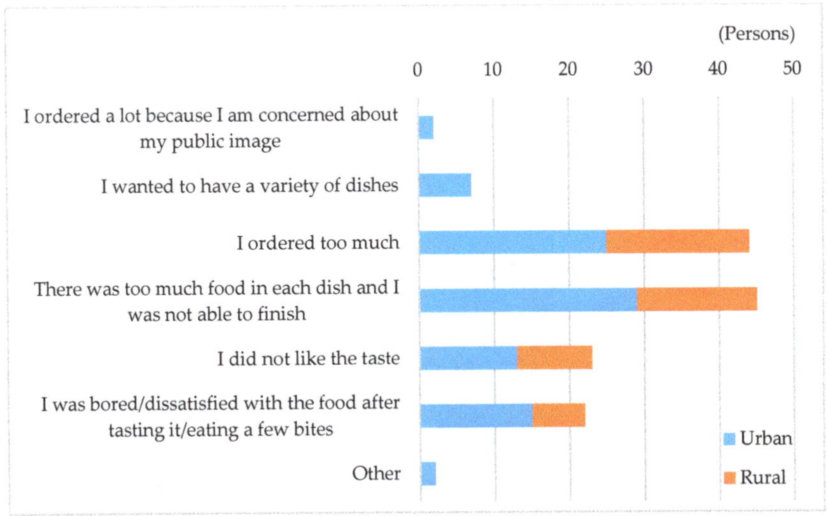

Figure 6. Reasons for food waste when eating out.

4.5. Waste Separation Habits in Households

Data obtained from respondents on methods to manage and dispose of food waste show that more than half of respondents (58%) did not separate food waste before disposal, while 24% separated food waste for sanitary reasons, and less than 20% actively separated and recycled food waste completely (Figure 7). Furthermore, the separation and reuse/recycling rate is higher in rural areas than urban areas for leftovers of cooked food/dishes and untouched food, in particular.

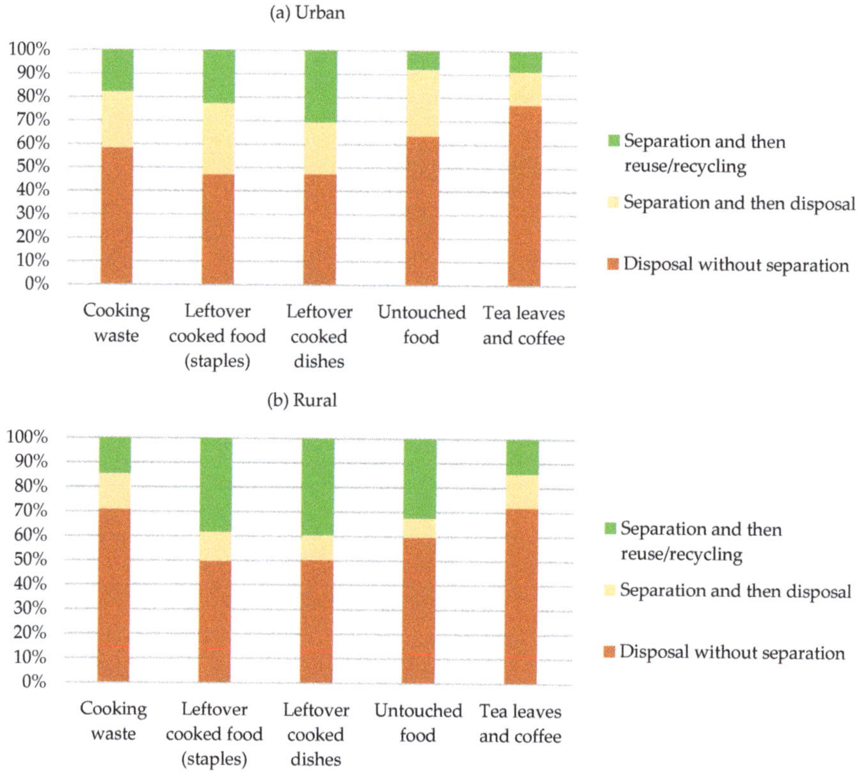

Figure 7. Methods of managing/disposing food waste in (**a**) urban and (**b**) rural areas.

The survey shows differences in the way that people in urban and rural areas separate and then reuse or recycle food waste (Figure 8). In urban areas, the majority of respondents share leftover food with other people, while in rural areas, food waste is used as animal feed and for other purposes. Follow-up interviews found that in urban areas, people look to acquaintances, such as neighbours who have pets like dogs or cats, or vendors in nearby markets who raise animals like pigs, for example, and share food that has been left over from meals, such as cooked rice, meat, and fish. In rural areas where most households raise livestock, it is easier to reuse leftovers directly as animal feed.

4.6. Outlook of Respondents on Reducing Food Waste

The majority of respondents in both rural and urban areas are quite willing to reduce food waste (Figure 9), which is evident in their efforts to reuse or recycle food waste. Only a few indicated that they have no plans to reduce food waste. Respondents were considering possible ways to reduce food waste, including limiting the amount of food cooked, reducing the amount of food ordered when eating out, regularly checking food in their refrigerators or cupboards, trying to avoid over-shopping or purchasing in bulk, checking "use-by" dates regularly, planning in advance for each meal, and figuring out how to make other dishes from leftover food (Figure 10).

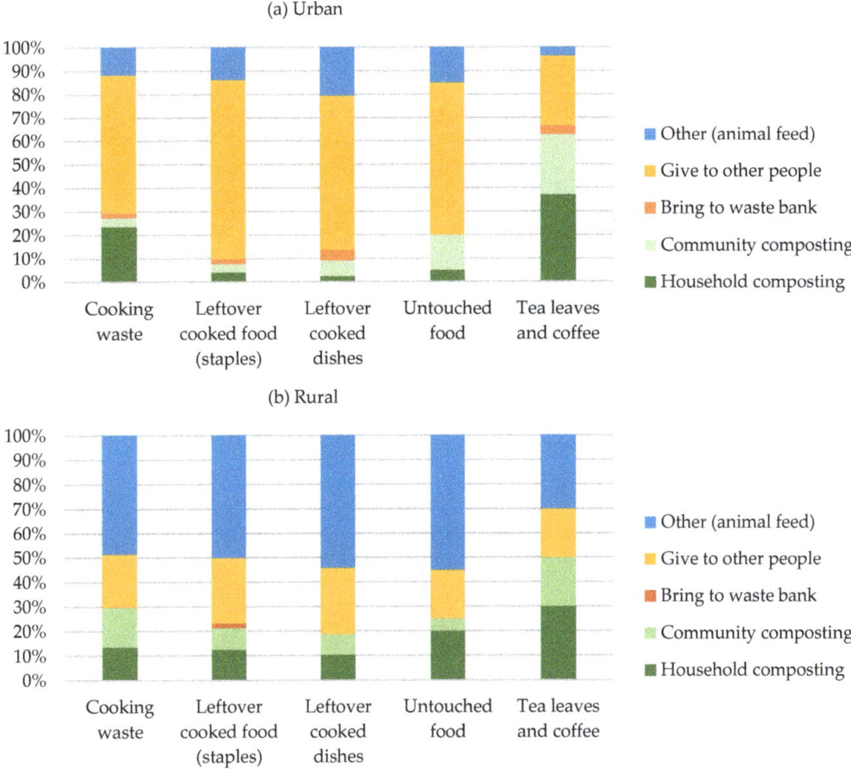

Figure 8. Reuse/recycling methods by respondents who answered that they separate, reuse, and recycle waste. (**a**) urban areas; (**b**) rural areas.

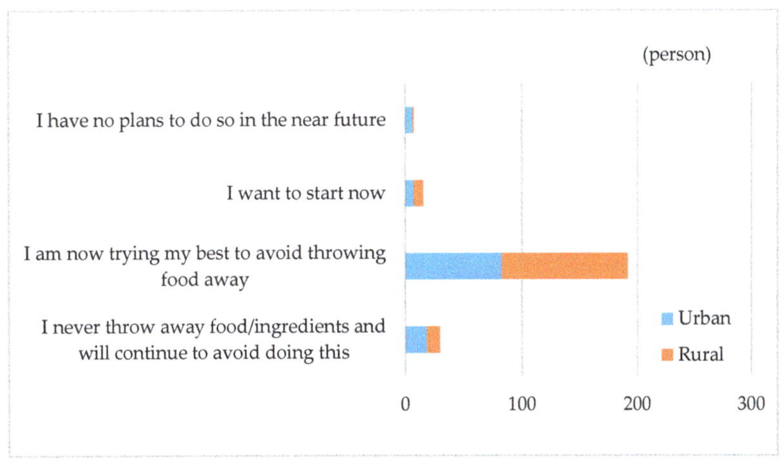

Figure 9. Respondents' efforts to reduce food waste.

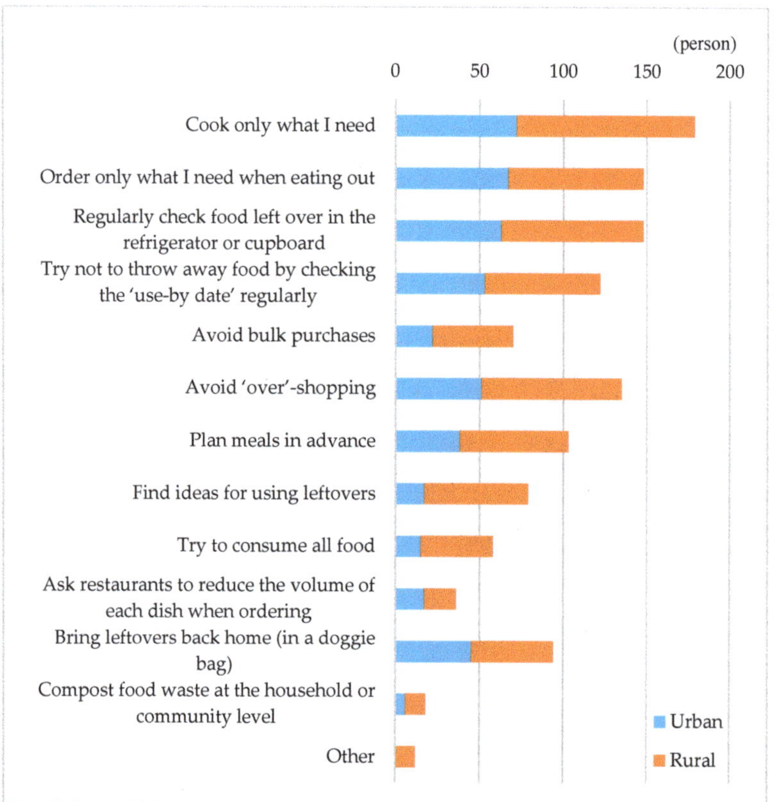

Figure 10. Respondents' intentions to reduce food waste.

Although earlier research suggests a tenuous correlation between people's attitudes in reducing food waste and actual behaviour [36,37], the results from this survey suggest the importance of identifying potential ways that consumers can apply promising methods to reduce food waste more easily than current practices.

5. Policy Recommendations and Potential Actions

The issue of food waste is considered to be a key to helping all countries achieve targets decided through international consensus, including the Paris Agreement and the 2030 Agenda (SDGs). A growing number of studies illustrate the ways in which households are embedded within broader social-economic contexts that create the paths through which food becomes waste and call for a more "holistic" policy approach that engages fully with the complexity of food waste across multiple scales [27,28,38]. Many unaddressed issues and challenges are obstacles that lie in the path to solutions to prevent and reduce the amount of food waste generated. Based on the results of the survey, some high-priority recommendations that may have implications for national and local policies are as follows:

(1) Development of comprehensive food waste policies to address the issue of food waste along the entire food supply chain

Food waste is still considered part of MSW in Vietnam. The country's policies related to food waste are disconnected and disjointed, and therefore are unable to cover the entire food supply chain.

However, the issue of food waste is not a simple end-of-life waste issue. The impact of waste policies on food waste reduction has been estimated as negligible by the European Commission [39]. The scholarly discussion on food waste has gradually shifted from a "waste" angle into a more holistic approach towards the development of a sustainable food system targeting the different stages of the food supply chain [40].

The European Commission has adopted a Circular Economy package which aims at helping European businesses and consumers transition toward more sustainable resource use [41]. To achieve the SDG 12.3 target, it is critical to develop a specific, comprehensive food waste policy covering the entire food supply chain based on the coordination of overlapping authority among related ministries. Several ministries are responsible for different stages of the food supply chain in relation to food waste issues. These include the Ministry of Agriculture and Rural Development in the agriculture production stage and food processing and manufacturing stage, the Ministry of Industry and Trade for food processing, wholesalers and retailers, and the Ministry of Natural Resources and Environment for issues that incorporate a waste management perspective. Policy coordination through the development of an inter-ministerial taskforce on food waste along the food supply chain of production, processing, storage, transportation, commercialisation, and household settings could help solve administrative challenges at the policymaking level.

(2)   Development of a practical food waste management system based on household separation and collection as well as "recycling loop" schemes to support the utilisation of recycled food waste products, especially in rural areas

In investigating food waste practices in households in Hanoi and relevant policies on food waste in Vietnam, this survey found the letter of the law lacking in options for the realistic implementation of policies, resulting in both a rise in the amount of food waste and an increase in MSW cross-contaminated with food waste sent to landfills. A case in point: although Vietnam set goals for ambitious waste management systems, recycling, and separation of waste at source in its National Strategy on Integrated SWM in 2009, the results of the survey indicated that waste is not being separated out at the household level and collection systems for separated waste have not been implemented. As a result, the target for 2015 has not been achieved, and other targets including those in the revised strategy still face an overabundance of challenges in view of the current state of SWM in the country.

Considering that the results of the survey showed that people in Hanoi often dine out and are willing to eat at home, and cooking waste generated during the preparation of meals covers more than 70% of the total, there is some indication that the household is a primary source of generated food waste. While the cooperation of residents in separating waste at source is vital, the development of practical food waste household separation and collection systems is also crucial in identifying appropriate solutions for food waste management as well as reducing the actual costs of waste management.

In the survey, residents expressed concern about the separate collection of waste. In this context, it is also important to establish a collection service that supports waste separation at source. It is necessary for the Hanoi government to fully understand the details of informational, educational, economic, and regulatory methods in national policies to accurately prioritise policy methods, identify effective policy measures in education and public awareness, and carefully design the entire waste management system. For example, collecting various types of waste on different days with high priority placed on fully-separated food waste and other recyclable waste, and setting different collection fees for recyclables and non-recyclables would act as a good economic incentive for household segregation [42].

There are no schemes in place that support the use of recycled products, even as regulations contain specific reference to integrated MSW facilities and suggestions for composting organic waste. A variety of technologies exist that can be leveraged in finding alternative paths for using food waste, such as animal feed processing, composting, and anaerobic digestion, with each demonstrating its own merits and demerits [40,43]. What Hanoi must consider is the development of a sustainable market mechanism that is predicated on local conditions to cover initial investment and running costs for food waste recycling technologies, production of commodities from recycled products, and creating demand for recycled products such as bio-fertiliser, animal feed, or biogas.

Furthermore, the results of the survey found that self-reported food waste generation in households in rural areas is relatively high, and less than 10% of food waste is reused/recycled as animal feed. In light of the high food waste generation rate and large population, it is important to strengthen food waste management in rural areas even with low waste collection and treatment rates.

(3)   Development of a shared platform for collaboration and intervention

The results of this survey demonstrate that in urban areas of Hanoi, people have adopted the habit of reusing leftover food directly and indirectly as animal feed. A shared platform would make it easier for people to identify others nearby who have pets or raise animals and supply them with food left over from their own meals, thereby encouraging collaboration among households. Furthermore, with this type of community intervention, such platforms for sharing food as well as leftovers would provide an option for connecting practices implemented by individuals, allow them to share their experiences with 3R activities, and eventually expand practices outside of smaller pilot areas. Also, digital platforms such as food rescue apps can be used as a mechanism for organising stakeholders to encourage their active participation in the process.

(4)   Reduction in portions and encouraging consumers to use surplus food at home

The survey shows that the top reasons cited for food waste when eating out were over-ordering and excessive portions. One way to prevent and encourage measures to reduce food waste in food courts and restaurants would be to cut portion sizes and encourage customers to take leftovers home. Previous studies have identified a better estimation of portion sizes as one of the most promising actions to avoid wasting food [44–46]. Serving small portions or offering a variety of options for portion sizes puts consumers in the driver's seat when deciding the size of their meals. Moreover, a relatively easy game-changer is repositioning social norms to encourage consumers to use doggy bags to take leftover food home.

(5)   Creation of uniform standards and a consensus on labels indicating food expiration dates

The results of the survey show that the prevailing reason for the generation of food waste at home is that food had passed its "use-by" date. Concerns about food safety and foodborne illnesses, together with a desire to eat fresh food are prominent reasons for the generation of food waste [47]. Instituting policies for uniform standards will help reduce confusion around expiration dates, such as when food has been manufactured/produced/packed, when it must be sold/used by/best-used-by or when it expires, and reduce food waste generated by households. The public must be kept informed not only about actual policies, but what implications these policies have on a day-to-day basis. Meanwhile, the review and elimination of unnecessary food safety standards has the potential to reduce unnecessary food waste. In addition, raising awareness on sustainable lifestyles, especially of the middle class, will help guide people in using their own sense to form a more nuanced assessment of edibility and may reduce food waste generation in households. The application of intelligent packaging indicating "use-by" dates in combination with a dynamic system that details the real state of the food inside packaging may also be considered as a way to significantly reduce food waste [48].

(6)   Promotion of sufficiency strategies to save food, reduce food waste and maintain healthy lifestyles

Policies aimed at changing consumption behaviours are designed according to efficiency, consistency, and sufficiency principles. Sufficiency strategies focus on the demand side by emphasising consumers' responsibilities to restrict the consumption of natural resources and the amount of food consumed [49]. Although sufficiency policies are not very popular due to the fear that they may impede individual well-being and quality of life, it would be a promising approach for improving the sustainability of the modern food system if conducted carefully and related to appropriate motives for changes [50]. In fact, sufficiency strategies are not one-way reductions, but keep amounts at "suitable" and "appropriate" levels. These strategies could help consumers who overconsume resources to not only reduce food waste and save natural resources, but maintain healthy lifestyles and well-being. The survey found that health is a key contributor to decisions on diet among consumer in Hanoi, and the government may be able to promote the alignment of government and citizens' actions based on sufficiency strategies to improve people's health and well-being while addressing food waste issues, to create a more sustainable society.

6. Conclusions

This paper uses Hanoi as a case study to clarify the current situation and trends in food consumption and food waste generation in households, examines ways in which consumers generate food waste, reviews relevant food waste policies and strategies already in place, and provides prescriptions for preventing and reducing food waste. This study found that (1) the level of food waste generation is high in Hanoi even when compared with developed countries, but is not being managed appropriately; (2) while the national government has put a number of laws, legislation, and strategies into effect and outlined ambitious targets for waste management, there is a lack of implementation at the local level; (3) both urban areas and rural areas in Hanoi are a major source of food waste; (4) there is a strong indication that people eat at home more often than dining outside or consuming ready-made meals, which might be a reason for the high generation of food waste by households and the high percentage of cooking waste in the total amount of food waste generated in households; (5) people rarely waste food when eating at home, while food waste tends to be generated more when dining out; (6) the main reasons for wasting food at home are that food has passed its "use-by date" and is seen to be of "deteriorating quality," while the top reasons for wasting food when eating out are that people over-order and that serving sizes are too large; (7) less than 20% of food waste is separated out for reuse and recycling, with the remainder is disposed and transported to landfills; (8) in urban areas, the majority of respondents offer leftover food to other people, while in rural areas, food waste is used as animal feed and for other purposes; and (9) most respondents are willing to reduce, reuse, and recycle food waste so interventions at the consumer level may hold promise in encouraging consumers to practice ways to reduce food waste.

Based on an analysis of the results of this survey, policy implications have been proposed for national and local policymakers and practitioners to consider as potential avenues to explore to reduce waste, including the (1) development of comprehensive food waste policies to address the issue of food waste along the entire food supply chain; (2) development of a practical food waste management system based on household separation and collection as well as "recycling loop" schemes to support the utilisation of recycled food waste products, especially in rural areas; (3) development of a shared platform for collaboration and intervention; (4) reduction in portions and encouraging consumers to use surplus food at home; (5) creation of uniform standards and a consensus on labels indicating food expiration dates; and (6) promotion of sufficiency strategies for saving food, reducing food waste, and maintaining a healthy lifestyle.

The issue of food waste is considered to be vital for all countries to achieve targets decided through international consensus, including the Paris Agreement and the 2030 Agenda (SDGs). The Vietnamese government is facing mounting pressure to establish a suitable waste management system. At the same time, Vietnam is an agriculture-based economy with a potentially high level of demand for bio-fertiliser, such as compost and fermented liquid manure by anaerobic digestion. Two decrees (Decree 109/ND-CP on Organic Agriculture issued in 2018 and Decree 84/ND-CP on Regulations on Fertiliser Management issued the following year) encourage the practice of organic agriculture and set out what constitutes an organic fertiliser. Collaboration between peripheral villages and towns could help drive a transition from conventional waste management to innovative resource management and to a sustainable food production-consumption system, although tremendous challenges remain.

It is clear that the data presented in this study is relatively uncertain due to the limited number of samples as well as limitations in self-reporting measures. However, this is the first attempt to link food waste to waste producers in urban and rural areas in Hanoi and to gain an understanding of the situation of food waste generation and the diversity of factors that can influence food waste behaviours at the household level in order to improve waste management systems and policies to reduce food waste. Future research should aim to validate these self-reporting measures with more objective techniques for data collection, and link food waste generation in households to both the food and waste systems to draw a picture of the overall image of food waste generation. Furthermore, the complexity of the food waste issue requires collaboration among different disciplines to bring the grainy picture of the reasons why consumers generate food waste into clearer focus. Finally, to achieve SDG target 12.3 of halving food waste at the consumer level by 2030, ensuring harmonised data collection on food waste remains an issue in Asia, and practical policies, strategies, and actions on source reduction is an area that can be considered for future study.

Supplementary Materials: The following is available online at http://www.mdpi.com/2071-1050/12/16/6565/s1, Questionnaire: Survey on Personal Lifestyles in Viet Nam.

Author Contributions: Conceptualization, C.L.; Data curation, T.T.N. and C.L.; Formal analysis, C.L. and T.T.N.; Methodology, C.L.; Validation, T.T.N.; Visualization, C.L.; Writing—original draft, C.L. All authors have read and agreed to the published version of the manuscript.

Funding: This research was supported by the Environment Research and Technology Development Fund (S-16) of the Environmental Restoration and Conservation Agency of Japan "Policy Design and Evaluation to Ensure Sustainable Consumption and Production Patterns in the Asian Region" (2016–2020), and the IGES SRF Fund 2020.

Acknowledgments: The authors would like to thank Phuong Anh Duong Thi and Thu Ha Nguyen Thi for their technical assistance with the questionnaire survey, and would also like to express gratitude to the referees for their useful comments.

Conflicts of Interest: The authors declare no conflict of interest.

References

1. Food and Agriculture Organization of the United Nations (FAO). *The state of Food and Agriculture*; FAO: Rome, Italy, 2019.
2. Beretta, C.; Stoessel, F.; Baier, U.; Hellweg, S. Quantifying food losses and the potential for reduction in Switzerland. *Waste Manag.* **2013**, *33*, 764–773. [CrossRef] [PubMed]
3. FAO. *Food Wastage Footprint & Climate Change*; FAO: Rome, Italy, 2015.
4. FAO. *Food Wastage Footprint: Full-Cost Accounting*; FAO: Rome, Italy, 2014.
5. FAO; IFAD; WFP. *The State of Food Insecurity in the World 2015. Meeting the International Hunger Targets: Taking Stock of Uneven Progress*; FAO: Rome, Italy, 2015.
6. United Nations. *Transforming Our World: The 2030 Agenda for Sustainable Development*; UN: New York, NY, USA, 2015.
7. Poore, J.; Nemecek, T. Reducing food's environmental impacts through producers and consumers. *Science* **2018**, *992*, 987–992. [CrossRef] [PubMed]
8. Gustavsson, J.; Cederberg, C.; Sonesson, U.; Van Otterdijk, R.; Meybeck, A. *Global Food Losses and Food Waste*; FAO: Rome, Italy, 2011.
9. Östergren, K.; Gustavsson, J.; Bos-Brouwers, H.; Timmermans, T.; Hansen, O.J.; Møller, H.; Anderson, G.; O'Connor, C.; Soethoudt, H.; Quested, T.; et al. *FUSIONS Definitional Framework for Food Waste*; EU: Goteborg, Sweden, 2014.
10. Schneider, F. Review of food waste prevention on an international level. *Proc. Institutions Civ. Eng.-Waste Resour. Manag.* **2013**, *166*, 187–203. [CrossRef]
11. Sahakian, M.; Shenoy, M.; Soma, T.; Watabe, A.; Yagasa, R.; Premakumara, D.G.; Liu, C.; Favis, A.M.; Saloma, C. Apprehending food waste in Asia: Policies, practices and promising trends. In *Routledge Handbook of Food Waste*; Routledge: London, UK, 2020; pp. 187–206.
12. Liu, C.; Mao, C.; Bunditsakulchai, P.; Sasaki, S.; Hotta, Y. Food waste in Bangkok: Current situation, trends and key challenges. *Resour. Conserv. Recycl.* **2020**, *157*, 104779.

13. Hansen, A. Meat consumption and capitalist development: The meatification of food provision and practice in Vietnam Geoforum Meat consumption and capitalist development: The meati fi cation of food provision and practice in Vietnam. *Geoforum* **2018**, *93*, 57–68. [CrossRef]

14. Rangkuti, F.Y.; Wright, T. *Indonesia Retail Foods: Indonesia Retail Report Update 2013*; Foreign Agricultural Service: Jakarta, Indonesia, 2013. Available online: https://apps.fas.usda.gov/newgainapi/api/report/downloadreportbyfilename?filename=Retail%20Foods_Jakarta_Indonesia_12-13-2013.pdf (accessed on 20 June 2020).

15. Aktas, E.; Sahin, H.; Topaloglu, Z.; Oledinma, A.; Huda, A.K.S.; Irani, Z.; Sharif, A.M.; van't Wout, T.; Kamrava, M. A consumer behavioural approach to food waste. *J. Enterprise Inf. Manag.* **2018**, *31*, 658–673. [CrossRef]

16. Goodman, D.S.; Robinson, R. *The New Rich in Asia: Mobile Phones, McDonald's and Middle Class. Revolution*; Routledge: London, UK, 2013.

17. Reardon, T.; Hopkins, R. The supermarket revolution in developing countries: Policies to address emerging tensions among supermarkets, suppliers and traditional retailers. *Eur. J. Dev. Res.* **2006**, *18*, 522–545. [CrossRef]

18. Lee, K.C.L. Grocery shopping, food waste, and the retail landscape of cities: The case of Seoul. *J. Clean. Prod.* **2018**, *172*, 325–334. [CrossRef]

19. General Statistics Office of Vietnam. *Statistical Yearbook of Vietnam 2018*; General Statistics Office: Hanoi, Vietnam, 2019.

20. Vietnam National Administration of Tourism. 2019. Available online: http://vietnamtourism.gov.vn/index.php/items/13462 (accessed on 24 January 2019).

21. World Bank Group. Climbing the Ladder: Poverty Reduction and Shared Prosperity in Vietnam; World Bank. 2018. Available online: http://documents1.worldbank.org/curated/en/206981522843253122/pdf/124916-WP-PULIC-P161323-VietnamPovertyUpdateReportENG.pdf (accessed on 12 July 2020).

22. World Bank Group. Ending Poverty, Investing in Opportunity; World Bank. 2019. Available online: http://documents1.worldbank.org/curated/en/156691570147766895/pdf/The-World-Bank-Annual-Report-2019-Ending-Poverty-Investing-in-Opportunity.pdf (accessed on 12 July 2020).

23. World Bank Group. Solid and Industrial Hazardous Waste Management Assessment; World Bank. 2018. Available online: http://documents1.worldbank.org/curated/en/352371563196189492/pdf/Solid-and-industrial-hazardous-waste-management-assessment-options-and-actions-areas.pdf (accessed on 20 July 2020).

24. Huong, L.T.; Kawai, K.; Thai, N.T. Physical Composition Analysis of Household Waste in Hanoi, Vietnam. Available online: http://www.gels.okayama-u.ac.jp/management/up_load_files/gakkan/2012_en/2012_en_3-14.pdf (accessed on 20 July 2020).

25. Kawai, K.; Huong, L.T.; Yamada, M.; Osako, M. Proximate composition of household waste and applicability of waste management technologies by source separation in Hanoi, Vietnam. *J. Mater. Cycles Waste Manag.* **2016**, *18*, 517–526. [CrossRef]

26. Aschemann-Witzel, J.; De Hooge, I.; Amani, P.; Bech-Larsen, T.; Oostindjer, M. Consumer-Related Food Waste: Causes and Potential for Action. *Sustainability* **2015**, *7*, 6457–6477. [CrossRef]

27. Roodhuyzen, D.M.; Luning, P.A.; Fogliano, V.; Steenbekkers, L.P. Putting together the puzzle of consumer food waste: Towards an integral perspective. *Trends Food Sci. Technol.* **2017**, *68*, 37–50. [CrossRef]

28. Schanes, K.; Dobernig, K.; Gözet, B. Food waste matters—A systematic review of household food waste practices and their policy implications. *J. Clean. Prod.* **2018**, *182*, 978–991. [CrossRef]

29. Leray, L.; Sahakian, M.; Erkman, S. Understanding household food metabolism: Relating micro-level material flow analysis to consumption practices. *J. Clean. Prod.* **2016**, *125*, 44–55. [CrossRef]

30. Sahakian, M.; Saloma, C.; Ganguly, S. Exploring the Role of Taste in Middle-Class Household Practices. *Asian J. Soc. Sci.* **2018**, *46*, 304–329. [CrossRef]

31. Soma, T. (Re)framing the Food Waste Narrative: Infrastructures of Urban Food Consumption and Waste in Indonesia. *Indonesia* **2018**, *105*, 173–190. [CrossRef]

32. Soma, T. Wasted infrastructures: Urbanization, distancing and food waste in Bogor, Indonesia. *Built Environ.* **2017**, *43*, 431–446. [CrossRef]

33. Yamane, T. *Statistics, An Introductory Analysis*, 2nd ed.; Harper and Row: New York, NY, USA, 1967.

34. Berkenkamp, C.; Phillips, J. Modeling the potential to increase food rescue: Denver, New York and Nashville. *Nrdc* **2017**, *41*, R-17-09-B. Available online: https://www.nrdc.org/sites/default/files/modeling-potential-increase-food-rescue-report.pdf (accessed on 20 July 2020).
35. WRAP. *Courtauld Commitment 2025: Annual review 2017–2018*; WRAP: Banbury, UK, 2019. Available online: https://www.wrap.org.uk/sites/files/wrap/Courtauld_2025_Annual_Review_2017_18.pdf (accessed on 20 July 2020).
36. Barr, S.; Gilg, A.W.; Ford, N.J. Differences Between Household Waste Reduction, Reuse and Recycling Behaviour: A Study of Reported Behaviours, Intentions and Explanatory Variables. *Environ. Waste Manag.* **2001**, *4*, 69–82.
37. Lober, D.J. Municipal solid waste policy and public participation in household source reduction. *Waste Manag. Res.* **1996**, *14*, 125–143. [CrossRef]
38. HLPE. *Food Losses and Waste in the Context of Sustainable Food Systems, A Report by the High Level Panel of Experts on Food Security and Nutrition of the Committee on World Food Security*; FAO: Rome, Italy, 2014.
39. Monier, V.; Shailendra, M.; Escalon, V.; O'Connor, C.; Gibon, T.; Anderson, G.; Hortense, M.; Reisinger, H. *Preparatory Study on Food Waste across EU 27, European Commission (DG ENV) Directorate C-Industry*; Final Report; EU: Paris, France, 2010; ISBN 978-92-79-22138-5.
40. Joshi, P.; Visvanathan, C. Sustainable management practices of food waste in Asia: Technological and policy drivers. *J. Environ. Manag.* **2019**, *247*, 538–550. [CrossRef]
41. Communication from the Commission to the European Parliament, the Council, the European Economic and Social Committee and the Committee of the Regions. In *Closing the Loop—An EU Action Plan for the Circular Economy*; European Commission; COM; EU: Brussels, Belgium, 2015. Available online: https://eur-lex.europa.eu/resource.html?uri=cellar:8a8ef5e8-99a0-11e5-b3b7-01aa75ed71a1.0012.02/DOC_1&format=PDF (accessed on 20 July 2020).
42. Onogawa, K.; Gamaralalage, P.J.; Liu, C. *Paradigm Shift from Incineration to Resource Management, and Town Development*; UNEP: Osaka, Japan, 2018. Available online: https://wedocs.unep.org/bitstream/handle/20.500.11822/30994/PSIR.pdf?sequence=1&isAllowed=y (accessed on 20 July 2020).
43. Li, Y.; Jin, Y.; Li, J.; Chen, Y.; Gong, Y.; Li, Y.; Zhang, J. Current Situation and Development of Kitchen Waste Treatment in China. *Procedia Environ. Sci.* **2016**, *31*, 40–49. [CrossRef]
44. Secondi, L.; Principato, L.; Laureti, T. Household food waste behaviour in EU-27 countries: A multilevel analysis. *Food Policy* **2015**, *56*, 25–40. [CrossRef]
45. Wansink, B.; van Ittersum, K. Portion size me: Plate-size induced consumption norms and win-win solutions for reducing food intake and waste. *J. Exp. Psychol. Appl.* **2013**, *19*, 320–332. [CrossRef] [PubMed]
46. Reynolds, C.; Goucher, L.; Quested, T.; Bromley, S.; Gillick, S.; Wells, V.K.; Evans, D.; Koh, L.; Kanyama, A.C.; Katzeff, C. Review: Consumption-stage food waste reduction interventions—What works and how to design better interventions. *Food Policy* **2019**, *83*, 7–27. [CrossRef]
47. Lanfranchi, M.; Calabrò, G.; De Pascale, A.; Fazio, A.; Giannetto, C. Household food waste and eating behavior: Empirical survey. *Br. Food J.* **2016**, *118*, 3059–3072. [CrossRef]
48. Poyatos-Racionero, E.; Ros-Lis, J.V.; Vivancos, J.L.; Martínez-Máñez, R. Recent advances on intelligent packaging as tools to reduce food waste. *J. Clean. Prod.* **2018**, *172*, 3398–3409. [CrossRef]
49. Schmidt, K.; Matthies, E. Resources, Conservation & Recycling Where to start fi ghting the food waste problem? Identifying most promising entry points for intervention programs to reduce household food waste and overconsumption of food. *Resour. Conserv. Recycl.* **2018**, *139*, 1–14.
50. Schäpke, N.; Rauschmayer, F. Going beyond efficiency: Including altruistic motives in behavioral models for sustainability transitions to address sufficiency. *Sustain. Sci. Pract. Policy* **2014**, *10*, 29–44. [CrossRef]

MDPI

St. Alban-Anlage 66

4052 Basel

Switzerland

Tel. +41 61 683 77 34

Fax +41 61 302 89 18

www.mdpi.com

Sustainability Editorial Office

E-mail: sustainability@mdpi.com

www.mdpi.com/journal/sustainability